Manufacturing Engineering and Management

Manufacturing Engineering and Management

Edited by **Jeff Hansen**

NY RESEARCH PRESS
New York

Published by NY Research Press,
23 West, 55th Street, Suite 816,
New York, NY 10019, USA
www.nyresearchpress.com

Manufacturing Engineering and Management
Edited by Jeff Hansen

International Standard Book Number: 978-1-63238-504-8 (Hardback)

Printed in China.

Contents

Preface

This book brings forth some of the most innovative concepts and elucidates the various applications of manufacturing engineering and management. It covers topics which deal with the core areas of this subject. Manufacturing engineering refers to a branch of engineering that deals with design, formation, planning, structuring and developing various systems, machines, tools, etc. The main objective of this field is to turn raw materials into finished goods in an efficient, economic and profitable way. This book is a compilation of chapters that discuss the most vital concepts and emerging trends in this area. Manufacturing engineering is an upcoming field that has undergone rapid development over the past few decades. The various studies that are constantly contributing towards advancing technologies and evolution of this discipline have been compiled in this book. It will prove to be immensely beneficial for students, engineers, managers and researchers related to this field.

Various studies have approached the subject by analyzing it with a single perspective, but the present book provides diverse methodologies and techniques to address this field. This book contains theories and applications needed for understanding the subject from different perspectives. The aim is to keep the readers informed about the progress in the field; therefore, the contributions were carefully examined to compile novel researches by specialists from across the globe.

Indeed, the job of the editor is the most crucial and challenging in compiling all chapters into a single book. In the end, I would extend my sincere thanks to the chapter authors for their profound work. I am also thankful for the support provided by my family and colleagues during the compilation of this book.

Editor

1

A method of meta-mechanism combination and replacement based on motion study

Yadong Fang

Department of Mechanical and Electrical Engineering, Xi'an Technological University, Xi'an, China

Lacking the effective methods to reduce labor and cost, many small- and medium-sized assembly companies are facing with the problem of high cost for a long time. In order to reduce costs of manual operations, the method of meta-mechanism combination and replacement is studied. In this paper, we mainly discuss assembling motion analysis, workpieces position information acquisition, motion library construction, assembling motion analysis by Maynard's operation sequence technique, meta-mechanism database establishment, and match of motion and mechanism. At the same time, the principle, process, and system realization framework of mechanism replacement are introduced. Lastly, problems for low-cost automation of the production line are basically resolved by operator motion analysis and meta-mechanism combination and match.

Keywords: production line; low-cost automation; motion analysis

1. Introduction

Motion Study, also known as motion research, element analysis, or work study, its main content is seeking the most economical and effective working method through a variety of analytical methods. It is an effective way to study and formulate correct and reasonable motion to shorten hours and improve efficiency by increasing valuable movement and shorten or eliminate waste motion (Du, 2013). Frank Bunker Gilbreth, an American engineer, established therbligs theory in 1912. The American Society of Mechanical Engineers added 'Discovery' motion and summarized 18 kinds of basic therbligs (Hua, 2012). In the process of analysis of therbligs, operation is observed and divided finely, and motion sequence is linked with hands, feet, eyes, and head movements in detail (Hirosea, Dokia, & Kondoa, 2013). Furthermore, operation is recorded and classified by therblig symbols, and the problems about operating sequence and method, such as single-hand waiting, unreasonable motion, and waste motion, are found and improved. The application researches in the country about motion study are mainly focused on the following four aspects (Wang, 2010):

*Email: fangyadong@xatu.edu.cn

- Motion analysis applied in production line balance.

Improvement measures based on motion analysis for production line in equipment or electronics manufacturing enterprise are proposed, and it improves the cycle time and balance ratio of production line (Gu & Zhou, 2009).

- Motion analysis applied in experimental teaching and assembly line study.

Wen Wei, Ding Wen-ying, and Dong Shao-hua choose reducer assembly line as the subject investigated, and eliminate bottleneck process of production line to improve its balance ratio. Gao Qi and Wang Shao-Hua research assembly line based on threbligs, and find out stasis, invalid, waste, imbalance, and disorder phenomenon in production process (Liao & Sun, 2009).

- Operation improvement and design by motion analysis.

Liu Shao-shan and Liu Feng-shan research therblig motion analysis and work improvement under lean production system to improve resource utilization rate. Lv Xiao-Ji and Chen Cang-ming take one automobile enterprises, for example, and eliminate various waste and imbalance problem in enterprise production process by motion analysis and line balance (Li & Gui, 2008).

- Operation process optimization by motion analysis.

Jiang De-yi and Liu Hong-xiao take snatch motion as instance, construct body mathematical model, and give optimal motion of snatch object in production activities. According to motion analysis for efficiency principle, Liu Hong-xiao optimizes male shirt shoulder sewing process (Yang, Wu, Zhu, Bao, & Wei, 2013).

However, there are few reports for the automation of the assembly line, especially the motion analysis and meta-mechanisms.

Since the automation technology appears, it develops crosswise with mathematics and computer technology. Increasingly mature hardware technologies, especially the rapidly developing computer technology, lay the foundations of automation technology (Wei, 1994). However, at the present stage, the automation technology not only solves the problems of production control for the single production cell, but also solves the problems of the entire production process as well as operation and management of business. The development of automation technology is divided into 4 stages, and it is shown in Table 1.

So far, for the automation of assembly enterprises all over the world, the four stages coexist, and they are information- and system-oriented (Jiang & Sui, 2012). But we know, at the fourth stage, computer-integrated manufacturing system (CIMS) costs too much, and the manufacturing period is too long; so in some small- and medium-sized enterprises, for the simple automation production lines, we can find a technology-intensive and low-cost way.

Therefore, the replacement of assembly operation with meta-mechanism and LCA have great potential and extensive foreground. Besides, they meet the requirements currently with advantages of low-cost investment, short period, quick effectiveness, and high automation.

Table 1. The development of automation technology.

The stage	Automatic level	Industrial mode of production	Technical means	Type of manufacturing: Discrete manufacturing	Type of manufacturing: Process manufacturing
First stage	Manual control	Labor intensive	Manual operation	Simple facility	Simple facility
Second stage	Unit automation	Equipment-intensive	CNC technology	CNC machine	Unit combination instrument
Third stage	Information automation	Information intensive	Computer-aided technology	CAD/CAM/FMS	DCS
Fourth stage	Decision automation	Knowledge intensive	Computer integrated technology	CIMS	CIPS

2. Common problems in the manual production lines

There are usually a lot of labor-intensive enterprises in the manufacturing production lines, in which the staffs are engaged in the simple and repetitive assembly motions, and easy to feel fatigued (Meng, 1994). Since the production efficiency and production quality are easily affected by workers, once lack of personnel or training, it will not be able to guarantee the delivery cycle (Son, 2005). In recent years, especially in the eastern coastal area, with the declining of the domestic manufacturing costs of parts and the increasing of the labor costs, reducing the costs of product's assembly become more and more important and evident, particularly in the industries of household appliances and electronics industries. The economic benefits brought by improving assembly efficiency are far more obviously increase than that brought by simply reducing manufacturing costs of parts. The applications of low-cost automation will make some small- and medium-sized assembly enterprises (or some large enterprises) run their own business with lower costs, and in the process, it can almost get the same profits as the profits of advanced automation enterprises. If we combine the low-cost automation with each production line, which can realize the combined operation of machines and operators, and the achievements will be much more valuable than trying to carry out the pure automation in a short term.

Although many enterprises pay much attention to the development and application of low-cost automation all over the world in recent years, but the executing process of low-cost automation in the production line is still faced with many problems (Yang, 2010). Mostly, three issues must be considered: firstly, the effective analysis of assembly motions; secondly, the replacement of similar movements by automation; and thirdly, the mechanisms must be easily replaced and just need low costs. In general, low-cost automation should have the following features:

(1) *Small investment*: The investment of computers and automation equipment generally accounts for 7–12% of all equipment investment, or less. The upper limit to the investment just likes this: the percentage of investment in a new factory should be lower than the general provisions. The percentage of some special projects should be far lower than the funds, which are approved by provinces and cities, and less than the enterprise's affordability. However, this part should account for more than 40% in CIMS and computer-integrated process system (CIPS).

(2) *Targeted investment*: Investment should increase output, improve quality, raise labor productivity, and reduce the consumption in the short term. Man should get benefits in a relatively short time (usually 1–2 years), and the lifetime of equipment should be 5–8 years.
(3) *The automation*: The most necessary part is automation. It includes reasonable classification between men's and machines' functions, and compact structures. Besides, the rate of function should be above 95%.
(4) Reliable and easy to maintain.
(5) Low-cost automation should be gradually improved, eventually progressed in the direction of the CIM and CIP.
(6) Use the mature technology as much as possible, invest least with quick effect, and reduce the investment risks.
(7) The maintenance cost must be low; the work should be reliable, and the online rate should be above 95%.

3. The principle and process of meta-mechanism replacement

Assembly enterprises are usually faced with these problems: whether input low-cost automation equipment in production line or not; how much automation equipment should be input; and what kind of low-cost automation equipment should be input, etc. But, even if enterprises have the ability to put simple automation into effect, they do not; or they spent a lot on expensive automation equipment, but it is difficult to get benefits in a short time, which will cause huge losses. In this article, through the analysis of operator motion in production line, man can establish a matching relation between operator motions and meta-mechanism combinations or automation equipment, implement low-cost automation in a production line, and then save a large part of the expenses.

The principle and operating steps about motion analysis of assembly and mechanism replacement have shown in Figure 1.

The process to the motion analysis and the method of meta-mechanism replacement includes the following aspects:

(1) Analyze the assembly motions. According to the category of products, use a video camera to do some real-time recording of the assembly operation, assembly operation is divided into parts by motion display devices, storage devices, and processing devices. At the same time, analyze the positional information about workpieces. Use the laser displacement sensor to collect the 3D data of workpieces in the assembling process from different positions.
(2) Use the algorithm of point cloud reconstruction and registration to rebuild the 3D data model of the workpieces (Wang, 2012), and then obtain the positional information of the workpieces.

The algorithm of point cloud reconstruction is shown as following:

Step1: Point cloud data to be spliced is inputted.
Step2: Initial triangle is determined.
  2.1 Any point is selected randomly as first vertex P_1 of initial triangle.
  2.2 Closet point P_2 from P_1 is searched, edge P_1P_2 is constructed.
  2.3 Optimum point is searched by traversing other points.

2.4 If $min(|P_1P_3|+|P_2P_3|)$, then goes to next step, else returns to 2.1.
2.5 If minimum interior angle of triangle$\Delta P_1P_2P_3$ isn't less than 35 degree, then goes to step 2.6, else returns to 2.1.
2.6 If maximum interior angle of triangle$\Delta P_1P_2P_3$ isn't more than 90 degree, then goes to next step, else returns to 2.1.
2.7 Initial triangle is determined, and point P_1, P_2 and P_3 are set as dead point. Besides, $\Delta P_1P_2P_3$ and three edges of initial triangle are put into directly triangle table and edge table.
Step3: The process of triangulation is realized.
3.1 Any edge E_1 is selected randomly (point P_4 and P_5).
3.2 Minimum envelope rectangle for triangles including edge E_1 is solved.
3.3 Optimum point P_6 is searched from active point set in minimum envelope rectangle.
3.4 If $min(|P_4P_5|+|P_4P_6|)$, then goes to next step, else returns to 3.1.
3.5 If minimum interior angle of triangle$\Delta P_4P_5P_6$ isn't less than 35 degree, then goes to step 2.6, else returns to 3.1.
3.6 If maximum interior angle of triangle$\Delta P_4P_5P_6$ isn't more than 90 degree, then goes to next step, else returns to 3.1.
3.6 P_6 is set as dead point, and P_4P_5 and P_4P_6 are added into edge table. Besides, edge E_1 is deleted, and$\Delta P_4P_5P_6$ is put into triangle table.

The algorithm of point cloud registration is shown as following (Zhang, 2013):

Step1: Original point cloud data sampling.
1.1 Workpiece is scanned progressively by laser displacement sensor.
1.2 Reflected light of workpiece is collected by laser collection device.
1.3 Laser signal is converted into workpiece 3D data.
Step2: Bilateral filtering denoising of point cloud is realized.
Step3: Point cloud is simplified.
3.1 Minimum enveloping hexahedron of point cloud is calculated.
3.2 By data traversing point, grid cell is divided, and data points are classified as grid cell.
3.3 Grid cell spatial location is set as mean value of data point set, and point cloud quantity and coordinate are reset.
Step4: Corresponding closest point of reference point cloud set X_1 is calculated, according to registration point cloud set X_2 to be matched.
Step5: Minimum optimal model for corresponding point mean distance is constructed.
Step6: Translation transformation and rotation transformation matrix are determined, and registration point coordinate value X_3 in new coordinate system are calculated.
Step7: If condition of convergence is satisfied, then point cloud registration is finished, else $X_2 = X_3$, and returns to 4.

(3) Set up the motion database. According to the products' category, store each motion of the operating personnel who assembles a product, the required time, and the variation of position of ready-to-assemble parts.
(4) Express the motion with Maynard's operation sequence technique (MOST). After each motion has been divided into therbligs, we can express the assembling operations by MOST. Then, classify and integrate it, such as moving, grabbing, body movements, pressure, pull, and rotation. We can inspect strictly, delete redundant motion, and change the existing procedure by canceling, merging, rearrangement, and simplification. Last, we can adjust the working methods to get the optimal operation.
(5) Set up the meta-mechanism repository. Sort the application of the existing meta-mechanism of assembly production line and automation equipment. Forming the meta-mechanism repository of the assembly line, such as object-clamping

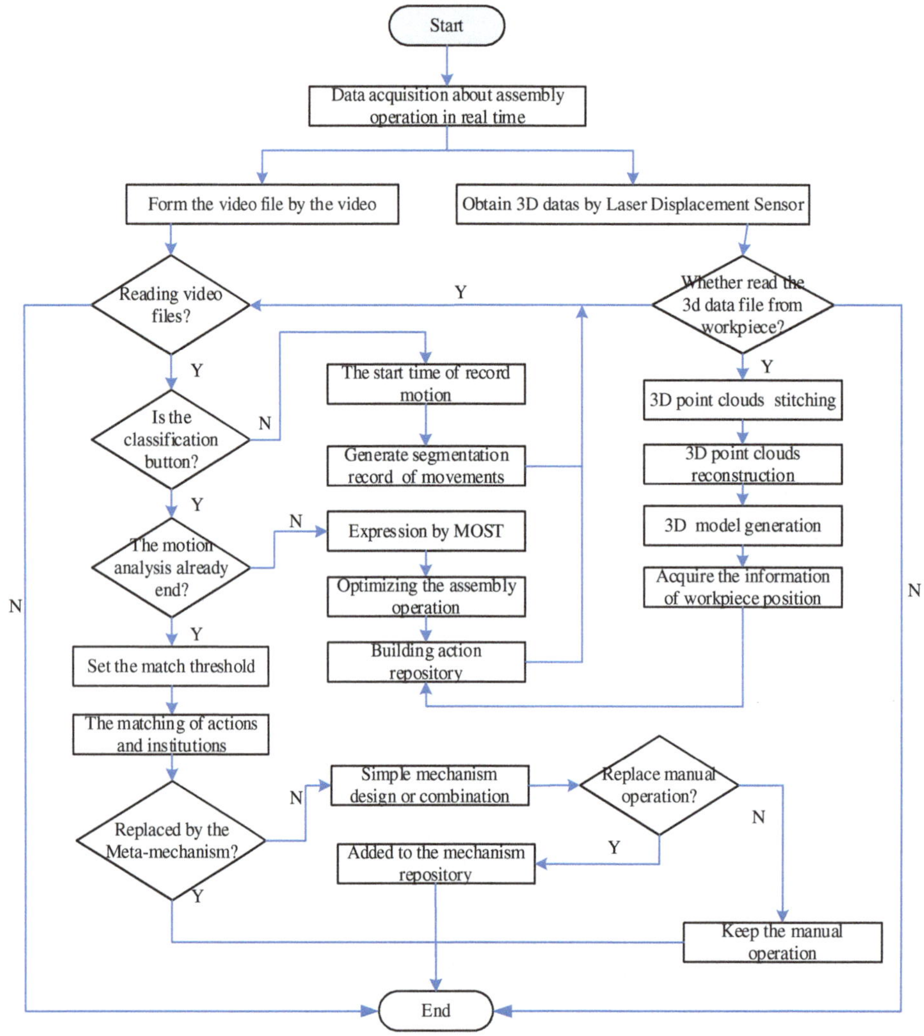

Figure 1. The principle of assembly motion analysis and mechanism substitution.

mechanism, moving mechanism, fixture-clamping mechanism, switches, positioning mechanism, locking mechanism, automatic halt mechanism, overload protecting mechanism, and so on, which usually includes 2D diagrams, key parameters, suitable operation, serial number, etc. The mechanism repository requires a lot of mechanisms and their combinations. The simple mechanisms and combinations which are commonly used can be divided into forms just like Figure 2.

(6) Match assembly motions with specific mechanisms. In view of each motion expressed by MOST in motion repository, search meta-mechanism repository and calculate the semantic similarity. If the similarity ratio is less than the setting threshold, combine mechanisms to form the project that can meet the requirements of assembly motion, which can contribute to evaluating and decision-making, and refer to the positional information about workpieces in the Step 2, then get the key technical parameters of mechanisms. The system interface has shown in Figure 3.

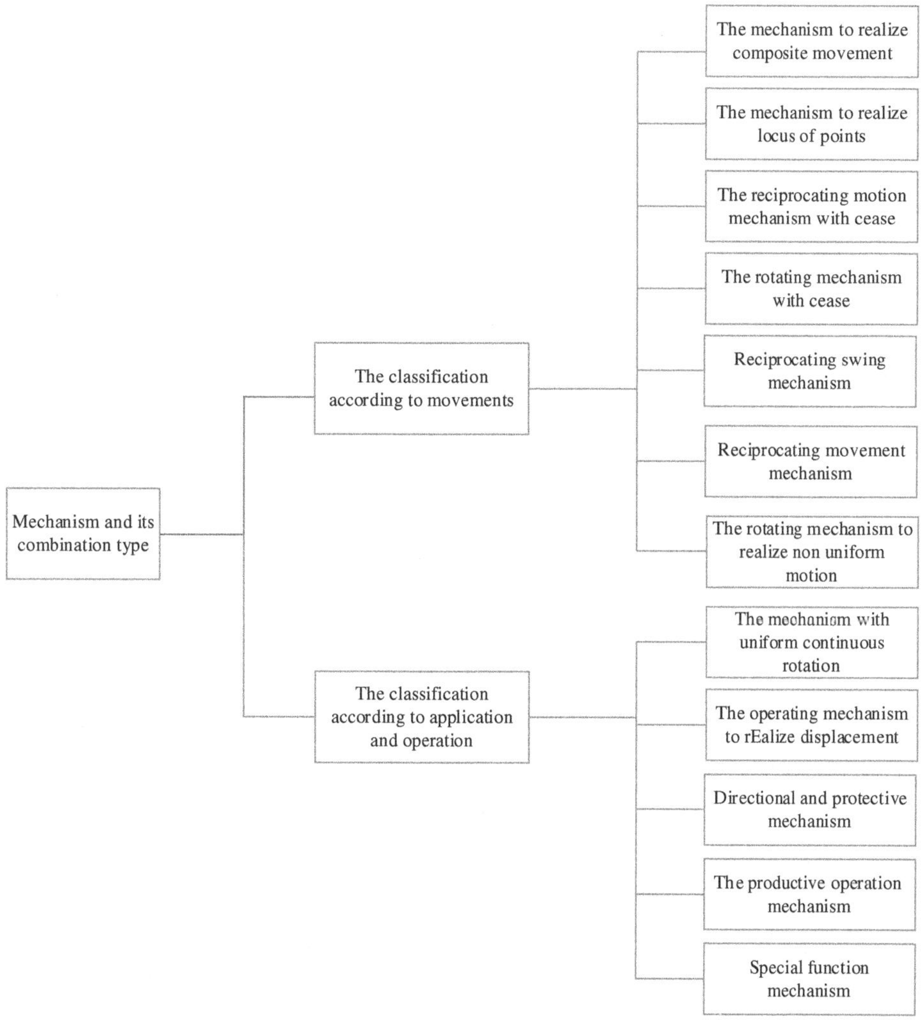

Figure 2. The classification of mechanisms and their combinations.

The realization of the above functions is based on the J2EE (Java 2 Platform Enterprise Edition) framework. Figure 4 shows that the system is divided into five layers: client layer, request to accept layer, application service layer, perception layer, and data storage layer from top to bottom. The function and effect of each layer are shown.

3.1. The client layer

When engineering technicians need data interaction, they use the server interface, which is provided by world web service and JavaBean (Enterprise Java Beans, EJB), to visit the system, and start the functions. The client layer includes two kinds of clients. One client is the thin client of Web service; the other is the specific client, which is based on Java and downloaded from the platform.

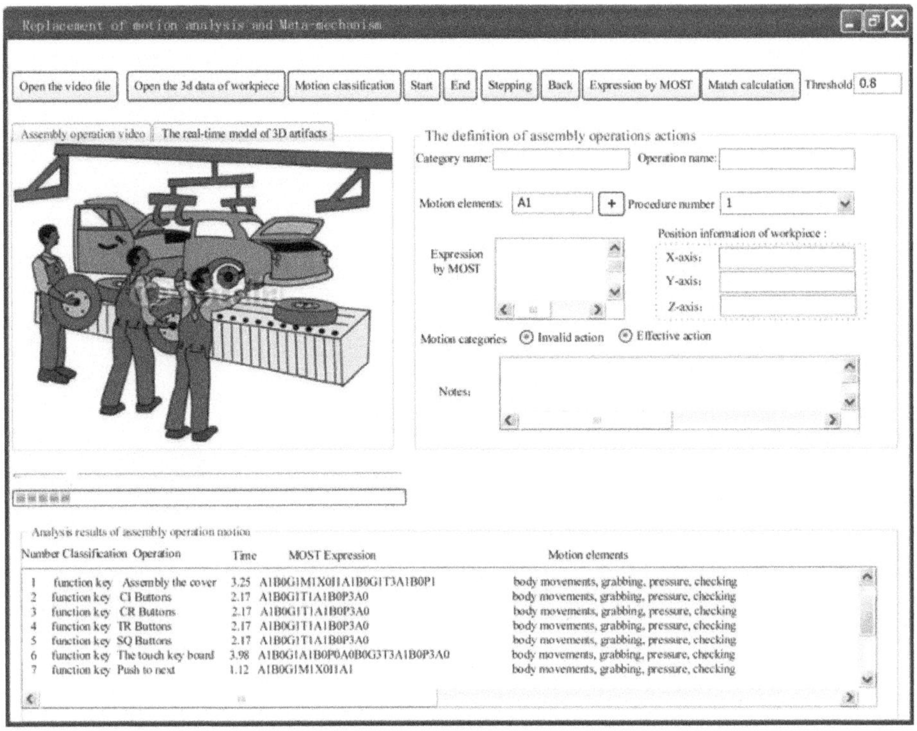

Figure 3. The interface of assembling motion analysis and meta-mechanisms.

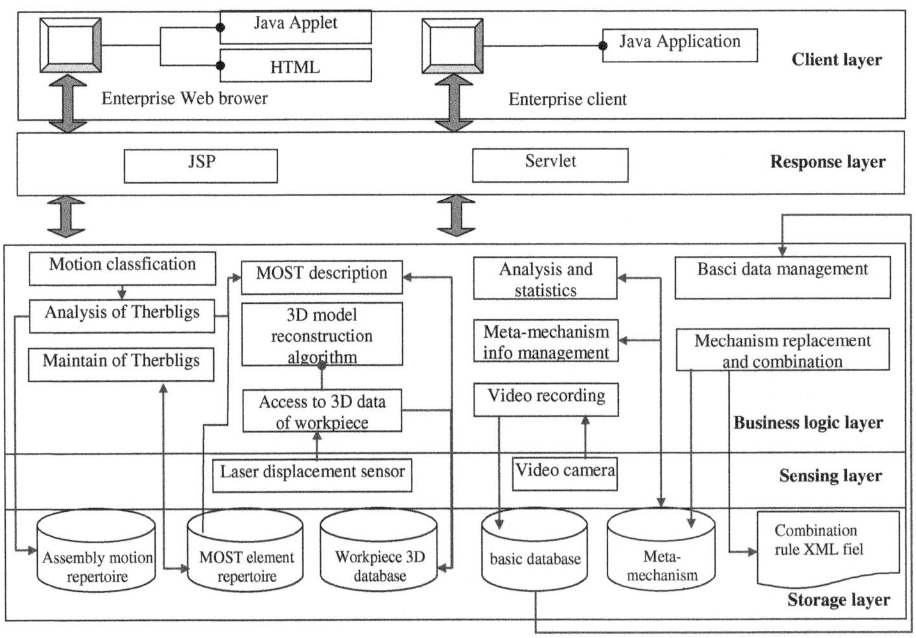

Figure 4. The framework of assembling motion analysis and meta-mechanisms.

3.2. Response layer

Response layer is used to receive the requests from the browser and give it to the bottom layer to handle, and send the result to the browser. The progress is made up of Java Server Page and Servlet that shows the information about the client layer. The request to accept layer can handle the data logically, for example, data calibration and the check of client browser. This layer cannot deal with complicated logical processing.

3.3. The business logic layer

The business logic layer is the core of the platform. It runs on the application server of Java and all logical processes are sealed in the EJB elements. The application server provides EJB elements with an optimized execution environment. The business processes are: engineering technicians use video cameras to record the assembling operation. They get 3D data of workpieces with a laser displacement sensor. Then, they use the algorithm of 3D reconstruction to reconstruct the 3D model of workpieces and get the changed information about positions of instruments. Changed positional information and the video files are synchronous. They utilize the classification of motions Bean to determine the sort of operations. They define and disassemble the assembly operations by the means of motion analysis. Then, get the standard database of MOST-elements. They use human–computer interaction to get the corresponding MOST-expression and its validity, and optimize the assembly operations. The standard database of MOST-elements can be controlled by the maintenance of therbligs. The information management module of meta-mechanism is in charge of the information maintenance of meta-mechanism in general assembling lines. Of course, this information is saved in meta-mechanism repository. The meta-mechanism combination and replacement are responsible for calculating the matched-degree between assembly operations and the typical meta-mechanisms. If the matched-degree meets the requirements, the mechanism will be replaced. After that, the operator will get changed positional 3D information of workpieces and the technical parameters. Otherwise, the operator will analyze XML, which is the rule of meta-mechanism combination, then the manual assembling operation is substituted by the valid combination of meta-mechanisms. Basic data management is responsible for personnel management, role management, user management, organizational management, and management of video files. The analytic and statistical model gets information of the assembly motion repository, foundation database, and meta-mechanism repository, so that it can analyze the operation of the assembling line and get reports and statistics of meta-mechanism replacement.

3.4. Sensing layer

The sensing layer involves two kinds of hardware: the video camera and laser displacement sensor. Video cameras record the assembly operation and send it to the database by Wi-Fi. The laser displacement sensor sends the collected 3D data to the server synchronously and stores the positional information of workpieces by the means of the reconstruction of a 3D model.

3.5. The data storage layer

The data storage layer provides the assembling motion analysis and the meta-mechanism combination and replacement with infrastructural support. It comprises assembling motion repository, MOST-element repository, 3D database of workpieces, foundation database, meta-mechanism repository, and XML files (the rule of meta-mechanism combination). The data storage layer is responsible for the storage of data, inquiring data, and backups. It maintains consistency and safety and provides the application service layer with data services. Assembling motion repository stores the motions' classifications, names, numbers, operating time, MOST-expressions, and therbligs. MOST-element repository stores the basic data table and consists of general motion table, controlling motion table, and instrument table. The 3D database of workpieces stores the changed positional information of workpieces. Fundamental database stores users, staff, roles, and motion video files. Meta-mechanism repository stores types, names, application, key parameters, and 2D images of the meta-mechanism that assembly line usually uses.

4. The mechanism replacement of a station in PS4 handles assembly line

There are some preconditions to realize meta-mechanism replacement of 'Installing R1/R2 buttons' workstation in PS4 handles line:

(1) Realize arrangement and classification of the meta-mechanism repository of productive operation in assembly line.

The meta-mechanism combination in an assembly line can be divided into five categories according to different ways of working: feeding mechanism, selecting mechanism, carrying mechanism, grasping mechanism, and assembly mechanism. The letters in parentheses express the corresponding acronym of each mechanism.

(2) Analyze the scope of application about meta-mechanism combination.

Analyze and research the way of motion and applicable object of meta-mechanism combination, thereby mainly confirm the range of application of meta-mechanism.

(3) Complete the matching process in MOST of commonly used productive mechanism.

Achieve the matching of MOST expression to each of the operating mechanism.

(4) The MOST expression of key motions.

Extract key motions of MOST from the each mechanism, and delete invalid motions, which can be easy to filtrate the meta-mechanism combination later.

(5) The index extracting of key motion of the operation mechanism.

Extract the key motion from the MOST expression (or MOST formula), which can be easy to improve the accuracy of matching.

After the above procedures, we can begin to replace the motion of 'Install R1/R2 buttons' workstation with corresponding meta-mechanism. The process can be achieved according to Table 2, and its detail description is shown as following steps:

(1) Observe the movement of objects.

Observing the operator's working process on 'Install R1/R2 buttons' workstation, and analyze the motion of the workpieces carefully, which can narrow down the range of selected mechanisms.

(2) Express by MOST.

The 'R1/R2 buttons' station is mainly composed of six motions, whose specific names and the MOST formulas are shown in Table 3.

(3) Select the key motions from MOST formulas.

The method of selecting key motions from MOST formulas is deleting the invalid motion sequences, and selecting the key motions, then we will get the key MOST motions as shown in Table 4.

(4) Extract the index from the key motions.

When the process of selecting key motions is finished, we can begin to extract the key index from these key motions. The results are shown in Table 5.

(5) Select the meta-mechanism.

When the above steps are completed, we can find the corresponding meta-mechanism or its combinations to replace the motions. Through the analysis we can obtain that six motions can be replaced by four different mechanisms, and the specific results are shown in Table 6.

Table 2. The replacement process of 'Install R1/R2 buttons' station.

Steps	Step name	Operation process
1	Observe the movement of objects	Observe the movement of objects, then grasp the mode of motion
2	Express by MOST	Analyze the operator's MOST motion on the 'Install R1/R2 buttons' workstation, and write down the working process
3	Select key motions from MOST formulas	Select key motions from MOST formulas, and must ensure they are effective ones
4	Extract the index from the key motion	Extract the index from the key motion which has got from the MOST formulas
5	Select the meta-mechanism	According to the MOST formulas, key MOST motion, its index and the motion-object, consult the corresponding meta-mechanism combination, then select the suitable mechanism-combination to replace them
6	Sort the mechanism combination	More than one motions often correspond to the different sets of mechanisms, so sometimes need to combine the multiple mechanisms to reduce input costs

Table 3. Express the motions of 'Install R1/R2 buttons' by MOST formulas.

Station name	No.	Motion name	MOST formula
Install R1/R2 buttons	1	Get a qualified product from the front workstation	A1B0G1T3A1B0P1A0
	2	Install the button 'R1' (right)	A1B0G1A1B0P3A0B0G0M1X0I3A0
	3	Get the button 'R2' and install a button spring in it (right)	A1B0G1(A1B0G3)A1B0P3A0
	4	Install the button 'R2' in the main holder(right)	A0B0G1A0B0P3A0B0G0M1X0I3A0
	5	Use adsorption-fixture to absorb a loudspeaker and install it	A1B0G1A1B0P1A1B0G3A1B0P3A1B0P1A1
	6	Push the product to the next workstation	A1B0G1M1X0I1A1(A1B0G1A1B0P1A1)

Table 4. Select the key motions of 'Install R1/R2 buttons' workstation.

Station name	No.	Motion name	Key motions
Install R1/R2 buttons	1	Get a qualified product from the front workstation	ABG/ABP
	2	Install the button 'R1' (right)	ABG/MXI
	3	Get the button 'R2' and install a button spring in it (right)	ABG/ABP
	4	Install the button 'R2' in the main holder (right)	ABG/MXI
	5	Use adsorption-fixture to absorb a loudspeaker and install it	ABG/ABP
	6	Push the product to the next workstation	ABG/MXI

It is observed that there are four different replacing mechanisms, and two of them are repeated, which can be achieved through a simple transformation on the primary mechanisms.

(6) Sort the mechanism combination.

Because the mechanism replacement involves four mechanisms, therefore, we can renew to combine and simplify them.

Table 5. Extract the index from key motions of 'Install R1/R2 buttons' work station.

Station name	No.	Motion name	Index of key motions
Install R1/R2 buttons	1	Get a qualified product from the front workstation	G1/T3
	2	Install the button 'R1' (right)	M1/X0/I3
	3	Get the button 'R2' and install a button spring in it (right)	G1/P3
	4	Install the button 'R2' in the main holder (right)	M1/X0/I3
	5	Use adsorption-fixture to absorb a loudspeaker and install it	G1/G3/P1/P3
	6	Push the product to the next workstation	M1/X0/I1

Table 6. The mechanism replacement of 'Install R1/R2 buttons' workstation.

Station name	No.	MOST formula	Replacement of meta-mechanism
Install R1/R2 buttons	1	A1B0G1T3A1B0P1A0	The stepping feeder-mechanism on the conveyor
	2	A1B0G1A1B0P3A0B0G0M1X0I3A0	The feeder-mechanism with cam-and-lever mechanism used by conveying billet
	3	A1B0G1(A1B0G3)A1B0P3A0	The feeder-mechanism with cam-and-lever mechanism used by conveying billet
	4	A0B0G1A0B0P3A0B0G0M1X0I3A0	The intermittent rotary conveying mechanism used in packaging cardboard
	5	A1B0G1A1B0P1A1B0G3A1B0P3A1B0P1A1	The intermittent rotary conveying mechanism used in packaging cardboard
	6	A1B0G1M1X0I1A1(A1B0G1A1B0P1A1)	The conveying mechanism with slider-crank

Note: Sometimes, in order to achieve the replacement of similar motions, we can simply change the mechanism structure.

5. Conclusions

Though the motion analysis is increasingly applied in enterprise production lines, it is still limited in follow aspects: relieving operational fatigue, improving operating process, measuring, and controlling the labor-hour. The motion analysis and the meta-mechanism replacement will improve the theory of motion analysis, expand the ranges of application of motion analysis, and find a good way to save manpower and costs.

Disclosure statement

No potential conflict of interest was reported by the author.

Funding

This work was supported by China Shaanxi Provincial Non-traditional Machining Key Laboratory Open Fund (ST-12010).

References

Du, Z. (2013). Motion analysis of application research in China. *Journal of Mechanical Management Development, 1* (1), 1–2.

Gu, T., & Zhou, B. (2009). The application of motion analysis in cylinder production line. *Journal of Mechanical Manufacturing, 8* (540), 1–2, 47.

Hirosea, K., Dokia, H., & Kondoa, A. (2013). Dynamic analysis and motion measurement of ski turns using inertial and force sensors. *Procedia Engineering, 60*, 355–360. Retrieved from http://www.sciencedirect.com/science/article/pii/S1877705813011338

Hua, J. (2012). *Optimization in the idea of lean production to the fixed telephone Bit Electronics*. Shandong province: Shandong University of Science and Technology.

Jiang, J., & Sui, R. (2012). Automatic assembly of combined checking fixture for auto-body components based on fixture elements libraries. *College of Mechanical and Electrical Engineering, 6* (30), 2–6.

Li, J., & Gui, W. (2008). The applications of methods research in G company operating standards decision. *Journal of Guilin University of Electronic Science and Technology, 28*, 2–10.

Liao, Y., & Sun, H. (2009). The application research of drug packaging assistive technology. *Journal of Packaging Engineering, 30*, 199–201.

Meng, X. (1994). *The modern mechanism manual* (pp. 152–229). Beijing: The Press of Mechanical Industry.

Son, C. (2005). Comparison of intelligent control planning algorithms for robot's part micro-assembly task. *Engineering Applications of Artificial Intelligence, 7*(19), 1–4.

Wang, F. (2010). *The technology and prospect of low cost automation concept.* Beijing: Tsinghua University.

Wang, J. (2012). *Technology research to the splicing and fusion of mass point-clouds model* (p. 10). Zhejiang: Zhejiang University of Technology.

Wei, C. (1994). Intelligent machine design with object-oriented design and constraints satisfaction mechanism. In *Proceedings of the third international conference on automation technology*, Taipei, China.

Yang, P. (2010). Drug packaging operation process analysis and optimization on the basis of motion research. *Journal of Packaging Engineering, 7*, 31.

Yang, Q., Wu, D. L., Zhu, H. M., Bao, J. S., & Wei, Z. H. (2013). Assembly operation process planning by mapping a virtual assembly simulation to real operation. *Computers in Industry, 64*, 869–879.

Zhang, F. (2013). *Research of point-clouds data processing with geometric method.* Xian: Northwestern University.

2

An extension for dynamic lot-sizing heuristics

Fabian G. Beck , Eric H. Grosse and Ruben Teßmann

Carlo and Karin Giersch Endowed Chair "Business Management: Industrial Management", Technische Universität Darmstadt, Hochschulstr. 1, 64289 Darmstadt, Germany

This paper presents an efficient procedure to extend dynamic lot-sizing heuristics that has been overlooked by inventory management literature and practice. Its intention is to show that the extension improves the results of basic heuristics significantly. We first present a comprehensive description of the extension procedure and then test its performance in an extensive numerical study. Our analysis shows that the extension is an efficient tool to improve basic dynamic lot-sizing heuristics. The results of the paper may be used in inventory management to assist researchers in selecting dynamic lot-sizing heuristics and may be of help for practitioners as decision support.

Keywords: EOQ; dynamic lot-sizing; heuristics

1. Introduction

Changes in the competitive environment, such as high time pressure and variations in the demand pattern of the customers, induce companies to simultaneously reduce the costs and increase the service level and quality of production and manufacturing processes. For this reason, the management of inventories is essential as inventories directly influence product and raw material availability as well as production lead time. It has a high cost impact and is therefore one of the main operational activities in the manufacturing industry. The management of inventories requires determining economic batch sizes by balancing inventory holding and setup/order costs with the objective of providing a high service level at minimal total costs. For its relevance in industry, it is not surprising that the lot-sizing problem has lost none of its attention in research and practice since the publishing of the first decision model to determine the economic order quantity (EOQ) (Harris, 1913).

The EOQ model is a simple, robust and efficient tool for companies to resolve the conflict between ordering and inventory costs (Dobson, 1988). While a plethora of extensions of the EOQ model exists and continues to be developed with a wide pallet of theoretical orientations from multi-stage models to incentives and productivity issues (see, for reviews, Brahimi, Dauzere-Prese, Najid, & Nordli, 2006; Glock, 2012; Glock, Grosse, & Ries, 2014; Jans & Degraeve, 2008), the basic case with deterministic time-varying demand, also known as dynamic lot-sizing problem, was solved more than 50 years ago (Wagner & Whitin, 1958). A typical dynamic lot-sizing decision is to determine the amount and timing of replenishment of items, with known but varying

*Corresponding author. Email: beck@bwl.tu-darmstadt.de

demand and fixed ordering and linear holding costs. Although the Wagner–Whitin (WW) algorithm derives optimal solutions for this problem, it is only very infrequently used in practice where heuristics are applied more commonly. One may ask why anyone would use a heuristic when a practical optimal algorithm is available. This is partly due to the fact that the algorithm is not commonly known in practice and that heuristics are simpler and faster to compute (Boe & Yilmaz, 1983). However, proper computer implementation, meanwhile, makes it possible that the WW algorithm runs in linear time (Karimi, Fatemi Ghomi, & Wilson, 2003; Saydam & Evans, 1990). Nevertheless, dynamic lot-sizing heuristics still provide value for practical applications, research and educational exercises (Simpson, 2001). For example, material requirements planning modules in enterprise resource planning (ERP) software, to the best of the authors' knowledge, do not provide the WW method (Bahl & Bahl, 2009). Instead, some of the heuristic methods that may lead to poor results are covered. This is consistent with the findings that poorer rules are well known in educational texts, while other more efficient rules have not received attention in research and practice (Simpson, 2001).

The paper at hand builds on this line of thought and presents an efficient extension algorithm for dynamic lot-sizing heuristics published in 1997 but inexplicably overlooked by inventory management literature. We test the extension for two basic and popular heuristics and illustrate its performance and robustness in an extensive numerical study. Thus, the intention of this paper is to introduce the algorithm and to show that it improves existing heuristics significantly. In addition, we present a practical approach as we implement the simulation study using the environment of MS Excel. This paper may support researchers in selecting heuristics for dynamic lot-sizing problems, and may be of help for practitioners to support inventory management decisions.

The remainder of the paper is structured as follows: The next section reviews related literature. Subsequently, formulations for the dynamic lot-sizing models and algorithms under study are presented in Section 3. This is followed by an extensive numerical analysis in Section 4 to evaluate the performance of the heuristics, and managerial implications are deduced from its results. Section 5 concludes the paper and provides suggestions for future research.

2. Literature review

Reviewing the overall literature on dynamic lot sizing is not within the scope of this paper due to the sheer number of works available. The reader may refer to Robinson, Narayanan, and Sahin (2009) for a review and Andriolo, Battini, Grubbström, Persona, and Sgarbossa (2014) as well as Glock et al. (2014) for recent overviews and surveys on dynamic lot sizing. In this paper, we present an overview of works that concentrated on the comparison and evaluation of dynamic lot-sizing heuristics and algorithms.

To find an optimal solution in the dynamic lot-sizing problem, dynamic programing was used (Wagner & Whitin, 1958). However, several authors developed heuristic solution procedures to solve the dynamic lot-sizing problem. A description and classification scheme of popular heuristics, such as least unit cost (LUC) (Gorham, 1968), Silver–Meal (SM) (Silver & Meal, 1973) and part-period balancing (PP) (DeMatteis, 1968), was given in the review paper of De Bodt, Gelders, and Van Wassenhove (1984). The authors noted that choosing a suitable heuristic for a specific application is not straightforward as cost structure and variability of demand determine how heuristics perform. Bitran, Magnanti, and Yanasse (1984) analytically derived worst case error bounds for LUC and PP. Early comprehensive overviews and performance evaluations of dynamic lot-sizing procedures, including PP, SM, and WW, were developed by Blackburn and Millen (1980), Axsäter (1982, 1985), Baker (1989) and Ritchie and Tsado (1986). Blackburn and Millen (1985) added other algorithms, such as Groff's rule (GR) (Groff, 1979), to previous performance evaluations. Another early classification scheme of basic heuristics was developed by Maes and Van Wassenhove (1988). The authors confirmed that the selection of heuristics depends on environmental characteristics and that universal conclusions about the performance of an algorithm are difficult to draw. Another comparison of dynamic lot-sizing techniques (including LUC, PP, SM and GR) with extensive numerical evaluation can be found in Zoller and Robrade (1988). A worst case and performance analysis of dynamic lot-sizing heuristics, including LUC, PP, and SM, was studied in Vachani (1992). Gupta, Keung, and Gupta (1992) compared the performance of dynamic lot-sizing heuristics, such as LUC and GR, in a multi-stage system. A comprehensive sensitivity analysis of popular dynamic lot-sizing heuristics, among them LUC and SM, was conducted by Pan (1994). Ganas and Papachristos (1997) presented an analytical evaluation and comparison of SM and PP. Kazan, Nagi, and Rump (2000) extended and compared the performance of WW and SM to account for unexpected changes in previously set production and setup schedules. Jeunet and Jonard (2000) evaluated the robustness of lot-sizing techniques, including WW, PP and LUC, taking into account uncertain environments. Simpson (2001) extensively evaluated WW, GR, SM and LUC, among others, in a computational study and concluded that there exist some inconsistencies in dynamic lot-sizing heuristics and that these algorithms are equivalent in performance. The author noted that, curiously enough, some algorithms that are known to generate poorer results are more popular in research and practice than other algorithms that outperform the well-known heuristics. A comparison of dynamic lot-sizing techniques (among them WW and SM), taking into account the so-called end-effect, was presented by Van Den Heuvel and Wagelmans (2005). Gafaar (2006) used SM as benchmark to test the applicability of a genetic algorithm to the dynamic lot-sizing problem with the special case of batch ordering. More recently, Jans and Degraeve (2007) presented a summary and comparison of meta-heuristic solution algorithms for the dynamic lot-sizing problem. Ho, Solis, and Chang (2007) compared the performance of SM, PP, and least total cost (LTC) for the special case of deteriorating inventory. Teunter, Bayindir, and Van Den Heuvel (2006) and Schulz (2011) extended some heuristics, such as SM and LUC, to take into account product returns and remanufacturing considerations. Another modification of dynamic lot-sizing heuristics, among them SM, LUC, and PP, was presented by Toy and Berk (2013), who studied special 'warm/cold' production processes. In a current study, Baciarello, D'Avino, Onori, and Schiraldi (2013) presented a comparison of basic lot-sizing heuristics (among them PP, SM, and GR) in a computational study using the results of WW as benchmark. A recent comparison of dynamic lot-sizing algorithms, including SM and WW, for the special case of convex production costs and setups, can be found in Kian, Gürler, and Berk (2014).

From the literature review, we can deduce that most studies compared and tested the same popular heuristics, such as LUC, SM, or GR, while other heuristics and possible extensions have mainly been overlooked (Simpson, 2001). One unrecognized algorithm that extends basic dynamic lot-sizing heuristics is addressed in this paper. Leinz, Bossert, and Habenicht (1997) introduced a procedure which improves the results of stop rules using a combination of stop rules and cost comparisons. Interestingly enough, this extension algorithm has not received, to the best of the authors' knowledge, any attention in research and practice since the publishing of the working paper (in German) in 1997. We present a formal description of the algorithm, in the following denoted as LBH and test its performance in an extensive numerical study.

3. Dynamic lot-sizing algorithms under study

In this chapter, we give a short description of the dynamic lot-sizing algorithms under study. For our comparative analysis, we choose WW, LUC and GR as the literature review showed that these algorithms are popular in dynamic lot sizing. Special attention is paid to the LBH extension algorithm and how it can improve basic heuristics (here: LUC and GR).

3.1. Definitions

The following definitions are made for the single-level dynamic lot-sizing problem:

- discrete time periods of equal duration over a finite planning horizon;
- demand is deterministic with time-varying feature;
- full demand occurs at the beginning of each period;
- all period demands and costs are non-negative;
- lead time is constant (fixed and known);
- order costs are fixed for every order;
- replenishment rate is infinite;
- stock is empty at the beginning and the end of the planning horizon;
- lot-splitting is not allowed;
- no shortages are allowed;
- all products are considered separately;
- holding costs only have to be determined for a stock which is carried over from one period to another;
- holding costs for carrying inventory during the consumption period are not considered.

The following terminology will be used throughout the paper:

l	last period of a preliminarily considered order cycle
l^*	final resulting last period which is calculated by a heuristic; the related demand d_{l^*} in period l^* is added to the order in period b
b	first period of an order cycle
$\tau_{b,l}$	order cycle range including period b to l
$\Delta\tau$	difference in the average order cycle range between a heuristic and WW
y_t	binary tag in period t
$c_{b,l}$	total cost per unit
c_O	fixed cost per order
c_H	holding cost per unit and period
C_O	total ordering costs
ΔC_0	approximated marginal savings in ordering costs
C_H	total holding costs
ΔC_H	approximated marginal-added holding costs
d_t	demand in period t

t	index for time periods $t = 1, \ldots, T$
x_t	amount ordered in period t
n_i	order frequency for algorithm i until the planning horizon T is reached
d	demand rate
q	order quantity in the EOQ-model
I_t	inventory at the end of period t
C	total costs
C_t	total costs in period t
C^P	preliminary total costs for the first order
C_a	total costs of two aggregated orders, $a \in \{\psi, \phi, \theta\}$
pp	part-period quotient
κ	relative difference in the cost variance between a heuristic and the WW algorithm
$\bar{\kappa}_{\text{erratic}}$	arithmetical mean of κ that is required for the calculations with erratic demand structure
T	end of planning horizon
X, Y	random variables
μ	expected value for uniform and normal distribution
σ^2	variance for uniform and normal distribution

3.2. Wagner–Whitin (WW)

The WW algorithm (Wagner & Whitin, 1958) leads to optimal solutions for the problem under study and is used in this paper to calculate reference values to evaluate the heuristic algorithms. It can be denoted as follows:

$$C = \sum_{t=1}^{T} C_t(x_t, I_t) = \sum_{t=1}^{T} (y_t(x_t) \cdot c_O + I_t \cdot c_H) \rightarrow \text{Min!}$$

subject to

$$I_t = I_{t-1} + x_t - d_t$$

$$I_t, x_t \geq 0, \quad 1 \leq t \leq T,$$

where

$$y_t(x_t) = \begin{cases} 0, & \text{if} \quad x_t = 0 \\ 1, & \text{if} \quad x_t > 0. \end{cases}$$

Without loss of generality, we take $I_0 = I_T = 0$ (Baker, 1989).

Because the WW algorithm is established in the literature, the reader may refer to Wagner and Whitin (1958), De Bodt et al. (1984) and Gupta and Keung (1990), among others, for a detailed description of the algorithm.

3.3. Least unit cost (LUC)

LUC (Gorham, 1968) is a heuristic algorithm that is based on one of the characteristics of the EOQ model, which says that the order quantity is equal, regardless of whether the unit costs or the total costs are minimized. A proof is given in Appendix 1. Because of that characteristic, the LUC heuristic calculates the unit costs $c_{b,l}$ from the beginning period $b = 1$ and adds order quantities as long as the unit costs are decreasing. When the costs are increasing, for the first time, the cumulated order size is ordered and the calculation of the unit costs starts again in the period $b = l^* + 1$ which is the first one not considered in the order before. The heuristic ends when the planning horizon is reached. The algorithm is expressed as follows:

$$c_{b,l} = \frac{c_O + c_H \cdot \sum_{t=b}^{l} (t-b) \cdot d_t}{\sum_{t=b}^{l} d_t}$$

The stop rule $c_{b,l^*-1} \geq c_{b,l^*} < c_{b,l^*+1}$ selects the final resulting last period l^* which leads to the following order quantity in period b:

$$x_b = \sum_{t=b}^{l^*} d_t$$

3.4. *Groff's rule (GR)*

One of the heuristic algorithms that has proven to lead to good results for the dynamic lot-sizing problem is GR (Groff, 1979; Simpson, 2001). It uses another feature of the optimal order quantity in the EOQ-model. For the optimal order quantity, the marginal holding costs and the marginal ordering costs are equal according to amount. A proof is given in Appendix 2. The heuristic starts in period $b = 1$ and ends when the planning horizon is reached. The order quantity is increased as long as the approximated marginal savings in ordering costs ΔC_O from adding the l-th period's demand exceed the approximated marginal holding costs ΔC_H. These two marginal costs are calculated as follows:

$$\Delta C_O = \frac{c_O}{(l-b-1)} - \frac{c_O}{(l-b)} = \frac{c_O}{(l-b-1) \cdot (l-b)}$$

$$\Delta C_H = \frac{d_l \cdot c_H}{2}$$

As soon as the stop rule $\frac{c_O}{(l^*-b) \cdot (l^*-b+1)} - \frac{d_{l^*+1} \cdot c_H}{2} \leq 0$ is reached, the final resulting last period l^* is selected and the order quantity for period b can be computed. The related order quantity is $x_b = \sum_{t=b}^{l^*} d_t$. The calculations are continued in period $b = l^* + 1$ until $l = T$ and the stop rule is not reached.

3.5. *Leinz–Bossert–Habenicht (LBH)*

The LBH extension algorithm can be described in five steps (Leinz et al., 1997).

Let $b = 1$ and $l = 1$:

- Step 1: Determine the first two order cycle ranges, τ_{b,l^*} and $\tau_{l^*+1,l^{**}}$ by starting in period b and using a stop rule such as the ones of LUC or GR. These results are cumulated to a preliminary order cycle range $\tau_{b,l^{**}} = \tau_{b,l^*} + \tau_{l^*+1,l^{**}} = l^{**} - b + 1$.
- Step 2: Set $\psi = \tau_{b,l^{**}}$. The associated costs can be calculated using
  $C_\psi = c_O + c_H \cdot \sum_{t=b}^{b+\psi-1} [(t-b) \cdot d_t]$.
  An attempt is made to decrease these costs by dividing the preliminary order cycle range into two parts which leads to $\psi - 1$ possible combinations of two cycles. The resulting costs can be determined as follows:
  $C_\phi = c_O + c_H \cdot \sum_{t=b}^{b+\phi-1} [(t-b) \cdot d_t] + c_O + c_H \cdot \sum_{t=b+\phi}^{b+\psi-1} [(t-(b+\phi)) \cdot d_t]$
  with $\phi = 1, 2, \ldots, \psi - 1$.
  After these calculations, choose $C^* = \min_\phi \{C_\phi\}$ with $\phi = 1, 2, \ldots, \psi$ so that $\tau^*_{b,b+\phi^*-1} = \phi^*$. The associated preliminary total costs for the first order can then be calculated with
  $C^P = c_O + c_H \cdot \sum_{t=b}^{b+\phi^*-1} [(t-b) \cdot d_t]$.
  If $\phi^* \neq \psi$ respectively $\phi^* > 1$ go to step 3 otherwise go to step 4.

- Step 3: Another attempt is made to decrease the total costs by dividing the first calculated order cycle range $\tau^*_{b,b+\phi^*-1}$ into two orders. To this end, the following costs have to be calculated:
  $C_\theta = c_O + c_H \cdot \sum_{t=b}^{b+\theta-1} [(t-b) \cdot d_t] + c_O + c_H \cdot \sum_{t=b+\theta}^{b+\phi^*-1} [(t-(b+\theta)) \cdot d_t]$
  with $\theta = 1, 2, \ldots, \phi^* - 1$.
  Determine $C^{**} = \min_\theta \{C_\theta\}$ with $\theta = 1, 2, \ldots, \phi^* - 1$.
- Step 4: Another case differentiation has to be made in this step.
  If $C^{**} \geq C^P$ or $\phi^* = \psi$ or $\phi^* = 1$, one gets the following order quantity $x_b = \sum_{t=b}^{b+\phi^*-1} d_t$.
  If $C^{**} < C^P$, one gets two different order quantities $x_b = \sum_{t=b}^{b+\theta-1} d_t$ and $x_{b+\theta} = \sum_{t=b+\theta}^{b+\phi^*-1} d_t$.
- Step 5: Set $b = b + \phi^*$ and $l = b + \phi^*$. Go back to step 1 and continue in this manner until the planning horizon is reached.

In the next section, an extensive simulation study is conducted to evaluate the performance of the algorithms.

4. Simulation study

4.1. Description

In the numerical simulation study, LUC, GR and the LBH extension based on LUC (LBH-LUC) as well as on GR (LBH-GR) are examined assuming different demand structures, i.e. constant, systematic and erratic demand (Leinz et al., 1997; Zoller & Robrade, 1988). The algorithms were programed using MS Excel and run on a laptop with Intel core i5 (2G) processer. The following assumptions for all tests are made:

- average demand is around 1000 [units] except all demands with trend structures (see I–VI below);
- *pp*-quotient $\left(pp = \frac{c_O}{c_H}\right)$ can attain the following values:
  $pp \in \{1000; 2500; 5000; 7500; 10,000; 15,000\}$;
- 100 periods of demand are considered ($T = 100$).

Using the assumptions above, six data-sets are created for the case of constant demand with $d_t = 1000$.

For the systematic demand, the trend and season are analyzed for eight systematic demand structures. The demand for each period will be calculated as follows:

(I)	linear increasing trend	$d_t = 1000 + 7.5 \cdot (t-1)$
(II)	progressive trend	$d_t = d_{t-1} \cdot \left[1 + \frac{5+5\cdot\left(\frac{t+99}{100}\right)}{1500}\right]$ with $d_1 = 1000$
(III)	saturation	$d_t = 2000 - \left[1000 \cdot e^{-\frac{(t-1)}{10}}\right]$
(IV)	linear decreasing trend	$d_t = 2000 + 7.5 \cdot (t-1)$
(V)	degressive increasing trend	$d_t = 1000 + \sqrt{5 \cdot d_{t-1} \cdot (t-1)}$ with $d_1 = 1000$
(VI)	degressive decreasing trend	$d_t = 2000 - \sqrt{5 \cdot d_{t-1} \cdot (t-1)}$ with $d_1 = 2000$
(VII)	additive constant season	$d_t = 1000 + s_t$
(VIII)	additive dynamic season	$d_t = 1000 + \frac{t}{25} \cdot s_t$

with $s_t = s_{t-5}; s_1 = -100; s_2 = 50; s_3 = 210; s_4 = 0; s_5 = -120$ for VII and VIII. In combination with the six different *pp*-values, there are 48 data sets to examine.
The erratic demand is simulated by a uniform and a modified normal distribution (MND). The uniformly distributed random variables have an expected value of $\mu = 1000$ and fluctuation ranges of $\{\pm 20\%; \pm 40\%; \pm 60\%; \pm 80\%; \pm 100\%\}$, which lead to the variances $\sigma^2 \in \{13,333;\ 53,333;\ 120,000;\ 213,333;\ 333,333\}$. The modified normal distributed random variables X are generated out of normal distributed random variables Y with an expected value $\mu = 1000$ and a variance $\sigma^2 \in \{10,000;\ 100,000;\ 500,000\}$. The latter assumptions make negative demand values in Y possible, so in case a negative demand occurs, this value is set to zero in X. By this alteration, the initially normally distributed values are actually no longer normally distributed, which also means, that the distribution of X is no longer symmetrical. In fact, there is the following change to the distribution $P_{\mathrm{MND}}(X = 0) = F_{X_{\mathrm{ND}}}(0)$, where $F_{X_{\mathrm{ND}}}(x)$ is the cumulative distribution function of the normal distribution (ND). The probability density function then shows a small rise at $x = 0$ and is no longer defined from $-\infty$ to $+\infty$ but is now defined from 0 to $+\infty$. With respect to the law of large numbers, we calculate 100 data sets for each erratic demand structure.

4.2. *Results*

The costs of the results obtained with a heuristic are compared to the results of the reference procedure WW (Leinz et al., 1997). The total cost variance is measured by the relative difference: $\kappa = \frac{C_{\mathrm{Heuristic}} - C_{\mathrm{WW}}}{C_{\mathrm{WW}}} \times 100[\%]$. Because of the law of large numbers, 100 relative differences κ_i with $i = 1, \ldots, 100$ for each erratic demand structure are created and the arithmetical mean is needed for further calculations. The arithmetical mean can be computed as follows: $\bar{\kappa}_{\mathrm{erratic}} = \frac{1}{100} \cdot \sum_{i=1}^{100} \kappa_i$.

In addition, the average order cycle range difference compared to WW can be calculated as follows: $\Delta\tau = \frac{T}{n_{\mathrm{Heuristic}}} - \frac{T}{n_{\mathrm{WW}}}$ [periods]. Note that results shown in Figures 1–8 are calculated using the arithmetical mean for all test sets with the specific indicated demand structures (in figure captions) and *pp*-values, i.e. $\bar{\kappa}$ and $\overline{\Delta\tau}$, respectively.

The results for the case of constant demand are summarized in Table 1. In this case, other values than $pp = 5000$ lead to no variances compared to WW and are thus not illustrated. However, our results show that *pp*-values of 5000 in combination with a fixed planning horizon lead to variance in total costs and order cycle range compared to WW. A rolling planning horizon, as common in practice, may be suitable to prevent this variance. As can be seen, LBH eliminates this variance as well.

Regarding systematic demand with trend (cf. Section 4.1, No. I to VI), our results show that LBH leads to substantial improvements in the basic heuristics as illustrated in Figure 1. As can be seen, GR generates the worst results in this case. However, even employing large *pp*-values, the cost variance is only slightly above 1%.

As for systematic seasonal demand (cf. Section 4.1, No. VII and VIII), we also observe that LBH performs well and, in this case, GR outperforms LUC. This relative advantage remains applying LBH as LBH-GR leads to better results than LBH-LUC.

Table 1. Results for constant demand.

Heuristic		GR	LUC	LBH-GR	LBH-LUC
$pp = 5000$	κ	.749	.749	0	0
	$\Delta\tau$	−.089	−.089	0	0

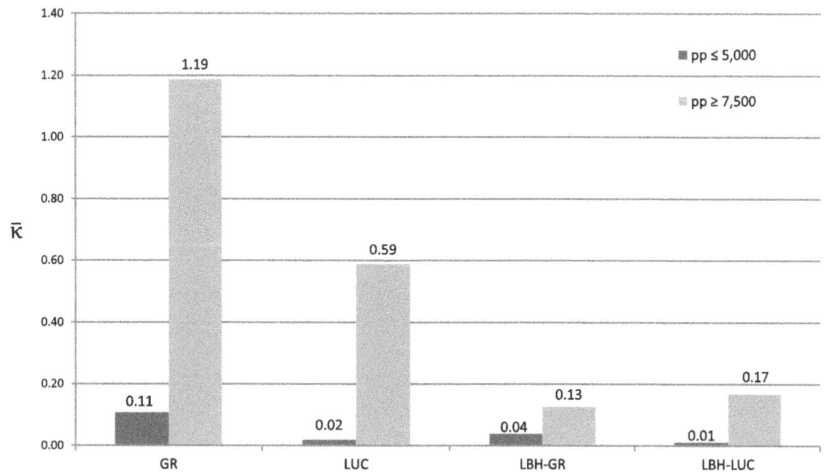

Figure 1. Results for systematic demand with trend structure (demand structures I–VI).

As can be seen, higher cost variances occur compared to the results of systematic demand with trend. This implies that inventory management for seasonal demand is a more challenging task in practice than for systematic demand with trend. However, as our results show, this situation could be enhanced significantly applying the LBH extension algorithm, in particular, LBH-GR as illustrated in Figure 2.

In the case of erratic demand, the results of our study show that higher *pp*-values lead to higher cost variances, except employing LUC, compare therefore Figure 3. Regarding LBH, the results show that average *pp*-values lead to the highest cost variances. As LUC calculates poor results with cost variance more than 6%, LBH-GR outmatches LBH-LUC accordingly.

All tested algorithms generate shorter average order cycle ranges compared to WW, as can be seen in Figure 4. One can observe a tendency that high variance in order

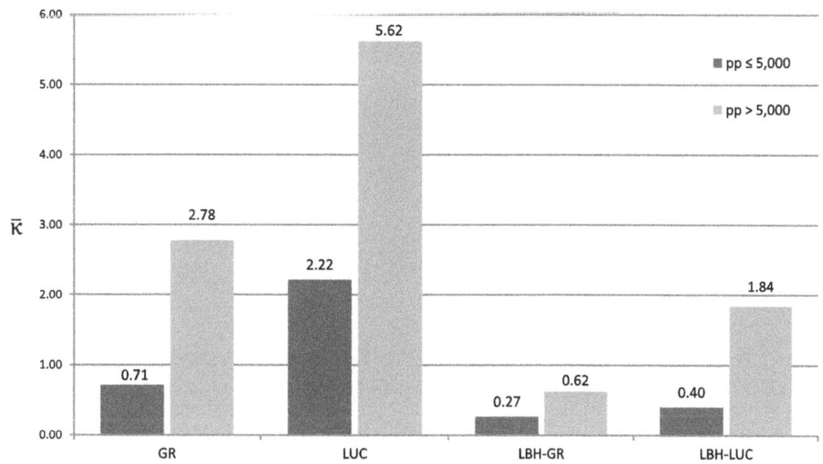

Figure 2. Results for systematic seasonal demand (demand structures VII–VIII).

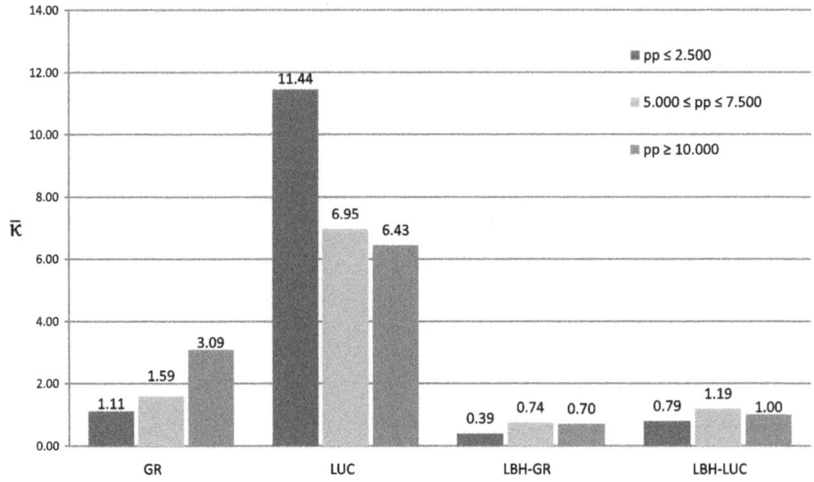

Figure 3. Results for erratic demand and varying *pp*-values.

cycle ranges may lead to high total costs variance. However, as can be seen for LUC (*pp* ≤ 2500) in Figures 3 and 4, for example, a low order cycle range variance may result in high total cost variance. The results in Figure 4 also show that the LBH extension algorithm generates longer order cycle ranges that are then closer to the results of WW compared to the results of the basic heuristics.

In case of $\sigma^2 < 60{,}000$, all heuristics lead to small cost differences apart from LUC, which generates high cost variances. As can be seen in Figure 5, a significant improvement of LUC is possible applying LBH with *pp* ≤ 5000. However, as the basic heuristic calculates poor results, the results of LBH-LUC are still quite far off the ones of WW. It is also notable that LBH-LUC leads to the same cost variances with different *pp*-values. Moreover, we can observe here, that LBH-GR generates the lowest cost

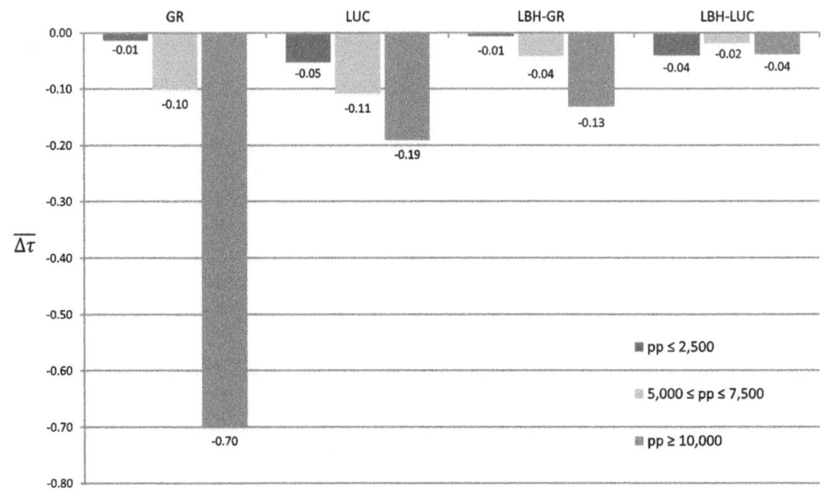

Figure 4. Order cycle range differences for erratic demand and varying *pp*-values.

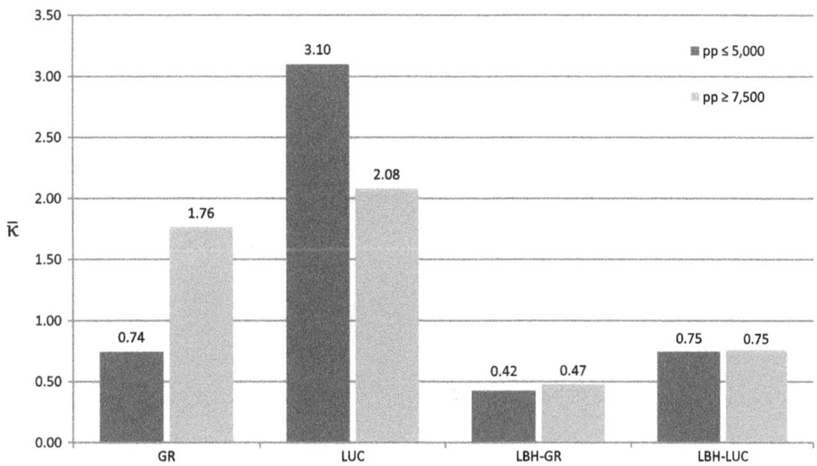

Figure 5. Cost variances for $\sigma^2 < 60{,}000$.

variances compared to WW with all *pp*-values. The additional costs of GR can be considerably reduced applying LBH even with higher *pp*-values.

Most procedures lead to shorter order cycle ranges compared to WW with low *pp*-values and $\sigma^2 < 60{,}000$. In turn, as illustrated in Figure 6, LBH-LUC generates longer order cycle ranges with high *pp*-values. We also observe the tendency that the closer the order cycle range generated by a basic heuristic and a heuristic extended with LBH gets to the order cycle range of WW, the better cost results are obtained (compare Figures 5 and 6). However, there are exceptions, such as LBH-GR with high *pp*-values. An explanation is probably that equal average order cycle ranges to WW are derived, but the order timing is different leading to cost variance of about .5%.

For $\sigma^2 > 200{,}000$, comparable results are obtained to the case of $\sigma^2 < 60{,}000$. However, cost variance is noticeably higher for $\sigma^2 > 200{,}000$, see therefore Figure 7. The only

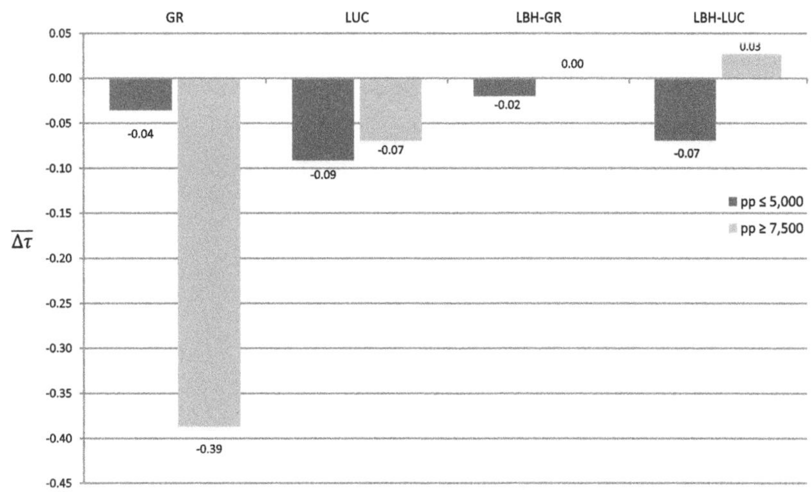

Figure 6. Order cycle ranges for $\sigma^2 < 60{,}000$.

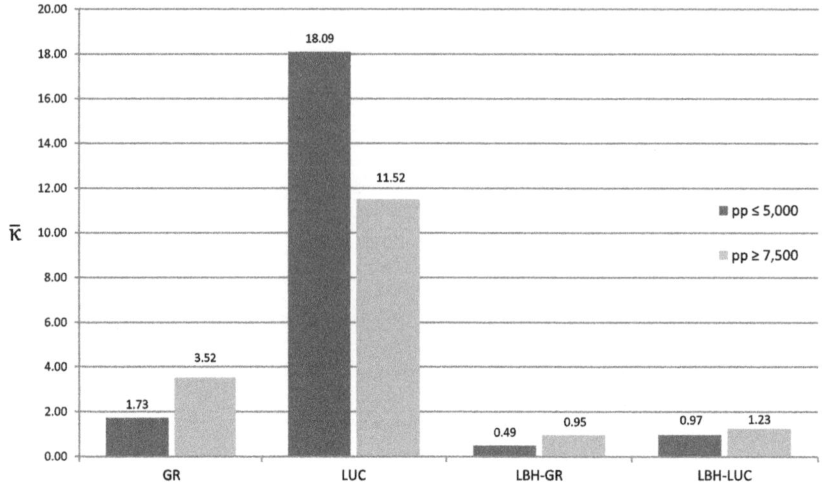

Figure 7. Cost variances for $\sigma^2 > 200{,}000$.

algorithm that leads to cost variances significantly lower than 1% compared to WW is LBH-GR with small *pp*-values. LBH-LUC is about 1% deviation. While GR generates acceptable cost variances with low *pp*-values, high *pp*-values lead in turn to significant poorer results. Obviously, LUC leads to the worst cost variances. Even though cost variance decreases with high *pp*-values, it is still very high. Particularly, in this scenario, one can see the strength of the LBH extension, which leads to considerable improvements in the results.

For the case of $\sigma^2 > 200{,}000$, we can again observe that average order cycle ranges calculated with all tested procedures are too short compared to WW, see Figure 8. Applying LBH leads to extended order cycle ranges. A peculiarity is that the cost results of LUC are better for high *pp*-values than for low *pp*-values (see Figure 7),

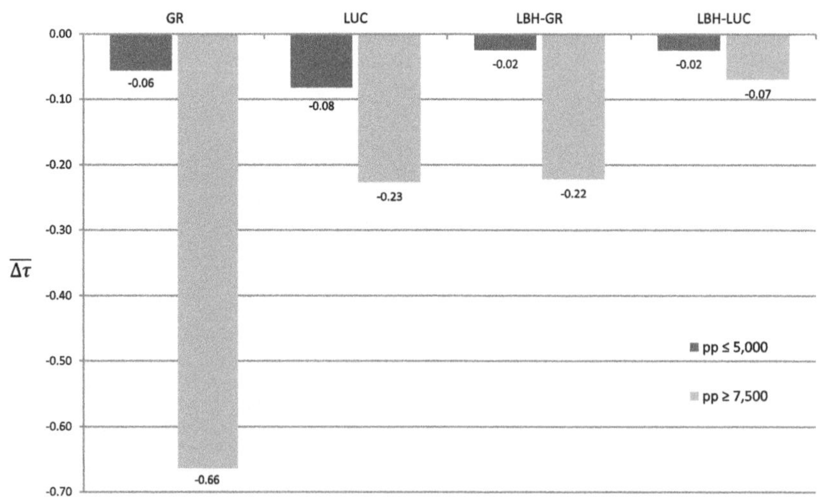

Figure 8. Order cycle ranges for $\sigma^2 > 200{,}000$.

although $\Delta\tau$ is nearer to zero, when *pp*-values are low (see Figure 8). However, we note that a certain variance in average order cycle ranges does not imply a certain goodness of an algorithm regarding cost performance (compare LUC with low *pp*-values).

The results of our study can be summarized as follows:

- LBH eliminates end-of-horizon effects for constant demand.
- All tested algorithms generate average order cycle ranges that are too short compared to WW.
- Average order cycle ranges that are close to WW can only be seen as an indication for good cost performance.
- LUC generates the highest cost variances compared to WW in most tested cases, particularly in the case of $\sigma^2 > 200{,}000$.
- High *pp*-values lead to higher cost variances for almost all tested scenarios in comparison to low *pp*-values. Only LUC leads to partly better results for high *pp*-values.
- The results of all tested heuristics can be significantly improved applying the LBH extension algorithm, and order quantities approximated the results of WW. However, poor results of a basic heuristic can only relatively be improved with LBH.

5. Conclusion

The intention of this paper was to present a comprehensive description of an extension for dynamic lot-sizing heuristics that has been overlooked by literature and practice. For this purpose, we conducted an extensive numerical study to evaluate its performance. The results of our study show that LBH leads to significant cost savings compared to the tested heuristics and may support researchers in improving existing dynamic lot-sizing heuristics. In addition, this paper may be of help for practitioners to support inventory management decisions. As ERP systems, to the best of the authors' knowledge, do not provide the WW algorithm, LBH represents a practical extension to improve the results of implemented procedures. As shown above for LUC and GR, LBH could be easy to integrate into ERP systems, as it may extend and improve any heuristic of the dynamic lot sizing problem significantly. The results of our numerical study showed, for example, that LBH-LUC reduced the cost variance of LUC compared to the optimal solution of WW from 11.4 to .79% (cf. Figure 3). We observed that LBH is adding about 20% computation time, whereas WW adds about 400% to the computation time of GR and LUC, which shows that LBH is suitable in practice. In addition, the results of this paper may be of help in education to advance existing heuristics in the literature.

This paper also has limitations. Generally, the improvements obtained with LBH depend on the results of the basic heuristic. The better the results of the heuristic, (i.e. near to the solution of WW) the better the result with LBH extension is. For example, the lot-for-lot heuristic (Orlicky, 1975) may be improved significantly applying LBH, but would still be far off the solution obtained with WW as LBH would solely consider two order periods, which means $\tau_{b,l^*} = 1$ and $\tau_{l^*+1,l^{**}} = 1$, for all order cycle ranges. In addition, we tested LBH solely within a range of possible heuristics. This and other limiting factors could be addressed in an extension of this paper.

References

Andriolo, A., Battini, D., Grubbström, R. W., Persona, A., & Sgarbossa, F. (2014). A century of evolution from Harris's basic lot size model: Survey and research agenda. *International Journal of Production Economics, 155*, 16–38.

Axsäter, S. (1982). Worst case performance for lot sizing heuristics. *European Journal of Operational Research, 9*, 339–343.

Axsäter, S. (1985). Performance bounds for lot sizing heuristics. *Management Science, 31*, 634–640.

Baciarello, L., D'Avino, M., Onori, R., & Schiraldi, M. M. (2013). Lot sizing heuristics performance. *International Journal of Engineering Business Management, 5*, 1–10.

Bahl, H. C., & Bahl, N. (2009). An empirical comparison of lot-sizing methods available in an ERP System. *California Journal of Operations Management, 7*, 77–83.

Baker, K. R. (1989). Lot-sizing procedures and a standard data set: A reconciliation of the literature. *Journal of Manufacturing and Operations Management, 2*, 199–221.

Bitran, G. R., Magnanti, T. L., & Yanasse, H. H. (1984). Approximation methods for the uncapacitated dynamic lot size problem. *Management Science, 30*, 1121–1140.

Blackburn, J. D., & Millen, R. A. (1980). Heuristic lot-sizing performance in a rolling-schedule environment. *Decision Sciences, 11*, 691–701.

Blackburn, J. D., & Millen, R. A. (1985). A methodology for predicting single-stage lot-sizing performance: Analysis and experiments. *Journal of Operations Management, 5*, 433–448.

Boe, W. J., & Yilmaz, C. (1983). The incremental order quantity. *Production and Inventory Management, 24*, 94–100.

Brahimi, N., Dauzere-Prese, S., Najid, N. M., & Nordli, A. (2006). Single item lot sizing problems. *European Journal of Operational Research, 168*, 1–16.

De Bodt, M. A., Gelders, L. F., & Van Wassenhove, L. N. (1984). Lot sizing under dynamic demand conditions: A review. *Engineering Costs and Production Economics, 8*, 165–187.

DeMatteis, J. J. (1968). An economic lot-sizing technique, I: The part-period algorithm. *IBM Systems Journal, 7*, 30–39.

Dobson, G. (1988). Sensitivity of the EOQ model to parameter estimates. *Operations Research, 36*, 570–574.

Gafaar, L. (2006). Applying genetic algorithms to dynamic lot sizing with batch ordering. *Computers & Industrial Engineering, 51*, 433–444.

Ganas, I. S., & Papachristos, S. (1997). Analytical evaluation of heuristics performance for the single-level lot-sizing problem for products with constant demand. *International Journal of Production Economics, 48*, 129–139.

Glock, C. H. (2012). The joint economic lot size problem: A review. *International Journal of Production Economics, 135*, 671–686.

Glock, C. H., Grosse, E. H., & Ries, J. M. (2014). The lot sizing problem: A tertiary study. *International Journal of Production Economics, 155*, 39–51.

Gorham, T. (1968). Dynamic Order Quantities. *Production and Inventory Management, 9*, 75–81.

Groff, G. K. (1979). A lot sizing rule for time phased component demand. *Production and Inventory Management, 20*, 66–74.

Gupta, Y. G., Keung, Y. K., & Gupta, M. C. (1992). Comparative analysis of lot-sizing models for multi-stage systems: A simulation study. *International Journal of Production Research, 30*, 695–716.

Gupta, Y. P., & Keung, Y. (1990). A review of multi-stage lot-sizing models. *International Journal of Operations & Production Management, 10*, 57–73.

Harris, F. W. (1913). How many parts to make at once. *Factory – The Magazine of Management, 10*, 135–136, 152.

Ho, J. C., Solis, A. O., & Chang, Y.-L. (2007). An evaluation of lot-sizing heuristics for deteriorating inventory in material requirements planning systems. *Computers & Operations Research, 34*, 2562–2575.

Jans, R., & Degraeve, Z. (2007). Meta-heuristics for dynamic lot sizing: A review and comparison of solution approaches. *European Journal of Operational Research, 177*, 1855–1875.

Jans, R., & Degraeve, Z. (2008). Modeling industrial lot sizing problems: A review. *International Journal of Production Research, 46*, 1619–1643.

Jeunet, J., & Jonard, N. (2000). Measuring the performance of lot-sizing techniques in uncertain environments. *International Journal of Production Economics, 64*, 197–208.

Karimi, B., Fatemi Ghomi, S. M. T., & Wilson, J. M. (2003). The capacitated lot sizing problem: A review of models and algorithms. *Omega, 31*, 365–378.
Kazan, O., Nagi, R., & Rump, C. M. (2000). New lot-sizing formulations for less nervous production schedules. *Computers & Operations Research, 27*, 1325–1345.
Kian, R., Gürler, Ü., & Berk, E. (2014). The dynamic lot-sizing problem with convex economic production costs and setups. *International Journal of Production Economics, 155*, 361–379.
Leinz, J., Bossert, B., & Habenicht, W. (1997). *Entwicklung eines Verfahrens zur dynamischen einstufigen Einprodukt-Bestellmengenplanung* [Development of a procedure for single-level single-item dynamic lot sizing]. Hohenheim: Arbeitspapier Nr.15 – Universität Hohenheim, Institut für Betriebswirtschaftslehre, Lehrstuhl für Industriebetriebslehre.
Maes, J., & Van Wassenhove, L. (1988). Multi-item single-level capacitated dynamic lot-sizing heuristics: A general review. *Journal of the Operational Research Society, 39*, 991–1004.
Orlicky, J. (1975). *Material requirements planning*. New York, NY: McGraw-Hill.
Pan, C.-H. (1994). Sensitivity analysis of dynamic lot-sizing heuristics. *Omega, 22*, 251–261.
Ritchie, E., & Tsado, A. K. (1986). A review of lot-sizing techniques for deterministic time-varying demand. *Production and Inventory Management, 27*, 65–79.
Robinson, P., Narayanan, A., & Sahin, F. (2009). Coordinated deterministic dynamic demand lot-sizing problem: A review of models and algorithms. *Omega, 37*, 3–15.
Saydam, C., & Evans, J. R. (1990). A comparative performance analysis of the Wagner–Whitin algorithm and lot sizing heuristics. *Computers & Industrial Engineering, 18*, 91–93.
Schulz, T. (2011). A new Silver–Meal based heuristic for the single-item dynamic lot sizing problem with returns and remanufacturing. *International Journal of Production Research, 49*, 2519–2533.
Silver, E. A., & Meal, H. C. (1973). A heuristic for selecting lot size quantities for the case of a deterministic time-varying rate and discrete opportunities for replenishment. *Production and Inventory Management, 12*, 64–74.
Simpson, N. C. (2001). Questioning the relative virtues of dynamic lot sizing rules. *Computers & Operations Research, 28*, 899–914.
Teunter, R. H., Bayindir, Z. P., & Van Den Heuvel, W. (2006). Dynamic lot sizing with product returns and remanufacturing. *International Journal of Production Research, 44*, 4377–4400.
Toy, A. Ö., & Berk, E. (2013). Dynamic lotsizing for a warm/cold process: Heuristics and insights. *International Journal of Production Economics, 145*, 53–66.
Vachani, R. (1992). Performance of heuristics for the uncapacitated lot-size problem. *Naval Research Logistics, 39*, 801–813.
Van Den Heuvel, W., & Wagelmans, A. P. M. (2005). A comparison of methods for lot-sizing in a rolling horizon environment. *Operations Research Letters, 33*, 486–496.
Wagner, H. M., & Whitin, T. M. (1958). Dynamic version of the economic lot size model. *Management Science, 5*, 89–96.
Zoller, K., & Robrade, A. (1988). Dynamic lot sizing techniques: Survey and comparison. *Journal of Operations Management, 7*, 125–148.

Appendix 1

To show: The minimum of the unit costs is equal to the minimum of the total costs in an EOQ model.
Proof:
The total cost function for an EOQ-model is given by

$$C(q) = C_H(q) + C_O(q) = c_H \cdot \frac{q}{2} \cdot T + c_O \cdot \frac{T \cdot d}{q} \rightarrow \text{Min!}$$

The first derivation is set to zero to find the critical point:

$$\frac{dC(q)}{dq} = c_H \cdot \frac{1}{2} \cdot T - c_O \cdot \frac{T \cdot d}{q^2} = 0$$

This leads to the optimal value for the order quantity which can be expressed as follows:

$$q = \sqrt{\frac{2 \cdot d \cdot c_O}{c_H}}$$

The unit costs can be denoted as follows:

$$c(q) = \frac{C(q)}{T \cdot d} = c_H \cdot \frac{q}{2 \cdot d} + c_O \cdot \frac{1}{q} \rightarrow \text{Min!}$$

Here, the first derivation is also needed and has to be set to zero:

$$\frac{dc(q)}{dq} = c_H \cdot \frac{1}{2 \cdot d} - c_O \cdot \frac{1}{q^2} = 0$$

The resulting optimal order quantity is

$$q = \sqrt{\frac{2 \cdot d \cdot c_O}{c_H}}$$

Appendix 2

To show: For the optimal order quantity, the marginal holding costs and the marginal ordering costs are equal according to amount.
Proof:
We need the total cost function of the EOQ-model again which is given by

$$C(q) = C_H(q) + C_O(q) = c_H \cdot \frac{q}{2} \cdot T + c_O \cdot \frac{T \cdot d}{q} \rightarrow \text{Min!}$$

Differentiation and setting equal to zero can be mathematically expressed by

$$\frac{dC(q)}{dq} = c_H \cdot \frac{1}{2} \cdot T - c_O \cdot \frac{T \cdot d}{q^2} = 0$$

And the transposing of the equation leads to

$$c_H \cdot \frac{1}{2} \cdot T = c_O \cdot \frac{T \cdot d}{q^2}$$

$$\left| \frac{dC_H(q)}{dq} \right| = \left| \frac{dC_O(q)}{dq} \right|$$

3

Developing location indicators for Agricultural Service Center: a Delphi–TOPSIS–FAHP approach

Morteza Zangeneh[a]*, Asadolah Akram[a], Peter Nielsen[b] and Alireza Keyhani[a]

[a]*Department of Agricultural Machinery Engineering, Faculty of Agricultural Engineering and Technology, School of Agriculture & Natural Resources, University of Tehran, Karaj, Iran;*
[b]*Department of Mechanical and Manufacturing Engineering, Aalborg University, Aalborg, Denmark*

The main goal of this paper is to construct objectives and attributes of the service centers location problem. This problem is often found within production management as a location problem, when designing for example supply chains or manufacturing layouts. The main contribution of this paper is constructing related location indicators. Since there was no similar literature in this discipline, so a Delphi survey applied to quantity expert's attitudes about location problem of Agricultural Service Center (ASC) and construct location selection attributes and also ASC objectives. A TOPSIS survey is done to rank extracted attributes to import in fuzzy analytical hierarchy process (AHP) study. Then a fuzzy AHP technique is applied to compute the weight of these most important attributes using four objectives, which obtained by Delphi technique too. In the simplest form with this assumption that all objectives have the same priority, the results illustrated that the service, cost, speed, and ASC profit are first to last important objectives, respectively. At the end of this paper, the multi-choice goal programming method recommended to use when the priorities of objectives are not the same. Finally, the weight of ASC location attributes computed to consider in ASC location problem.

Keywords: Agricultural Service Centers (ASC); location; objectives; attributes; service

1. Introduction

Each member of supply chain needs to perform specific functions or activities in value addition process. Performance of supply chain could be improved if supply chain is integrated and the concerned activities are properly coordinated (Shukla, Garg, & Agarwal, 2014). Energy, labor, input costs, markets conditions, crop yielding and prices, machinery service availability, pest and disease infection, environmental conditions, and even economic and political strategies change the agricultural productivity in the supply chain (Lak & Almasi, 2011). Among these dynamic variables, providing agricultural services in the right way can have a strategic role to improve the agricultural productivity. Several services can be given to the whole of agricultural supply chain. These problems can be seen as analog to facility location problems in supply chains or design problems inside physical manufacturing systems. Some services are for farms, specifically. In following section, four types, including input supply, mechanization services,

*Corresponding author. Emails: mzangeneh@ut.ac.ir, morteza@m-tech.aau.dk

advisory services, and financial services, which are considered in this research, are introduced.

A wide range of location problems effect today's firm and organizational objectives and decision-makings. The interest and service level, economic profits, quality of service, security, environmental protection, availability, and optimum performance are some of the many criteria and objectives of the facility location problem; and hence, different methods and approaches have been employed in the literature to maximize the achievement of these objectives (Hosseini-Nasab, Tavana, & Yousefi, 2014).

1.1. Input supply service

Unfortunately after market liberalization, a majority of the farm input supply companies remains concentrated in urban areas or central rural zones. Due to these changes millions of poor farmers in rural areas do not have access to agricultural inputs on time and in right way, such as improved seeds, chemical fertilizers which are needed to help them improve their productivity. So poor development and weak performance of rural agricultural input markets explain the current low productivity of small holder farmers (Dorward & Chirwa, 2011).

1.2. Agricultural mechanization service

The manufacture, distribution, repair, maintenance, management and utilization of agricultural tools, implements, and machines are covered under mechanization services (Lak & Almasi, 2011). The important point in mechanization services is that how supply mechanization services to the farmer in an efficient and effective manner. If mechanization is implemented in the right way, it will have a considerable effect on agricultural productivity improvement (Lak & Almasi, 2011). In many countries, agricultural mechanization has made a significant contribution to the agricultural and rural development. After applying farm mechanization levels of production have increased, soil and water conservation measures were constructed, the profitability of farming improved, the quality of rural life enhanced, and development in the industrial and service sectors was stimulated so mechanization services are highly required to be supplied in an effective way (Bishop, 1997). Inns (1995) mentioned that agricultural mechanization development depends on the farmers' satisfaction and capability to identify opportunities for achieving sustainable benefits by improved and/or increased use of power and machinery, selecting the most worthwhile opportunity and carrying it through to successful implementation. Lack of consideration to the necessity of development in mechanization of agricultural sector, insufficient cooperation between industrial and agricultural sector, and unrealistic selection of goals and objectives and perhaps, more importantly misuse and poor management of resources could all be counted toward the considerable fall back in the agricultural sector (Bagheri & Moazzen, 2009). Sims and Kienzle (2009) have done three mechanization supply chain case studies in three countries. They presented five elements for mechanization supply chain (see Figure 1).

The role and activity of machinery hire service stakeholders should include following aspects: Coordination with other stakeholders; Business management; Quality control; Operator training; Maintenance and servicing; and closely fallow farmers' needs (Sims & Kienzle, 2009). The previous papers did not consider to efficient manner of supplying and distributing the mechanization services for customers. In current research the strategy of providing mechanization services with all other agricultural services is

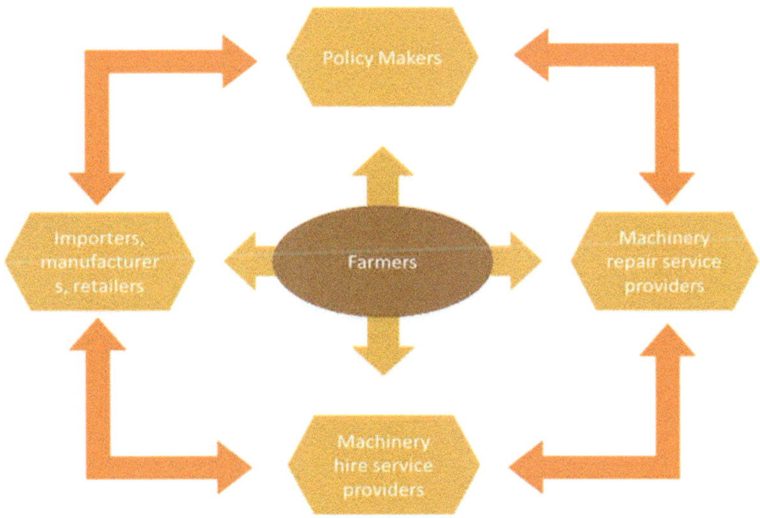

Figure 1. Farm power and machinery supply chain stakeholders (Sims & Kienzle, 2009).

considered. Therefore, supplying high quality mechanization services in coordination with the needs of farms by Agricultural Service Centers (ASCs) can lead to the improvement of productivity in whole supply chain of agricultural production.

1.3. Agricultural advisory service

Agricultural consulting services are known as activities that make new knowledge available to the agricultural producers and assist them to improve their farming and management skills. The services may include: Sharing the information, training and advice of farmers, testing new technologies on farms, and developing farm management tools. The basic indicators for success of a demand-driven advisory service system in agriculture are: (a) farmers have access to agricultural advisory services; (b) farmers use the advisory services; (c) advices lead to income's increase from the agricultural production; and also (d) competition among agricultural advisers (Chipeta, 2006). Altogether, the advisory is a critical need to the agricultural supply chain and the access to this service must be easy to encourage the farmers to use it. This can be obtained by locating service centers in accurate places. The demands from farmers in many cases are different. The advisory service demands which are formulated by various stakeholders of agricultural supply chain can be seen in Figure 2.

1.4. Financial service

Access to external financial resources in agriculture is constrained. This is due to low enterprise profitability in agriculture, accumulated debts, high inflation, risk and uncertainty, and collateral problems (Johan, Swinnen, & Gow, 1999). The limited access to financial resources may lead to several problems in agricultural production such as: disturb the time of farming operations (input supply, planting, trading, and etc.), decrease the input quality, etc. So increasing the access to financial services including loan and insurance can improve the productivity in agriculture.

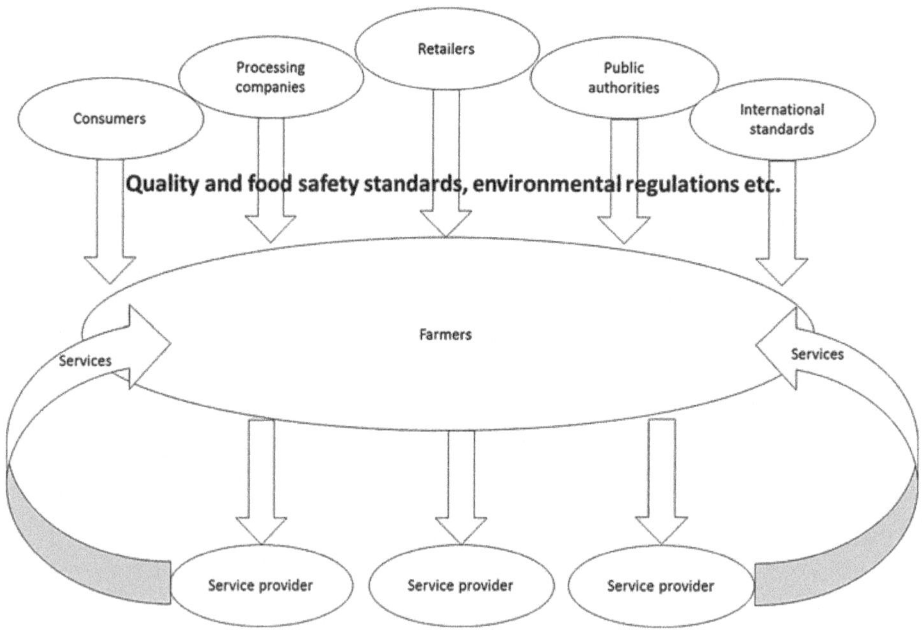

Figure 2. Demands are formulated by various stakeholders (Chipeta, 2006).

1.5. Location problem of ASCs

Agriculture is the only major sector that uses the land surface as an essential input into its production function. This wide geographical dispersion of agricultural production has the important economic consequence; transportation becomes essential. Output must be transported for consumption by others and inputs; such as modern seeds, fertilizer, pesticides, or machinery, and all required agricultural services must be transported to the farm to raise output (Timmer, Falcon, & Pearson, 1983). Therefore, to success in providing and distributing agricultural services for improvement of productivity in this sector, the right location of such service centers should be selected. In this regard, one of the initial steps to solve this location problem is finding related location objectives and attributes to use in multi-criteria decision-making (MCDM) location models. In following section the multi-attribute decision-making (MADM) and multi objective decision-making (MODM) techniques is reviewed and appropriate decision-making technique for ASC location problem is selected.

In a complementary study from the authors of this paper, a MADM approached developed, to use the attributes and objectives extracted in this study, and best candidate locations for establishing service centers for agricultural production identified (Zangeneh, Nielsen, Akram, & Keyhani, 2014).

1.6. MADM and MODM decision-making techniques

There are various techniques for ranking alternatives. In many real-world problems, the decision-maker likes to pursue more than one target or consider more than one factor or measure, which called MODM problem and MADM problem. There are many decision-making problems that their information is spatial and known as location decisions problems. Facility location is a branch of operation's research related to locating or positioning at least a new facility among several existing in order to optimize at least one objective function (Farahani, Steadieseifi, & Asgari, 2010).

There are many techniques which are used to solve the MADM problems. The most popular ones are as follows: lexicographic, permutation method, simple additive weighting (SAW), elimination and choice expressing reality (ELECTRE), technique for order preference by similarity to ideal solution (TOPSIS), linear programming techniques for multidimensional analysis of preference (LINMAP), and interactive SAW method and MDS with the ideal point (Hwang & Yoon, 1981a).

Also there are many techniques which are used for the MODM problems. The most popular ones are as follows: metric LP methods, bounded objective method, lexicographic method, goal programming (GP), goal attainment method, method of Zionts–Wallenius, the methods as step method (STEM) and related method, sequential multi-objective problem solving and sequential information generator for multi-objective problems method, GP STEM, and C-constraint method and adaptive search method (Hwang & Masud, 1979; Szidarovszky, Gershon, & Duchstein, 1986; Ulungu & Teghem, 1994; Zionts, 1979). Also, papers such as the work by Do et al. could be considered relevant as they combine simulation under uncertainty with heuristics to find suitable solutions.

The analytic hierarchy process (AHP) is widely used by authors to solve MADM problems. García et al. (2014) generated a multi-criteria and multi-attribute assessment model that allows selecting the ideal location for warehouses for perishable agricultural products. In another paper, Akıncı, Özalp, and Turgut (2013) determined suitable lands for agricultural use in Turkey by AHP. As García et al. (2014) newly reviewed the literatures, no papers has been seen, which focused on ASC location problem. A good paper on implementation of Fuzzy AHP and Delphi method in MADM problems is Cho and Lee (2013). They identified four decision areas and further prioritized the sixteen factors under a hierarchy model structured by Fuzzy AHP approach.

Location and the number of ASCs in each region is a key parameter to ensure the success of agricultural servicing. So the main scope of this study is constructing indicators for location selection of ASCs to improve the performance of agricultural supply chain viewpoint of productivity, quality, and competitiveness. Secondly, to give primary weight to the objectives and location attributes, Fuzzy AHP method is used. Also specifying the final weight of location attributes using GP method is provided at last step of this study. In other words an integrated approach which combines MADM and MODM is developed. First to third phase belongs to MADM and fourth phase is a MODM problem. Also GP is used to generalize proposed approach and make it utilizable for different/future real world applications. The managers can use developed approach to import their ideal values of agricultural supply chain goals to the model and get appropriate weight for the ASC location attributes. The results of current research are used in facility location problem of ASCs in another paper of the authors.

2. Material and methods

In this research, the FAHP has been used to assess the location indicators (both objectives and attributes) of ASCs. Using the FAHP method, the ASC location attributes is prioritized based on objectives, which assumed that they have same importance. But another phase added to the framework to consider the possible difference between objectives using multi-choice goal programming (MCGP), i.e. in the case that the priorities of objectives are not the same. So this research has four main phases for assessing the location attributes (see Figure 3):

(1) Constructing initial location attributes and refining them and also location objectives using Delphi method.
(2) Ranking initial ASC location attributes using TOPSIS technique.
(3) Developing the AHP hierarchal model containing the assessment of objectives and attributes, and determining the weights of objectives and attributes through FAHP.
(4) Prioritizing the attributes through GP.

2.1. *Phase I*

First of all, the location problem of ASCs is defined. Often the ASC location problem occur in developing countries, but may occurs in development ones where the strategic decision be taken to improve the productivity of their agricultural supply chain. However, the problem is also well known within production, where placing central maintenance or service facilities is often a critical issue when determining the cost of supporting an ongoing manufacturing environment. Services, however, play a particularly critical role in the agricultural sector. Tractor and farm machineries are needed, but in developing countries considerable number of them are old and their efficiency has been reduced, so need to be replaced. Also other services such as advisory, input supply

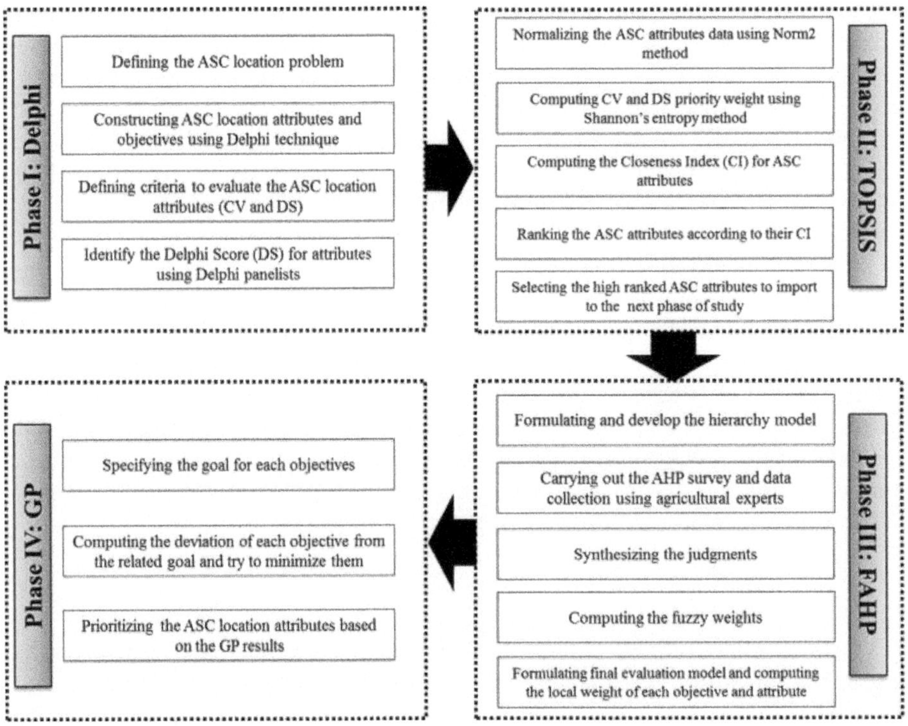

Figure 3. The schematic process of overall research framework.

and financial services can improve the efficiency of agricultural production. In some developing countries (for example Iran) governments have been concluded that establishing ASCs and keep them update in the aspects of tractors, machineries, and also other agricultural services will be improve the efficiency. So, the main issue for this policy is selecting the location of these centers. This problem has many aspects, but it is MADM problem and an AHP can solve it, primarily. It is common that in similar location problems, desired attributes be collected from literatures and similar studies. But in current problem no similar research study for the attributes found, so Delphi method was used to construct ASC location attributes.

2.1.1. Delphi method

Delphi is a technique of popular survey method which extract consensus of ideas among a set of experts or panelists by maintaining the unanimity among them. Delphi technique has been used for various purposes like setting goals, finding problems, developing system models, decision-making, etc. (Prusty, Mohapatra, & Mukherjee, 2010). In current research several rounds of Delphi survey is needed to construct the objectives and attributes of ASC location problem. In the following sections the process of Delphi survey has been described.

2.1.2. Delphi survey process

2.1.2.1. Design and administration of the initial questionnaire. There is no similar research about location of ASC in literatures, so the initial questionnaire was basically a brainstorming session among the selected Delphi panelists. Designer team was included four people who designed the open questions of first questionnaire and contributed in different steps of this research. The aim of this step of Delphi study was construction of the location selection indicators (both objectives and attributes) of ASC location problem. Each Delphi panelist was asked to answer two main questions as follows:

(1) What location can be appropriate to establish ASC? In other words what attributes should be considered to select a place for an ASC?
(2) What advantages can be achieved if an ASC located in best possible location (based on attributes you mentioned in previous question)? In other words what are the objectives of ASC location selection?

First question was designed to collect and construct ASC location attributes and the second to determine main objectives of ASC facility location problem. Eight experts replied to the first round questionnaire. The composition of the panelists in this round is given in Table 1.

The second questionnaire was designed with the objective of prioritizing the attributes raised in the first questionnaire and sent to the Delphi panelists. Attributes were rated with respect to priority criterion. The rating scale (see Table 2) was in the range of 1–9, with '1' representing 'not important' and '9' representing 'very highly important' meant that how much an attribute is important for each Delphi panelists participated in the survey to select best location of ASCs.

There are several methods to calculate the score of each attributes to determine the most important attributes to use in Fuzzy AHP method. The number of items for pairwise comparisons should be reduced in AHP to simplify judgment process and

Table 1. Composition of experts in the first round.

Participant type	Educations	Office	Number
Professor	Ph.D. with average background of 12 years in the field of Agricultural Engineering	University of Tehran, University of Zanjan	3
Ph.D. student	M.Sc. in the field of Agricultural Engineering	University of Tehran	3
Government expert	B.Sc. and M.Sc. with average background of 14 years in the field of Agricultural Mechanization Engineering	Agricultural Department of Zanjan and Alborz provinces	2

ensure the accuracy of results. So here coefficient of variation and Delphi score are used simultaneously in a TOPSIS ranking survey to reduce the number of items.

If the absolute dispersion is defined as the standard deviation, and the average is mean, the relative dispersion is called the coefficient of variation (CV) or coefficient of dispersion. The CV is attractive as a statistical tool because it apparently permits the comparison of varieties free from scale effects, i.e. it is dimensionless. The CV is defined as the ratio of the standard deviation σ to the mean μ (Brown, 1998) (Equation (1)):

$$\mathrm{CV} = \frac{\sigma}{\mu} \tag{1}$$

The lower value of CV has preference to select one attribute as ASCs location indicator, i.e. Delphi panelist opinions is similar to each other and average of them is high about an attribute, illustrate that all panelists want to consider that attribute in study.

Another method to analysis the Delphi panelist's scores to select best indicators is Delphi score, which proposed by Linstone and Turoff (2002) (Equation (2)):

$$\text{Delphi Score} = \frac{(\text{Lowest Score} + \text{Highest Score}) + (4 \times \text{Average Score})}{6} \tag{2}$$

Against the CV higher values of Delphi score is preferred. Each attribute which have the most Delphi score can be selected to include in ASCs location selection process.

2.2. *Phase II*

In this phase, the most important attributes will be selected according to their Delphi score and CV. Then the hierarchy model will be formulated, and the AHP survey can be run. Here, the TOPSIS technique is applied to rank attributes according to the value of CV and DS.

Table 2. Linguistic variables for the rating of attributes.

Linguistic variable	Value
Not important	1
Very low	2
Medium low	3
Low	4
Fair	5
Medium	6
High	7
Medium high	8
Very high	9

TOPSIS is a MCDM method which is initially developed by Hwang and Yoon (Hwang & Yoon, 1981b). The technique is based on the idea that the optimal solution should have the shortest distance from the positive ideal solution and the farthest from the negative ideal solution (Oztaysi, 2014). In current research, the maximum value of each attribute is considered as positive ideal solution and the minimum value as negative ideal solution. This value is mentioned for positive attributes where higher values of them are preferred for the location of ASC, e.g. population of candidate location, while for negative attributes is vice versa.

2.3. Phase III

The main aim of current research is estimating the weights of objectives and attributes from the FAHP model which is developed. So the AHP questionnaire distributed to the agricultural professors and experts to respond to judge about relative weight of each pairwise comparison between objectives and attributes. In the production environment this can easily be changed to managers and production technology experts.

2.3.1. FAHP

AHP is a structured technique for organizing and analyzing complex decisions based on paired comparisons of both projects and criteria (Saaty, 1986). The strength of the AHP method lies in its ability to construct complex, multi-person, multi-attribute, and multi-period problem hierarchically (Chin, Xu, Yang, & Lam, 2008). The judgments of conventional AHP method usually have ambiguity problem because the verbal attitudes of decision-makers evaluation process contain vague and multiplicity of meaning (Lee, 2010). Thus, in order to overcome the verbal ambiguity of linguistic variables, fuzzy set theory has applied to the judgment as an extension of AHP method. Fuzzy set theory was first introduced by Zadeh (1976) to deal with the uncertainty.

2.4. Phase IV

In this research, it is assumed that the importance of all objectives is the same, but for applying this approach to any agricultural region, the managers must consider their agricultural conditions and also their strategic decision to use ASC in their region. If the managers prefer to stress some objectives, in other words the managers follows specific level of each objective function to improve the performance of their ASC, so they strictly recommended prioritizing the objectives and then use GP technique (Chang, 2007) to compute the final weight of ASC location attributes. Then they can use these weights to define the process of MADM to find best location of ASCs. Traditional GP can be used in this case when there is only one goal level in each objective function, but MCGP is another version of GP which considers several maximum or minimum goals in (Chang, Chen, & Zhuang, 2014).

3. Results and discussion

3.1. The results of Delphi survey

The answers made by the Delphi panelists to the initial questionnaire covered a wide range of issues. According to the results, the panelists mentioned number of machinery available in the region, distances, future work condition of service centers in future, etc. Similarly, García et al. (2014) constructed attributes for perishable product warehouse which relatively is an ASC including: accessibility, security, needs of the agricultural product warehouse, social acceptance that the warehouse may have on the environment where it is supposed to be built, the costs of product transportation, wages, and salaries of workers and managers. After constructing the attributes various scores were obtained in the second questionnaire against each ASC location attributes. The title and scores of both evaluation methods (i.e. CV and DS) methods have been shown in Table 3.

Based on the main objectives of development of ASC establishment policy, four objectives was made for ASC location problem shown in Table 4, by consulting agricultural experts using Delphi method (the second question in first round questionnaire is related to the objectives). It can be said that the objectives constructed here is almost the main objectives of agricultural supply chain. The quality is the first and the most important objective in any supply chain management. All actions are done to improve the quality. So this study strictly focused on quality of services in accomplishing our research. The quality of services in agricultural operations such as mechanization and inputs can directly affect the quality and quantity of production. Competitiveness of agricultural production directly related to their quality and price. The farmers always try to reduce their costs and compete with other markets. Here the location attribute is

Table 3. Score of ASC location attributes.

No.	Attribute	CV	DS
C1	Surplus of tractor and machinery which concentrated in region	0.763	4.20
C2	Lack of tractor and machinery	0.393	6.50
C3	Easy reach of farmers to the ASC	0.447	5.73
C4	Easy reach to spare parts, repair, customer service center, and gas station	0.225	6.90
C5	Easy reach to main roads and byroads	0.553	5.27
C6	Possibility to develop the workspace in future	0.525	5.27
C7	Minimum distance to the covered regions	0.435	5.80
C8	Easy reach to its villages and covered regions	0.313	6.23
C9	Easy reach to water, electricity, gas, and telephone with minimum cost	0.484	5.67
C10	Existence of crop rotation with least fallow (irrigated be more than dry farming)	0.384	5.93
C11	Social acceptance by the farmers and bring their adoption	0.518	5.27
C12	Good climate for developing ASC operations and be safe from natural disasters	0.503	4.20
C13	Tractor and machinery being old and farmers could not be able replace them	0.479	5.20
C14	Maximum distance to other service centers or similar centers	0.334	5.60
C15	Possibility to extend new crop production in future	0.506	3.30
C16	The price of land for cheap extend	0.838	3.13
C17	Easy reach to required labor	0.569	4.23
C18	Enough security	0.594	4.63
C19	Visibility for farmers	0.859	4.27
C20	Minimum distance to residential areas	0.929	3.00
C21	Existence of progressive farmers	0.566	5.07
C22	Existence of good educations and tendency to modernism by farmers	0.536	4.70
C23	Good soil, good climate, less slope, big plot lands, and appropriate wind	0.516	5.00
C24	Existence of crowded population and vast cultivated area	0.316	6.60
C25	Minimum distance of villages to each other and large number of them (population being concentrated)	0.472	5.43
C26	Maximum number of crop types cultivated	0.503	5.47

Table 4. ASC location selection objectives made after first round Delphi survey.

Objectives	Objective Function (Max/Min)	Abbreviations
Quality of agricultural services	Max	Service quality (O1)
Cost of agricultural services paid by farmers instead of delivering services	Min	Service cost (O2)
Speed of service	Max	Service speed (O3)
Profitability of ASCs	Max	ASC profit (O4)

defined, which can minimize the agriculture service costs and consequently product cost. Locations which are near to their customers can reduce their transportation cost, so according to the objective of service cost, location attributes could be selected, which minimized the distance between service centers and the farms/farmers. Time is a critical factor in agricultural commodities, and extremely can influence the quality and price of agricultural production because of being perishable, dependence on whether, etc. The location and also capacity of service centers affects the service time. Therefore the attributes reflects these items can improve the location selection results. Service centers must be sustainable to provide appropriate services to the farms, so their profitability can be considered in this case.

3.2. *Refinement of attributes by TOPSIS*

Attributes were analyzed to refine and reduce their size, and also by selecting the most important ones. If too many factors be included in AHP model, then the number of pairwise comparisons will be increased, so the accuracy of AHP judgments may reduce. Also it is necessary to make them mutually exclusive and nonredundant. So to find five best attributes after the second round Delphi survey, a TOPSIS survey designed to rank ASC attributes, which have highest Delphi score and lowest CV simultaneously. Then they were selected as alternatives in the AHP model to make judgment easy and possible for experts and increase their accuracy. The results of TOPSIS technique are illustrated in Table 5 and Figure 4.

The highest rank ASC attributes has been introduced in Table 6.

3.3. *Process of the AHP survey*

The hierarchy model of ASC location problem formulated and developed as shown in Figure 5.

To find the weight of the ASC location selection attributes, an AHP questionnaire was designed with a 9-point scale and pairwise comparison format. The questionnaire was distributed to 10 agricultural experts. Approximately, 3 of 10 participants had more than 10 years of work experience in agricultural sector and also have related higher academic educations. Other experts almost are professor and PhD student of agricultural engineering. They were asked to make a pairwise comparison judgment and give the relative importance amongst the performance objectives and attributes. The pairwise judgment is conducted from the first level to the third level. Finally, each participant individually expressed their preference between each pair of elements.

Table 5. The results of TOPSIS survey.

Attribute	Normalized data		Distance to positive ideal solution	Distance to negative ideal solution	Closeness index to ideal solution	Rank
	CV	DS				
C1	0.208	0.666	0.345	0.116	0.252	23
C2	0.055	1.594	0.063	0.396	0.863	3
C3	0.072	1.240	0.160	0.304	0.655	7
C4	0.018	1.796	0.000	0.458	1.000	1
C5	0.110	1.047	0.220	0.244	0.526	15
C6	0.099	1.047	0.218	0.249	0.534	13
C7	0.068	1.269	0.152	0.313	0.673	6
C8	0.035	1.466	0.093	0.372	0.800	4
C9	0.084	1.212	0.170	0.293	0.632	9
C10	0.053	1.328	0.133	0.332	0.714	5
C11	0.096	1.047	0.217	0.250	0.535	12
C12	0.091	0.666	0.321	0.182	0.362	20
C13	0.082	1.020	0.222	0.251	0.530	14
C14	0.040	1.183	0.172	0.305	0.639	8
C15	0.092	0.411	0.391	0.158	0.287	22
C16	0.252	0.370	0.433	0.042	0.089	25
C17	0.116	0.676	0.321	0.168	0.343	21
C18	0.126	0.810	0.287	0.186	0.394	19
C19	0.264	0.687	0.357	0.102	0.223	24
C20	0.309	0.340	0.458	0.000	0.000	26
C21	0.115	0.969	0.242	0.225	0.482	17
C22	0.103	0.833	0.276	0.203	0.423	18
C23	0.096	0.943	0.245	0.228	0.482	16
C24	0.036	1.644	0.045	0.414	0.903	2
C25	0.080	1.114	0.196	0.272	0.582	11
C26	0.091	1.128	0.194	0.271	0.582	10
Shannon's weight	0.726	0.273				

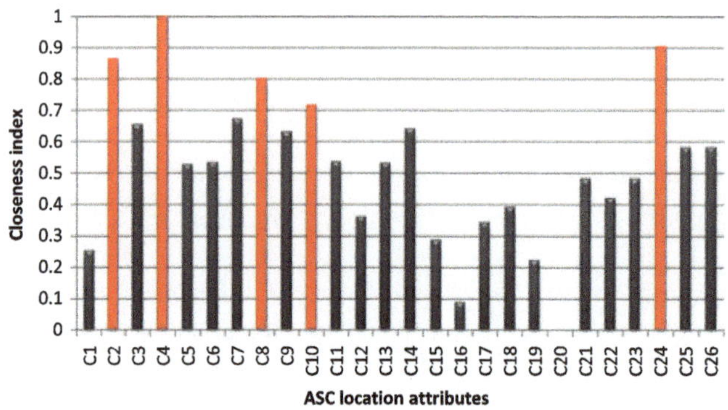

Figure 4. The value of closeness index in TOPSIS ranking method.

Table 6. The most important attributes based on Delphi survey.

No.	Attribute	New abbreviation
C2	Lack of tractor and machinery	Machinery lack (C1)
C4	Easy reach to spare parts, repair, customer service center, and gas station	Easy support (C2)
C8	Easy reach to its villages and covered regions	Easy reach (C3)
C10	Existence of crop rotation with least fallow (irrigated be more than dry farming)	Crop rotation (C4)
C24	Existence of crowded population and vast cultivated area	Population and area (C5)

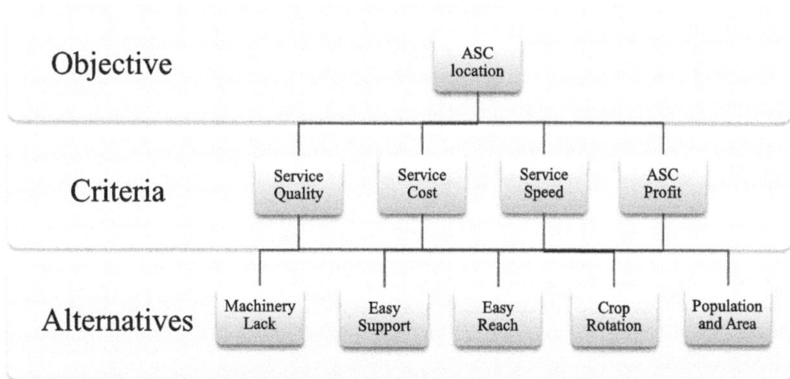

Figure 5. AHP model for location selection of the ASCs.

3.4. *FAHP model development*

3.4.1. Computing the pairwise comparison matrices of objectives and attributes using fuzzy numbers

By using triangular fuzzy numbers, the fuzzy pairwise comparisons matrix for the main objectives is constructed as follows in Table 7.

First, via pairwise comparison, the fuzzy synthetic extent values of the attributes were calculated in order to assess the priority weights of the main objectives. The values of the fuzzy synthetic extent of the four objectives were denoted $S_{\text{Service Quality}}$, $S_{\text{Service Cost}}$, $S_{\text{Service Speed}}$, and $S_{\text{ASC Profit}}$, respectively:

$$S_{\text{Service Quality}} = (1.600, 5.661, 19.000) \otimes (6.169, 16.841, 72.000)^{-1} = (0.022, 0.336, 3.080)$$

$$S_{\text{Service Cost}} = (1.616, 4.521, 19.000) \otimes ((6.169, 16.841, 72.000))^{-1} = (0.022, 0.268, 3.080)$$

$$S_{\text{Service Speed}} = (1.476, 3.233, 18.000) \otimes ((6.169, 16.841, 72.000))^{-1} = (0.021, 0.192, 2.918)$$

Table 7. Pairwise comparison matrix for objectives.

	O1	O2	O3	O4	Normalized weight
O1	(1.000, 1.000, 1.000)	(0.200, 1.415, 6.000)	(0.200, 1.711, 6.000)	(0.200, 1.533, 6.000)	0.257
O2	(0.166, 0.706, 5.000)	(1.000, 1.000, 1.000)	(0.200, 1.245, 7.000)	(0.250, 1.570, 6.000)	0.252
O3	(0.166, 0.567, 5.000)	(0.142, 0.785, 5.000)	(1.000, 1.000, 1.000)	(0.166, 0.880, 7.000)	0.245
O4	(0.166, 0.652, 5.000)	(0.166, 0.636, 4.000)	(0.142, 1.135, 6.000)	(1.000, 1.000, 1.000)	0.244
CR	0.001				

Table 8. Pairwise comparison matrix for service quality.

Service quality	C1	C2	C3	C4	C5	Normalized weight
C1	(1.000, 1.000, 1.000)	(0.125, 0.322, 4.000)	(0.142, 0.792, 4.000)	(0.142, 1.086, 5.000)	(0.142, 0.569, 5.000)	0.189
C2	(0.250, 3.101, 8.000)	(1.000, 1.000, 1.000)	(1.000, 3.282, 8.000)	(1.000, 3.702, 8.000)	(1.000, 2.907, 6.000)	0.225
C3	(0.250, 1.261, 7.000)	(0.125, 0.304, 1.000)	(1.000, 1.000, 1.000)	(0.125, 1.584, 5.000)	(0.125, 0.977, 6.000)	0.195
C4	(0.200, 0.920, 7.000)	(0.125, 0.270, 1.000)	(0.200, 0.630, 8.000)	(1.000, 1.000, 1.000)	(0.166, 0.812, 3.000)	0.190
C5	(0.200, 1.756, 7.000)	(0.166, 0.343, 1.000)	(0.166, 1.022, 8.000)	(0.333, 1.231, 6.000)	(1.000, 1.000, 1.000)	0.199
CR	0.007					

Table 9. Pairwise comparison matrix for service cost.

Service cost	C1	C2	C3	C4	C5	Normalized weight
C1	(1.000, 1.000, 1.000)	(0.142, 0.344, 2.000)	(0.142, 0.468, 3.000)	(0.166, 0.892, 4.000)	(0.200, 0.753, 4.000)	0.186
C2	(0.500, 2.905, 7.000)	(1.000, 1.000, 1.000)	(0.142, 1.492, 4.000)	(0.250, 2.334, 8.000)	(0.250, 1.614, 6.000)	0.214
C3	(0.333, 2.134, 7.000)	(0.250, 0.670, 7.000)	(1.000, 1.000, 1.000)	(0.333, 1.231, 6.000)	(0.333, 0.719, 6.000)	0.205
C4	(0.250, 1.120, 6.000)	(0.125, 0.428, 4.000)	(0.166, 0.812, 3.000)	(1.000, 1.000, 1.000)	(0.166, 0.748, 2.000)	0.192
C5	(0.250, 1.326, 5.000)	(0.166, 0.619, 4.000)	(0.166, 1.390, 3.000)	(0.500, 1.335, 6.000)	(1.000, 1.000, 1.000)	0.200
CR	0.01					

Table 10. Pairwise comparison matrix for service speed.

Service speed	C1	C2	C3	C4	C5	Normalized weight
C1	(1.000, 1.000, 1.000)	(0.111, 0.455, 3.000)	(0.111, 0.483, 6.000)	(0.200, 1.157, 3.000)	(0.200, 0.950, 4.000)	0.193
C2	(0.333, 2.197, 9.000)	(1.000, 1.000, 1.000)	(0.250, 0.821, 9.000)	(0.333, 2.099, 5.000)	(0.250, 1.876, 6.000)	0.209
C3	(0.166, 2.069, 9.000)	(0.111, 1.216, 4.000)	(1.000, 1.000, 1.000)	(0.200, 1.586, 9.000)	(0.142, 1.290, 6.000)	0.207
C4	(0.333, 0.863, 5.000)	(0.200, 0.476, 3.000)	(0.111, 0.630, 5.000)	(1.000, 1.000, 1.000)	(0.333, 0.835, 2.000)	0.207
C5	(0.250, 1.052, 5.000)	(0.166, 0 533, 4.000)	(0.166, 0.774, 7.000)	(0.500, 1.196, 3.000)	(1.000, 1.000, 1.000)	0.191
CR	0.1					

Table 11. Pairwise comparison matrix for ASC profit.

ASC profit	C1	C2	C3	C4	C5	Normalized weight
C1	(1.000, 1.000, 1.000)	(0.200, 1.453, 7.000)	(0.200, 2.065, 7.000)	(0.200, 2.432, 8.000)	(0.166, 1.000, 4.000)	0.207
C2	(0.142, 0.688, 5.000)	(1.000, 1.000, 1.000)	(0.200, 1.048, 8.000)	(0.333, 2.334, 6.000)	(0.166, 0.679, 4.000)	0.201
C3	(0.142, 0.484, 5.000)	(0.125, 0.954, 5.000)	(1.000, 1.000, 1.000)	(0.250, 1.231, 8.000)	(0.142, 0.357, 5.000)	0.196
C4	(0.125, 0.411, 5.000)	(0.166, 0.428, 3.000)	(0.125, 0.812, 4.000)	(1.000, 1.000, 1.000)	(0.166,0.328, 1.000)	0.183
C5	(0.250, 1.000, 6.000)	(0.250, 1.472, 6.000)	(0.200, 2.794, 7.000)	(1.000, 3.047, 6.000)	(1.000, 1.000, 1.000)	0.210
CR	0.1					

$$S_{\text{AMSC Profit}} = (1.476, 3.424, 16.000) \otimes ((6.169, 16.841, 72.000))^{-1} = (0.021, 0.203, 2.594)$$

The degree of possibility of S_i $(i \neq j)$ was determined.

$$V(S_{\text{Service Quality}} \geq S_{\text{Service Cost}}) = 1$$

$$V(S_{\text{Service Quality}} \geq S_{\text{Service Speed}}) = 1$$

$$V(S_{\text{Service Quality}} \geq S_{\text{AMSC Profit}}) = 1$$

$$V(S_{\text{Service Cost}} \geq S_{\text{Service Quality}}) = \frac{3.080 - 0.022}{(3.080 - 0.022) + (0.336 - 0.268)} = 0.978$$

$$V(S_{\text{Service Cost}} \geq S_{\text{Service Speed}}) = 1$$

$$V(S_{\text{Service Cost}} \geq S_{\text{AMSC Profit}}) = 1$$

$$V(S_{\text{Service Speed}} \geq S_{\text{Service Quality}}) = \frac{2.918 - 0.022}{(2.918 - 0.022) + (0.336 - 0.192)} = 0.952$$

$$V(S_{\text{Service Speed}} \geq S_{\text{Service Cost}}) = \frac{2.918 - 0.022}{(2.918 - 0.022) + (0.268 - 0.192)} = 0.974$$

$$V(S_{\text{Service Speed}} \geq S_{\text{AMSC Profit}}) = \frac{2.918 - 0.021}{(2.918 - 0.021) + (0.203 - 0.192)} = 0.996$$

$$V(S_{\text{AMSC Profit}} \geq S_{\text{Service Quality}}) = \frac{2.594 - 0.022}{(2.594 - 0.022) + (0.336 - 0.203)} = 0.950$$

$$V(S_{\text{AMSC Profit}} \geq S_{\text{Service Cost}}) = \frac{2.594 - 0.022}{(2.594 - 0.022) + (0.268 - 0.203)} = 0.975$$

$$V(S_{\text{AMSC Profit}} \geq S_{\text{Service Speed}}) = 1$$

The minimum degree of possibility was stated as follows:

$$\acute{d}(\text{Service Quality}) = \min(1.000, 1.000, 1.000) = 1.000$$

$$\acute{d}(\text{Service Cost}) = \min(0.978, 1.000, 1.000) = 0.978$$

$$\acute{d}(\text{Service Speed}) = \min(0.952, 0.974, 0.996) = 0.952$$

$$\acute{d}(\text{AMSC Profit}) = \min(0.950, 0.975, 1.000) = 0.950$$

Accordingly, the weight vector was derived as $\acute{W} = (1.000, 0.978, 0.952, 0.950)$ and the normalized weight was given by $W = (0.257, 0.252, 0.245, 0.244)^T$. By applying the same calculation process to the remaining attributes, priority weights could be calculated.

From the calculated results listed in Table 7, it can be concluded that, by comparing the weights of objective in Level 2, service quality and service cost are more important than service speed and ASC profit objective.

The local weights of the relative importance are computed to the attributes against the objectives; and also the global weights, which are the relative importance attributes against the goal. To derive the global weight of each attribute, its local weight was multiplied by the local weight of each corresponding objectives. In this study, all consistency ratios range from 0.001 to 0.1 and overall CR is 0.04, which fell within the acceptance level of 0.10 as recommended by Saaty (1994). This shows that the survey respondents have assigned their weights consistently after examining the priorities of success attributes of location selection for ASCs.

3.4.2. Local weights based on the synthesized judgments

Tables 8–11 hierarchically display local weights according to the Level 3, with the CR using a 5 × 5 matrix. In the Level 3 attributes of service quality, 'easy support' (local *W*: 0.225) and 'population and area' (local *W*: 0.199) attributes rank first and second, respectively. Also 'easy reach' (local *W*: 0.195), 'crop rotation' (local *W*: 0.190), and 'machinery lack' (local *W*: 0.189) rank the third to fifth, respectively. To visualize this result, Figures 6 and 7 show the local weights of the attributes under service quality and service cost objectives in a radiated diagram.

Similarly from Figures 8 and 9 the priority of attributes under 'Service Speed' and 'ASC Profit' can be seen.

3.4.3. Global weights based on the synthesized judgments

To determine the overall prioritization of the five attributes with respect to the goal of best location of ASCs, the global weights of the attributes were calculated. Table 12 shows the prioritization of the four objectives and five attributes in terms of the global weights based on the synthesized judgments of evaluators.

From the Level 2 objective weights, we can derive the prioritization of four objectives. Table 12 shows that 'Service Quality' (global *W*: 0.257) is most important

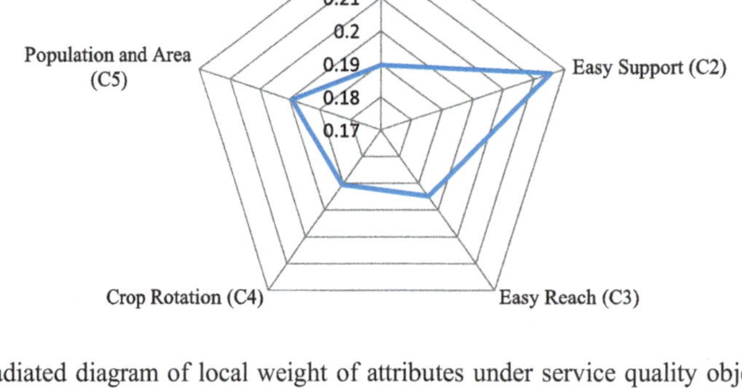

Figure 6. Radiated diagram of local weight of attributes under service quality objectives.

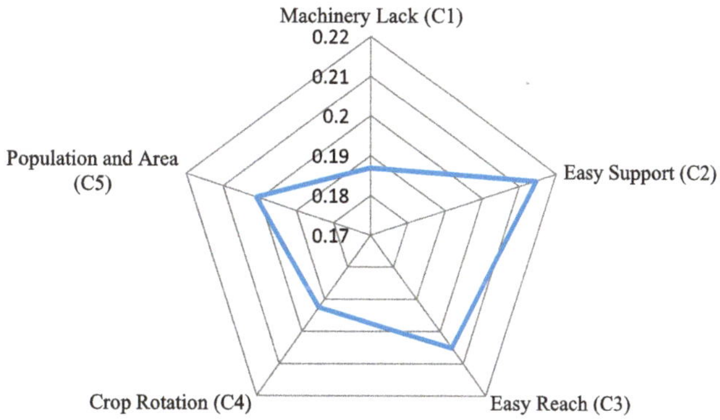

Figure 7. Radiated diagram of local weight of attributes under service cost objectives.

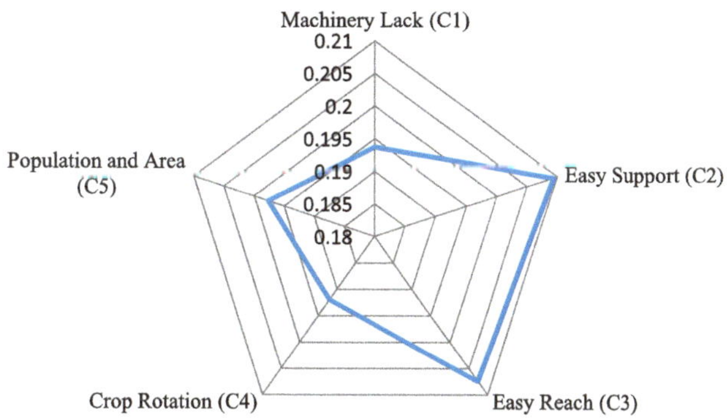

Figure 8. Radiated diagram of local weight of attributes under service speed objectives.

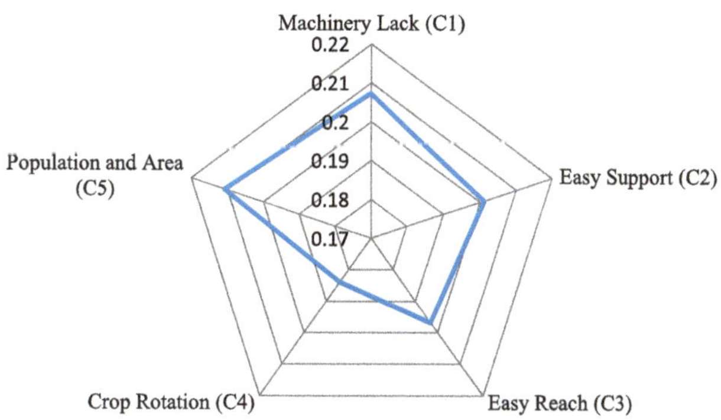

Figure 9. Radiated diagram of local weight of attributes under ASC profit objectives.

criterion for assessing the ASC location attributes. 'Service Cost' (global *W*: 0.252), 'Service Speed' (global *W*: 0.245) and 'ASC Profit' (global *W*: 0.244) are the second to fifth most important objectives, respectively, in this level.

For Level 3 perspective, the global weight of each attributes was calculated by multiplying the local weight of each attribute under each criterion to global weight of

Table 12. Global weight and prioritization of objectives and attributes.

Objectives	Global W	Prioritizations	Attributes	Global W	Prioritizations
Service quality (O1)	0.257	1	Easy support (C2)	0.212	1
Service cost (O2)	0.252	2	Population and area (C5)	0.202	2
Service speed (O3)	0.245	3	Easy reach (C3)	0.201	3
ASC profit (O4)	0.244	4	Machinery lack (C1)	0.194	4
			Crop rotation (C4)	0.189	5

related objectives and calculating sum of them for each attribute. As can be seen in Table 12 the attribute 'Easy Support' (global W: 0.212) has relatively higher value than other attributes. 'Population and area' (global W: 0.202) and 'Easy reach' (global W: 0.201) with less difference are the second and third important attributes. Also 'Machinery lack' (global W: 0.194) and 'Crop rotation' (global W: 0.189) are the less important attributes for location selection of ASC.

The attributes selected in this research are interested to use in facility location problem of ASCs. Service centers should be located on places which can reach to their required inputs to provide the services for farms. Easy support attribute means that candidate location is better to have access to fuel stations, agricultural input markets, human workers, etc. The second important attribute is population and area, which directly influence on the service cost and centers profit. Locations which are near to more customers can reduce their costs and get more profits. Accessibility to service centers is almost related to the distance between farmers and centers. But in agricultural areas the accessibility may related to centrality of centers location among villages, because farmers of around areas needed to travel between their farms/home and the center several times in each season. Easy reach to service center will increase the service lead time and center profit. Transportation and also distance are key factors in location problems. Since in this research, we have assumed same priority for objectives, the easy reach ranked third, but if in a situation the service cost and service speed be more important than quality and profit for service centers, then the easy reach will be the most important location attribute, because the transportation directly affect the cost and speed of service. The number of available tractors and machineries in any region can be considered as an attribute, because the service centers should provide mechanization services to the farms which have no machinery or needs some machinery, therefore the centers can be located there. The last attribute is crop rotation which means that service centers can locate on regions that several types of crops are cultivated. Agricultural operations are completely seasonal; it means that some operations in one month of a year are needed and its facilities will remain useless until next season. This phenomenon is not favorable for service centers economically. As our expert team selected the work for the facilities should be available in all times of season. It can be obtained where several crops cultivated in the region.

In this research, we assumed that all objectives have same importance in normal case, but if any manager/researcher want to use this approach with other assumption about objectives and make priority, they are recommended to use the GP method introduced through the paper. They can use some targets/goals for the objectives and then prioritize the location attributes of service centers. It can be very useful to manage the agricultural systems to be efficient.

4. Conclusions

The main contribution of this paper was to identify the list of objectives and attributes for assessing the best location for establishing ASCs. The paper elaborated upon a hierarchically assessment model and identified important objectives, and their weights to evaluate the relative importance of the ASC location selection problem. This paper contributes to experts by applying the TOPSIS and FAHP model with the Delphi method, which is considered one of the most structured techniques for industrial and managerial decision-making. Based on literature review and Delphi study, a TOPSIS, and an AHP model was generated by incorporating four decision objectives. In addition, the fuzzy approach has been used to make the conventional AHP model more sophisticated with much precise judgment. The results of the FAHP method indicated that agricultural service quality is the most important objective to assess the ASC location attributes. It can be concluded that the service quality is the main expectation from agricultural services. Further, several location attributes are adopted for ASC location and found that support for ASC is the main attribute. Tractors and agricultural machineries have too depreciations because of their working conditions. So these centers need continual repairs both in working season and off-season for next working season. This paper also proposed MCGP to prioritize the objective functions and define targets for them and then rank the attributes based on them. This method can be used to rank ASC location attributes, and this is another contribution of current research, because it is generalizable to other regions and other conditions. The studied problem closely relates to such problems found in manufacturing systems. The proposed methods and the conclusions reached can also be used for solving, e.g. maintenance facility placement problems in manufacturing or supply chain design problems. In future studies, extended research is needed to investigate the effect of current situation of services in agricultural supply chain on priority of location problem objectives in model developed in this paper into a variety of different areas. The main limitation of developed approach is related to the experts participated in this research. Since the methodology used in this paper relies on knowledge and experience of experts, thus it is so critical to use elite experts in such studies.

Acknowledgments

The financial support provided by University of Tehran is duly gratefully acknowledged. Thanks and appreciations of authors also go to colleagues in Aalborg University and people who have willingly helped us with their abilities.

ORCID

Peter Nielsen http://orcid.org/0000-0002-4882-7942

References

Akıncı, H., Özalp, A. Y., & Turgut, B. (2013). Agricultural land use suitability analysis using GIS and AHP technique. *Computers and Electronics in Agriculture, 97*, 71–82.

Bagheri, N., & Moazzen, S. A. (2009). Optimum strategy for agricultural mechanization development in Iran. *Journal of Agricultural Technology, 6*, 225–237.

Bishop, C. (1997). *A guide to preparing an agricultural mechanization strategy*. Rome: Food and Agriculture Organization of the United Nations.

Brown, C. E. (1998). *Applied multivariate statistics in geohydrology and related sciences*. Berlin: Springer, Verlag Berlin Heidelberg.

Chang, C. T. (2007). Multi-choice goal programming. *Omega, 35*, 389–396.

Chang, C. T., Chen, H. M., & Zhuang, Z. Y. (2014). Integrated multi-choice goal programming and multi-segment goal programming for supplier selection considering imperfect-quality and price-quantity discounts in a multiple sourcing environment. *International Journal of Systems Sciences, 45*, 1101–1111.

Chin, K. S., Xu, D.-L., Yang, J. B., & Lam, J. P.-K. (2008). Group-based ER–AHP system for product project screening. *Expert Systems with Applications, 35*, 1909–1929.

Chipeta, S. (2006). *Demand driven agricultural advisory services*. Lindau: Neuchatel Group.

Cho, J., & Lee, J. (2013). Development of a new technology product evaluation model for assessing commercialization opportunities using Delphi method and fuzzy AHP approach. *Expert Systems with Applications, 40*, 5314–5330.

Dorward, A., & Chirwa, E. (2011). The Malawi agricultural input subsidy programme: 2005/06 to 2008/09. *International Journal of Agricultural Sustainability, 9*, 232–247.

Farahani, R. Z., Steadieseifi, M., & Asgari, N. (2010). Multiple criteria facility location poblems: A survey. *Applied Mathematical Modeling, 34*, 1689–1709.

García, J. L., Alvarado, A., Blanco, J., Jiménez, E., Maldonado, A. A., & Cortés, G. (2014). Multi-attribute evaluation and selection of sites for agricultural product warehouses based on an analytic hierarchy process. *Computers and Electronics in Agriculture, 100*, 60–69.

Hosseini-Nasab, H., Tavana, M., & Yousefi, M. (2014). A new heuristic algorithm for the planar minimum covering circle problem. *Production & Manufacturing Research, 2*, 142–155.

Hwang, C. L., & Masud, S. A. M. (1979). *Multiple objective decision making – Methods and applications: A state-of-the-art survey*. Berlin: Springer.

Hwang, C. L., & Yoon, K. (1981a). *Multiple attribute decision making – Methods and applications: A state-of-the-art survey*. New York, NY: Springer.

Hwang, C. L., & Yoon, K. P. (1981b). *Multiple attribute decision making methods and aplications*. New York, NY: Springer-Verlag.

Inns, F. M. (1995). *Selection, testing and evaluation of agricultural machines and equipment theory*. Rome: FAO Agricultural Services Bulletin.

Johan, F., Swinnen, M., & Gow, H. R. (1999). Agricultural credit problems and policies during the transition to a market economy in Central and Eastern Europe. *Food Policy, 24*, 21–47.

Lak, M. B., & Almasi, M. (2011). An analytical review of parameters and indices affecting decision making in agricultural mechanization Australia. *Journal of Agricultural Engineering, 2*, 140–146.

Lee, S. H. (2010). Using fuzzy AHP to develop intellectual capital evaluation model for assessing their performance contribution in a university. *Expert Systems with Applications, 37*, 4941–4947.

Linstone, H. A., & Turoff, M. (2002). *The Delphi method, techniques and applications*. Addison Wesley.

Oztaysi, B. (2014). A decision model for information technology selection using AHP integrated TOPSIS-Grey: The case of content management systems. *Knowledge Based Systems*. Retrieved from http://dx.doi.org/10.1016/j.knosys.2014.02.010

Prusty, S. K., Mohapatra, P. K. J., & Mukherjee, C. K. (2010). GOS tree (goal-objective-strategy tree) approach to strategic planning using a fuzzy-Delphi process: An application to the Indian Shrimp Industry. *Technological Forecasting & Social Change, 77*, 442–456.

Saaty, T. L. (1986). Axiomatic foundation of the analytic hierarchy process. *Management Science, 32*, 841–855.

Saaty, T. L. (1994). How to make a decision: The analytic hierarchy process. *Interfaces, 24*, 19–43.

Shukla, R. K., Garg, D., & Agarwal, A. (2014). An integrated approach of Fuzzy AHP and Fuzzy TOPSIS in modeling supply chain coordination. *Production & Manufacturing Research, 2*, 415–437.

Sims, G. B., & Kienzle, J. (2009). *Farm equipment supply chains; guidelines for policy-makers and service providers: Experiences from Kenya, Pakistan and Brazil*. Rome: FAO.

Szidarovszky, F., Gershon, M. E., & Duchstein, L. (1986). *Techniques for multi-objective decision making in systems management*. Amsterdam: Elsevier Science Publishers B.V.

Timmer, C. P., Falcon, W. P., & Pearson, S. R. (1983). *Food policy analysis*. Baltimore: The Johns Hopkins University Press.

Ulungu, E. L., & Teghem, J. (1994). Multi-objective combinatorial optimization problems: A survey. *Journal of Multicriteria Decisions Analysis, 3*, 83–104.

Zadeh, L. A. (1976). A fuzzy-algorithmic approach to the definition of complex or imprecise concepts. *International Journal of Man-Machine Studies, 8*, 249–291.

Zangeneh, M., Nielsen, P., Akram, A., & Keyhani, A. (2014). Agricultural service center location problem: Concept and a MCDM solution approach. In *APMS 2014 International Conference Advances in Production Management Systems "Innovative and knowledge-based production management in a global-local world" Ajaccio*. 20–24 September 2014.

Zionts, S. (1979). A survey of multiple criteria integer programming methods. *Annals of Discrete Mathematics, 5*, 389–398.

4

An enhancement to the Bees Algorithm with slope angle computation and Hill Climbing Algorithm and its applications on scheduling and continuous-type optimisation problem

B. Yuce[a]*, D.T. Pham[b], M.S. Packianather[c] and E. Mastrocinque[d]

[a]*School of Engineering, BRE Institute of Sustainable Construction, Cardiff University, Newport Road, The Parade, Queen's Building, Cardiff CF24 3AA, UK;* [b]*School of Mechanical Engineering, University of Birmingham, Edgbaston, Birmingham B15 2TT, UK;* [c]*Institute of Mechanical and Manufacturing Engineering, Cardiff University, Queen's Buildings, The Parade, Cardiff CF24 3AA, UK;* [d]*School of Management, Royal Holloway, University of London, Egham Hill, Egham, Surrey TW20 0EX, UK*

This paper focuses on improvements to the Bees Algorithm (BA) with slope angle computation and Hill Climbing Algorithm (SACHCA) during the local search process. First, the SAC was employed to determine the inclination of the current sites. Second, according to the slope angle, the HCA was utilised to guide the algorithm to converge to the local optima. This enabled the global optimum of the given problem to be found faster and more precisely by focusing on finding the available local optima first before turning the attention on the global optimum. The proposed enhancements to the BA have been tested on continuous-type benchmark functions and compared with other optimisation techniques. The results show that the proposed algorithm performed better than other algorithms on most of the benchmark functions. The enhanced BA performs better than the basic BA, in particular on higher dimensional and complex optimisation problems. Finally, the proposed algorithm has been used to solve the single machine scheduling problem and the results show that the proposed SAC and HCA-BA outperformed the basic BA in almost all the considered instances, in particular when the complexity of the problem increases.

Keywords: the Bees Algorithm; slope angle computation; Hill Climbing Algorithm; benchmark functions; single machine scheduling

1. Introduction

Optimisation algorithms play a big role in the area of operational research, artificial intelligence, mathematics, computer science, robotics and many other areas, and are utilised to find the optimum solution for highly complicated problems. Several optimisation algorithms have been proposed to solve a variety of problems. Many real-life problems are not easy and are hard to solve in polynomial time which are called NP-Hard (Woeginger, 2003). NP-Hard-type problems are generally solved with stochastic-based optimisation algorithms such as genetic algorithm, ant colony optimisation, tabu search, simulated annealing and the Bees Algorithm (BA). Every algorithm has advantages and disadvantages depending on the type of problem which it is solving. Although there is no one best algorithm to solve all types of optimisation problems (Wolpert & Macready, 1997), the

*Corresponding author. Email: yuceb@cardiff.ac.uk

BA has been used successfully to solve many problems including mechanical design, job shop scheduling, supply chain optimisation, robot path planning, chemical engineering problems, assembly problems and several other applications (Ang, Ng, & Pham, 2013; Ang, Ng, Pham, & Soroka, 2013; Castellani, Pham, & Pham, 2012; Lien & Cheng, 2014; Mastrocinque, Yuce, Lambiase, & Packianather, 2013; Pham, Castellani, & Fahmy, 2008; Pham, Afify, & Koç, 2007; Pham, Castellani, & Ghanbarzadeh, 2007; Pham, Darwish, Eldukhri, & Otri, 2007; Pham, Koc, Lee, & Phrueksanant, 2007; Pham, Otri, Afify, Mahmuddin, & Al-Jabbouli, 2007; Pham, Soroka, Koç, Ghanbarzadeh, & Otri, 2007; Yuce, Mastrocinque, Lambiase, Packianather, & Pham, 2014). Despite these several successful applications of the BA, a careful examination shows that the algorithm needs to be more robust and sensitive to a particular problem. Several attempts have been made in the past to improve the BA using methods such as early neighbourhood search and efficiency-based recruitment (Pham, Packianther, Imaguliyev, & Yuce, 2012), adaptive neighbourhood search and site abandonment strategy (Yuce, Packianather, Mastrocinque, Pham, & Lambiase, 2013), combination of adaptive enlargement and reduction of the search neighbourhood (Azfanizam, Pham, & Faieza, 2014) developing new local search strategies (Ahmad, 2012), tuning the algorithms parameters (Otri, 2011), introducing new parameters (Pham & Ghanbarzadeh, 2007), combining with other optimisation algorithms (Sholedolu, 2009) novel initialisation based on the patch concept and Levy flight distribution (Hussein, Sahran, & Abdullah, 2014). In this paper, another such improvement is proposed to address the insensitive movement of the BA based on a determination of promising sites and convergence to the related sites with deterministic approach using a hybrid BA combining with hill climbing. The first step is to define a measurement technique to determine the direction along which promising solutions can be found. This is based on the steepness angle mimicking the direction along which a scout bee performs its figure-of-eight waggle dance during the recruitment of forager bees. The second step is to develop a hybrid algorithm combining BA and a Hill Climbing Algorithm (HCA) based on the threshold value of the steepness angle. The proposed enhanced BA has been tested on continuous-type benchmark functions and compared with other optimisation techniques, and finally has been applied to solve a well-known manufacturing problem such as the single machine scheduling problem, in order to show its benefit in solving a real-world problem.

This paper is organised as follows: basic version of the BA is presented in Section 2. The improved version of the BA follows in Section 3. The algorithm tests conducted on benchmark functions and results are given in Section 4. The single machine scheduling problem, its solution by the proposed algorithm and results are discussed in Section 5. Finally, the conclusions are given in Section 6.

2. The basic BA

The BA is a type of swarm-based algorithm, which was proposed by Pham, Ghanbarzadeh, Koc, Otri, Rahim, and Zaidi (2005, 2006). Swarm-based algorithms mimic the swarm behaviour of animals in nature such as ant colonies, bee colonies, bird flocking, animal herding, bacterial growth, fish schooling and many other (Beni & Wang, 1989). These behaviours have been implemented in artificial intelligence and swarm intelligence.

BA is also a type of swarm intelligence which has been inspired from foraging behaviour of the honey bees. A swarm of honey bees consists of a queen bee and thousands of worker bees which are allocated to do different works in a colony such as cleaning their nest, foraging and constructing comb (Seeley, 1995). In the nature when a scout bee (forager) finds a food source (nectar), it returns to the hive and performs special ritual which is a communicational movement, known as the waggle dance; this informs the colony about the nectar (Gould & Gould, 1988; Seeley, 1995). The waggle dance contains information about direction, distance and quality of the flower patch found by the bee (Talbi, 2009). After the waggle dance, the colony decides the number of bees needed to be assigned for that particular food source as promising patches should have more bees than less promising ones. The recruited bees evaluate the related patch and look around of the patch as well as share this information with their peers in the hive. The same process happens for all patches emulating local search. After the recruitment stage, the scout bees will continue searching for other promising patches in a random way emulating the global search and whenever they find another patch, this information will be given to the colony randomly (Von Frisch, 1955). This behaviour of honey bees was computationally modelled as an optimisation algorithm. The parameters and the pseudo-code of the algorithm are given below in Table 1 and Figure 1, respectively.

According to Figure 1, the BA has following steps: the first step is placing the '*n*' scout bees on the search space, and then in the next step, fitness values of the visited patches are evaluated. Step 3 is related to the selection of the best patches with respect to their fitness value where these selected best patches will be split into two groups containing more scout bees to the elite patches '*e*', and less scout bees to the non-elite best patches '*m–e*' in step 4. The next step 5 covers the neighbourhood search in the patches given in steps 3 and 4, and so according to neighbourhood search, the patches' fitness values will be evaluated. In step 6, the remainder bees, which are created in initial population, will be recruited for the random search to find better random solutions. In step 7, the random patches' fitness values will be evaluated and this process will continue until one of the stopping criteria is met.

The procedure described above equips the algorithm to combine exploitative neighbourhood search with explorative global search enabling effective location of the globally optimal solution to a problem (Yuce, 2012). Due to the powerful search capability of the algorithm, the basic BA was successfully utilised to solve different types of optimisation problems such as job scheduling (Pham, Koc et al., 2007), forming manufacturing cells (Pham, Afify et al., 2007), data clustering (Pham, Otri et al., 2007), tuning a fuzzy logic controller (Pham, Darwish et al., 2007), training neural networks (Pham, Koc, Ghanbarzadeh, & Otri, 2006), finding the multiple feasible solutions to preliminary design problems (Pham, Castellani et al., 2007) and design of mechanical

Table 1. The parameter for the basic BA.

Parameter	Symbols
Number of scout bees in the selected patches	*n*
Number of best patches out of the selected patches	*m*
Number of elite patches out of the selected best patches	*e*
Number of recruited bees in the elite patches	*nep*
Number of recruited bees in the non-elite best patches	*nsp*
The size of neighbourhood for each patch	*ngh*
Number of iterations	*MaxIter*
Difference between fitness values of consecutive iterations	*Error*

Generate the initial population size as *n*, set the best patch size as *m*, set the elite patch size as *e*, set the number of forager bees recruited to the elite sites as *nep*, set the number of forager bees around the *m-e* non-elite best patches as *nsp*, set the neighbourhood size as *ngh*, set the maximum iteration number as *MaxIter*, and set the error limit as *Error*.
i= 0
Step 1: Generate initial population *n*.
Step 2: Evaluate Fitness Value of initial population.
Sort the initial population based on the fitness result.
While $i \leq MaxIter$ or $FitnessValue_i - FitnessValue_{i-1} \geq Error$
    $i= i+1;$
    Steps 3, 4: Select the elite patches and non-elite best patches for neighbourhood search.
    Step 5: Recruit the *nep* forager bees to the *e* elite patches and *nsp* forager bees to *m-e* non-elite best patches.
    Evaluate the fitness value of each patch.
    Sort the results based on their fitness.
    Step 6: Allocate the rest of the *n-m* bees for global search to the non-best locations.
    Step 7: Evaluate the fitness value of non-best patches.
    Sort the overall results based on their fitness.
Run the algorithm until termination criterion is met.
End

Figure 1. Pseudo-code for the basic BA.

components (Pham, Soroka et al., 2007). The improvements made to the basic BA with slope angle computation and Hill Climbing Algorithm (SACHCA) is given in the following section.

3. The BA with SACHCA

This study presents a new enhancement to the basic BA based on the SACHCA. Even though the BA has both local and global search capability, it still has some weakness such as a high level of randomness, computational time and blind search in the local search process (Yuce, 2012). The HCA is an iterative single element-based local search algorithm, also known as gradient ascent/descent algorithm. Although the local minimum of an optimisation problem can be found by the HCA, the global optimum is not guaranteed (Grosan & Abraham, 2011).

The SACHCA-based improved BA is concerned with the process of locating the best sites. SAC is employed to determine the inclination of the current sites. In general, the promising location will be far from the current position if the slope angle is close to 90°. The current location will be close to a local optimum point if the slope angle is close to 0°. The direction of the local optimum can be determined according to the slope angle orientation.

The slope angle is computed using the first-order numerical derivation. The numerical derivation of each site is calculated from its neighbourhood. The two end points in the vicinity of the neighbourhood are used to compute the numerical derivation. The central difference method is utilised for numerical derivation (see Equation (1)).

$$\text{Slope-angle} = F'(X) = \frac{F(X + \frac{\Delta X}{2}) - F(X - \frac{\Delta X}{2})}{\Delta X} \tag{1}$$

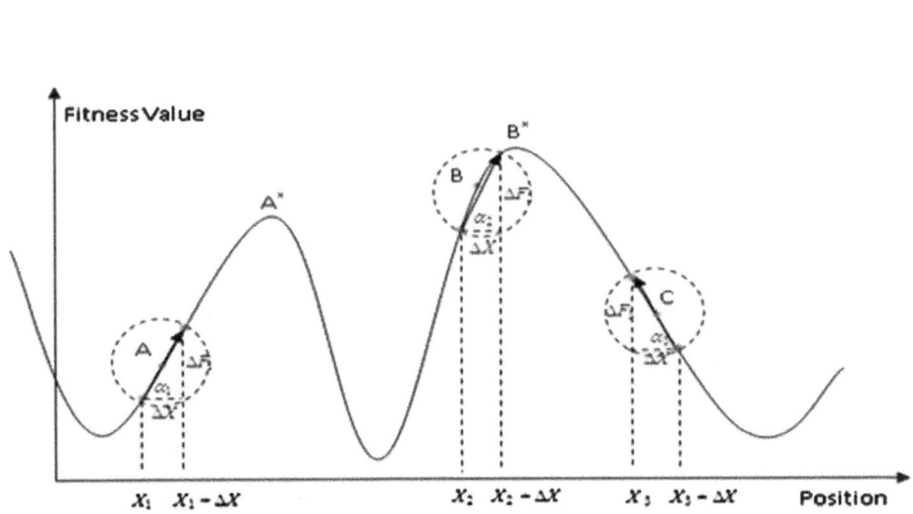

Figure 2. Slope angle for each of the best selected sites.

If the slope angle is very steep, then the promised location is far from the selected site but when the slope angle is close to zero, the promised location is very close to the selected site, as shown in Figure 2.

There are three selected (best) sites shown in Figure 2. Around site A, the angle direction is towards site A^x, which is a local optimum. The direction of sites B and C are towards the site B^x, which is another local optimum. At the end of the searching process, all the local optima will be sorted and the biggest will be selected as the global optimum. In the case of Figure 2, the site B^x was selected as global optimum. The local search process was accomplished with the use of the HCA, as shown in Equation (2).

$$X(i+1) = X(i) + h\nabla F(X_i) \tag{2}$$

where i is the iteration number, $X(i)$ is the current position, $X(i+1)$ is the next position, h is the incremental size and $\nabla F(X_i)$ is the gradient of the current position. Further, the pseudocode of the algorithm is given in Figure 3.

4. Benchmark functions tests

To measure the performance of the proposed algorithm, it was tested on 10 continuous-type benchmark functions. These functions are given in Table 2 (Ahmad, 2012; Pham & Castellani, 2009; Pham, Packianther, Imaguliyev, & Yuce, 2012). The BA requires a number of parameters to be set manually as given in Table 1. The other parameters of the proposed algorithm are given in Table 3. Several experiments were conducted in order to tune the algorithm parameters and to find the combination of them giving the best trade-off between the number of iterations and optimum that was found. All the experiments were carried out with the maximum number of iterations fixed to 5000 and $ngh = 1$. Five different combinations of the BA parameters were tested on benchmark functions according to Table 4. The tests were repeated 100 times for each function and the average number of iterations and the average optimum results obtained with the SACHCA-BA are given in Table 5. The combination number 3 gave the best results in terms of number of iterations and optimum found, and for this reason, it has been

```
Generate the initial population size as n, set the best patch size as m, set the elite patch size as e, set the number of
forager bees around of elite as nep, set the number of forager bees around of non-elite best patches as nsp, set the
step size for HC algorithm as h, set the angle limit as angle_limit, set the "number of the waiting time for HC"
algorithm as HC_time_limit, set the neighbourhood size as ngh, set the maximum iteration number as
MaxIter, and set the error limit as Error.
i= 0, time= 0; slope_angle(1 : m) = 0; Generate initial population; Evaluate Fitness Value of initial population;
Sort the initial population based on the fitness result.
While i ≤ MaxIter or FitnessValue_i - FitnessValue_(i-1) ≥ Error
        i= i+1;
        Select the elite patches and non-elite best patches for neighbourhood search.
        Recruit the forager bees to the elite patches and non-elite best patches.
        Evaluate the fitness value of each patch; Sort the results based on their fitness.
        For   k=1:m
        Calculate angle(k)
        While slope_angle(k) > angle_limit and time ≤ HC_time_limit
        X(i+1,k)=X(i,k)+h∇F(X(i,k)).
        Evaluate Fitness value for each position
        End
        Record all the local optimum sites found and sort them (end of the neighbourhood search).
        Allocate the rest of the bees for global search to the non-best locations;
        Evaluate the fitness value of non-best patches;
        Sort the fitness values and positions;
Run the algorithm until termination criterions are met.
End
```

Figure 3. Pseudo-code of improved BA with SACHCA strategy.

selected for the successive experiments. Then, the results were utilised for further analysis and comparison.

The first comparison were carried out between the proposed technique, the basic BA and other well-known optimisation techniques such as particle swarm optimisation (PSO), evolutionary algorithm (EA) and artificial bee colony (ABC) whose parameters values can be found in Table 6 (Ahmad, 2012). The performance of the algorithm was assessed according to the accuracy and the average evaluation numbers (Tables 7 and 8). Experimental results for BA, PSO, EA and ABC were extracted from another source (Ahmad, 2012).

Further, *T*-test was utilised to measure the statistical significance of the proposed algorithm and the basic BA and results are given in Table 9.

4.1. Discussion the results of benchmark functions

In this study, the neighbourhood search in the BA was identified as the focus for improving the BA by utilising SACHCA. The accuracy of the proposed algorithm was computed with average absolute differences of the best results. According to this, the

Table 2. Test functions (Ahmad, 2012; Pham & Castellani, 2009).

No	Function name	Interval	Function	Global optimum
1	Goldstein & Price (2D)	[−2, 2]	$\min F = [1 + (X_1 + X_2 + 1)^2(19 - 14X_1 + 3X_1^2 - 14X_2 + 6X_1X_2 + 3X_2^2)]$ $[30 + (2X_1 - 3X_2)^2(18 - 32X_1 + 12X_1^2 + 48X_2 - 36X_1X_2 + 27X_2^2)]$	$X = [0, -1]$ $F(X) = 3$
2	Schwefel (2D)	[−500, 500]	$\min F = \sum_{i=1}^{2} [-X_1 \sin(\sqrt{\lvert X_i \rvert})]$	$X = [0, 0]$ $F(X) = -837.658$
3	Schaffer (2D)	[−100, 100]	$\min F = 0.5 + \frac{(\sin(\sqrt{X_1^2 + X_2^2}))^2 - 0.5}{1 + 0.001(X_1^2 + X_2^2)^2}$	$X = (0, 0)$ $F(X) = 0$
4	Rosenbrock (10D)	[−1.2, 1.2]	$\min F = \sum_{i=1}^{10} [100(X_i^2 - X_{i+1})^2 + (1 - X_i)^2]$	$X = [1, 1, 1, 1, 1, 1, 1, 1, 1, 1]$ $F(X) = 0$
5	Sphere (10D)	[−5.12, 5.12]	$\min F = \sum_{i=1}^{10} X_i^2$	$X = [0, 0, 0, 0, 0, 0, 0, 0, 0, 0]$ $F(X) = 0$
6	Ackley (10D)	[−32, 32]	$\min F = -20e^{-0.2\sqrt{\frac{\sum_{i=1}^{10} x_i^2}{10}}} - e^{\frac{\sum_{i=1}^{10} \cos(2\pi X_i)}{10}} + 20 + e$	$X = [0, 0, 0, 0, 0, 0, 0, 0, 0, 0]$ $F(X) = 0$
7	Rastrigin (10D)	[−5.12, 5.12]	$\min F = 100 + \sum_{i=1}^{10} (X_i^2 - 10\cos(2\pi X_i))$	$X = [0, 0, 0, 0, 0, 0, 0, 0, 0, 0]$ $F(X) = 0$
8	Martin & Gaddy (2D)	[0, 10]	$\min F = (X_1 - X_2)^2 + (\frac{X_1 + X_2 - 10}{3})^2$	$X = [5, 5]$ $F(X) = 0$
9	Easom (2D)	[−100, 100]	$\min F = -\cos(X_1)\cos(X_2)e^{-((X_1 - \pi)^2 + (X_2 - \pi)^2)}$	$X = [\pi, \pi]$ $F(X) = -1$
10	Griewank (10D)	[−600, 600]	$\min F = \frac{1}{4000} \sum_{i=0}^{i=10} (x_i - 100)^2 - \prod_{i=0}^{i=10} \cos\left(\frac{x_i - 100}{\sqrt{i+1}}\right)$	$X = [100, 100, 100, 100, 100, 100, 100, 100, 100, 100]$ $F(X) = 0$

Table 3. The test parameter for the proposed SACHCA.

Parameters (symbols)	Symbols
Number of scout bees in the selected patches	*n*
Number of best patches in the selected patches	*m*
Number of elite patches in the selected best patches	*e*
Number of recruited bees in the elite patches	*nep*
Number of recruited bees in the non-elite best patches	*nsp*
The size of neighbourhood for each patch	*ngh*
Number of Iterations	*MaxIter*
Difference between fitness values of consecutive Iterations	*Error*
Angle limit	*angle_limit*
Step size for HC algorithm	*h*
Number of waiting time for HC algorithm	*HC_time_limit*

Table 4. Combinations of parameters values tested.

	Parameters									
Combination	*n*	*m*	e	*nep*	*nsp*	*h*	*ngh*	*MaxIter*	*angle_limit*	*HC_time limit*
1	10	5	1	3	2	.01	1	5000	.5	1000
2	25	10	2	6	4	.05	1	5000	.5	1000
3	50	15	5	12	8	.1	1	5000	.5	1000
4	100	30	10	24	16	.5	1	5000	.5	1000
5	200	60	20	50	30	1	1	5000	.5	1000

more accurate results are closer to zero. The proposed algorithm performed significantly better than the other methods on most of the test functions such as, 2D-Goldstein&Price,2D-Schewel, 2D-Schaffer, 10D-Rosenbrock, 10D-Sphere, 2D-Martin&Gaddy and 2D-Easom where the average absolute difference of these functions was found to be 0 (see Table 6). However, the average absolute difference for 10D-Ackley was found to be .0013 and for 10D-Griwank was .003. Further, all the results for SACHCA were found to be better than the results of the basic BA. This behaviour was verified with the number of evaluations given in Table 8 and with the t-test given in Table 9.

As shown in Table 7, the average numbers of evaluations for 2D-Goldstein & Price 2D-Schewel, 2D-Schaffer, 10D-Rosenbrock, 10D-Sphere and 10-Rastrigin were found with proposed algorithm as 250,049, 1049, 118,049, 60,195.5, 11,292.3 and 49,678, respectively, where the average number of evaluation for the basic BA were found as 554,000, 1140, 121,088, 935,000, 285,039 and 885,000. However, the results found from 2D-Martin Gaddy and 2D-Easom using the proposed algorithm were not better than the basic BA. Further, the results found for the 10D-Ackley and 10D-Griwank were better for the proposed algorithm compared with the basic BA. The proposed algorithm performed better overall, especially for higher dimensional functions. According to *T*-test results (Table 9), the results found with the enhanced BA were significant compared with the basic BA in half of the cases when the confidence level was selected to be 95% ($\alpha < .005$).

Table 5. SACHCA-BA results for the different combinations of parameters values tested on continuous type functions.

		Combinations									
		Combination 1		Combination 2		Combination 3		Combination 4		Combination 5	
No	Function	Avg. Iter	Avg. Opt	Avg. Iter	Avg. Opt	Avg. Iter	Avg. Opt	Avg. Iter	Avg. Opt	Avg. Iter	Avg. Opt
1	Goldstein & Price (2D)	5000	3.610	5000	3.381	1427	3.000	1869	3.000	2756	3.000
2	Schwefel (2D)	5000	−837.163	5000	−837.428	6	−837.658	426	−837.658	1568	−837.658
3	Schaffer (2D)	5000	.632	5000	.258	675	.000	2503	.000	2846	.000
4	Rosenbrock (10D)	5000	.293	5000	.179	344	.000	745	.000	1956	.000
5	Sphere (10D)	5000	.635	5000	.458	65	.000	305	.000	1420	.000
6	Ackley (10D)	5000	.389	5000	.162	5000	.001	5000	.001	5000	.001
7	Rastrigin (10D)	5000	.575	5000	.348	284	.000	532	.000	2763	.000
8	Martin & Gaddy (2D)	5000	.415	5000	.326	8	.000	268	.000	1235	.000
9	Easom (2D)	5000	−.548	5000	−.759	1429	−1.000	2126	−1.000	3154	−1.000
10	Griewank (10D)	5000	.765	5000	.421	5000	.000	5000	.000	5000	.000

Note: Avg_Iter, average iterations; Avg_Opt, average optimum.

Table 6. The test parameters for the EA, PSO and ABC (Ahmad, 2012).

EA				
Parameters		Crossover		No crossover
Population size			100	
Evaluation cycles (max number)			5000	
Children per generation			99	
Crossover rate		1		0
Mutation rate (variables)		.05		.8
Mutation rate (mutation width)		.05		.8
Initial mutation interval width α (variables)			.1	
Initial mutation interval width ρ (mutation width)			.1	

PSO				
Parameters			Value	
Population size			100	
PSO cycles (max number) T			5000	
Connectivity			See below	
Maximum velocity			See below	
C1			2	
C2			2	
w_{max}			.9	
w_{min}			.4	
Velocity of the each connectivity (connectivity, u)			Max particle velocity u	
Connectivity (number of neigbourhood)	(2, .005)	(2, .001)	(2, .05)	(2, .1)
	(10, .005)	(10, .001)	(10, .05)	(10, .1)
	(20, .005)	(20, .001)	(20, .05)	(20, .1)
	(100, .005)	(100, .001)	(100, .05)	(100, .1)

ABC	
Parameters	Value
Population size	100
ABC cycles (max number)	5000
Employed bees n_e	50
Onlooker bees n_e	49
Random scouts	1
Stagnation limit for site abandonment *stlim*	50× dimension

5. Application on single machine scheduling problem

5.1. Single machine scheduling problem

The proposed algorithm has been tested on a well-known manufacturing problem such as the single machine scheduling problem.

A single machine scheduling problem is a well-studied optimisation problem where a set of n-jobs with given deterministic processing times p_i and a common due date d have to be processed on a machine according to some constraints. The goal is to find a schedule for the n-jobs which minimises the sum of all the penalties which can occur due to the constraints. This is a challenging optimisation problem and is, therefore, chosen to test the performance of the proposed BA.

There are many solution techniques which could be applied to solving single machine scheduling problems. If the objective is to minimise the average flow time in Equation (7), then the optimal schedule will have the shortest processing time. If the objective is to minimise the maximum tardiness in Equation (5), then the optimal schedule will have the earliest due date. The Hodgson's algorithm gives an optimal solution if the objective is to minimise the number of jobs with tardiness greater than zero. The lateness in Equation (6) calculates any deviation from the due date, where a positive value for lateness indicates tardiness and a negative value is earliness. In this study, the goal will be to find a schedule for the n-jobs which jointly minimises the sum of earliness and tardiness penalties according with Equation (3).

$$\text{SUM-P} = \sum_{i=1}^{n} \alpha_i E_i + \beta_i T_i \tag{3}$$

where α_i and β_i are the earliness and tardiness penalties for job i per unit time, and E_i and T_i are the jobs' earliness and tardiness for job i, respectively, derived from

$$E_i = \max(0, d - C_i) \tag{4}$$

$$T_i = \max(0, C_i - d) \tag{5}$$

$$\text{Lateness} = C_i - d \tag{6}$$

$$\text{Flow time} = \text{End date} - \text{Start date} \tag{7}$$

where d is due date and C_i is the completion time of the job i.

5.2. *Experimental procedure*

The data-sets available in website (http://people.brunel.ac.uk/~mastjjb/jeb/orlib/schinfo.html) were used to find the common due dates for different jobs assigned to a single machine. A data-set is composed by nj (number of jobs) with given deterministic processing times p_i and a common due date d. Say that the jobs have to be processed on one machine and that for each of the jobs, an individual earliness E_i and tardiness T_i penalty is given, which is incurred, if a job is finished before or after the common due date d, respectively.

The format of the data file is as follows:

For each problem in turn: nj

for each job i ($i = 1, \dots, nj$) in turn: p_i, E_i, T_i are given.

The common due date d is calculated by:

$$d = round[\text{SUM\_}P \times rf] \tag{8}$$

where round[X] gives the biggest integer which is smaller than or equal to X, SUM_P denotes the sum of the processing times of the nj jobs and the parameter rf(restrictive factor) is used to calculate more or less restrictive common due dates.

The proposed SACHCA-BA was implemented in Matlab code on a Pentium i7(2.93 GHz) PC and was used to solve the single machine scheduling problem. The data-sets available in (http://people.brunel.ac.uk/~mastjjb/jeb/orlib/schinfo.html) were used to find the common due dates for different jobs assigned to a single machine. There are seven different data-sets with different numbers of jobs (nj = 10, 20, 50, 100,

Table 7. Accuracy of proposed algorithm compared with other well-known optimisation techniques.

Function	PSO		EA		ABC		BA		SACHCA-BA	
	Avg. Abs. Dif.	Std. Dev.	Avg. Abs. Dif.	Std. Dev.	Avg. Abs. Dif.	Std. Dev.	Avg. Abs. Dif.	Std. Dev.	Avg. Abs. Dif.	Std. Dev.
1. Goldstein & Price	.0000	.0000	.0000	.0000	.0000	.0001	.0000	.0003	.0000	.0000
2. Schwefel	4.7376	23.4448	4.7379	23.4448	.0000	.0000	.0000	.0005	.0000	.0000
3. Schaffer	.0000	.0000	.0009	.0025	.0000	.0000	.0000	.0003	.0000	.0002
4. Rosenbrock	.5998	1.0436	61.5213	132.6307	.0965	.0880	44.3210	112.2900	.0000	.0001
5. Sphere	.0000	.0000	.0000	.0000	.0000	.0000	.0000	.0003	.0000	.0000
6. Ackley	.0000	.0000	.0000	.0000	.0000	.0000	1.2345	.3135	.0013	.0032
7. Rastrigin	.1990	.4924	2.9616	1.4881	.0000	.0000	24.8499	8.3306	.0000	.0038
8. Martin& Gaddy	.0000	.0000	.0000	.0000	.0000	.0000	.0000	.0003	.0000	.0000
9. Easom	.0000	.0000	.0000	.0000	.0000	2.0096	.0000	.0003	.0000	.0002
1. Griewank	.0008	.0026	.0210	.0130	.0052	.0078	.3158	.1786	.0003	.0005

Note: Abs. Dif., absolute difference; Std. Dev., standard deviation.

Table 8. Average evaluation of proposed algorithm compared with other well-known optimization techniques.

	PSO		EA		ABC		BA		SACHCA-BA	
Function	Avg. Eval.	Std. Dev.	Avg. Eval.	Std. Dev.	Avg. Eval.	Std. Dev.	Avg. Eval.	Std. Dev.	Avg. Eval.	Std. Dev.
1.Goldstein & Price	3262	822	2002	390	2082	435	504.0000	211.0000	250,049	0
2. Schwefel	84,572	90,373	298,058	149,638	4750	1197	1140	680	1049	0
3. Schaffer	28,072	21,717	219,376	183,373	21,156	13,714	121,088	174,779	118,049	0
4. Rosenbrock	492,912	29,381	500,000	0	497,728	16,065	935,000	0	60,195.5	16,092.3
5. Sphere	171,754	7732	36,376	2736	13,114	480	285,039	277,778	11,292.3	2343
6. Ackley	236,562	9119	50,344	3949	18,664	627	910,000	0	15,426	5402.6
7. Rastrigin	412,440	67,814	500,000	0	207,486	57,568	885,000	0	49,678.12	17.341.2
8.Martin & Gaddy	1778	612	1512	385	1498	329	600	259	1300.8	672.3
9. Easom	16,124	15,942	36,440	28,121	1542	201	5280	6303	250,049	0
10. Griewank	290,466	74,501	490,792	65,110	357,438	149,129	4300,000	0	87,777.8	38,285.3

Note: Avg. Eval., average evaluation; Std. Dev., standard deviation.

Table 9. The statistical significance between the improved BA and the basic BA.

		Significance between the basic BA and the enhanced BA	
No.	Function	Significant ($\alpha < .05$)	α
1	Goldstein & Price (2D)	Yes	1.58E−08
2	Schwefel (2D)	No	.4553
3	Schaffer (2D)	No	.2599
4	Rosenbrock (10D)	Yes	.027
5	Sphere (10D)	Yes	6.16E−09
6	Ackley (10D)	Yes	6.08E−06
7	Rastrigin (10D)	Yes	.0098
8	Martin & Gaddy (2D)	Yes	.0018
9	Easom (2D)	No	.1040
10	Griewank (10D)	No	.9567

200, 500 and 1000) and four value of restrictive factor (rf = .2, .4, .6 and .8). Each data-set contains 10 instances. Therefore, the algorithm was run on 280 instances and the results were compared with those obtained using the basic BA. The parameter values in common of the basic BA and the SACHCA-BA were selected according to Table 3.

The percentage offset between the basic BA and the SACHCA-BA results are calculate for each value of nj and rf as follows:

$$\%\text{Offset} = \frac{\text{SACHCA BA} - \text{BA}}{\text{BA}} 100 \tag{9}$$

5.3. *Discussion the results of single machine scheduling*

In order to evaluate the performance of the proposed algorithm, it has also been implemented on the single machine scheduling problem. Moreover, the results of the proposed algorithm were comparing with the results of basic BA using the offset between the respective results given in Equation (10). Further, the average offset of 10 instances for each couple of nj and rf are given in Table 10. The results show that the average offset is negative in all the cases, which means that, the SACHCA-BA outperformed compare to the basic BA in solving the single machine scheduling problem, shown in Figure 4.

$$\text{Average } \% \text{ Offset} = \frac{1}{10} \sum_{k=1}^{10} \%\text{Offset}_k \tag{10}$$

Figure 4 shows the trends of the average offset for each couple of nj and rf. The performance of the SACHCA-BA improves compared with the BA when the rf and the nj increase, that is, when the problem becomes more complex. Lower is the value of rf and more stable is the average offset when nj increases. For instance, when rf = .2, the average offset is .00% for nj = 10, and −6.33% when nj = 1000. On the contrary, when

Table 10. The average offset between the SACHCA-BA and the basic BA.

nj	*rf* = .2	*rf* = .4	*rf* = .6	*rf* = .8
10	.00%	−.37%	−2.25%	−3.07%
20	−.97%	1.79%	−1.12%	−2.00%
50	−1.72%	−2.42%	−9.52%	−11.05%
100	−1.32%	−2.77%	−9.26%	−20.80%
200	−1.96%	−4.17%	−9.52%	−21.51%
500	−5.00%	−11.15%	−13.59%	−19.22%
1000	−6.33%	−24.04%	−35.75%	−45.33%

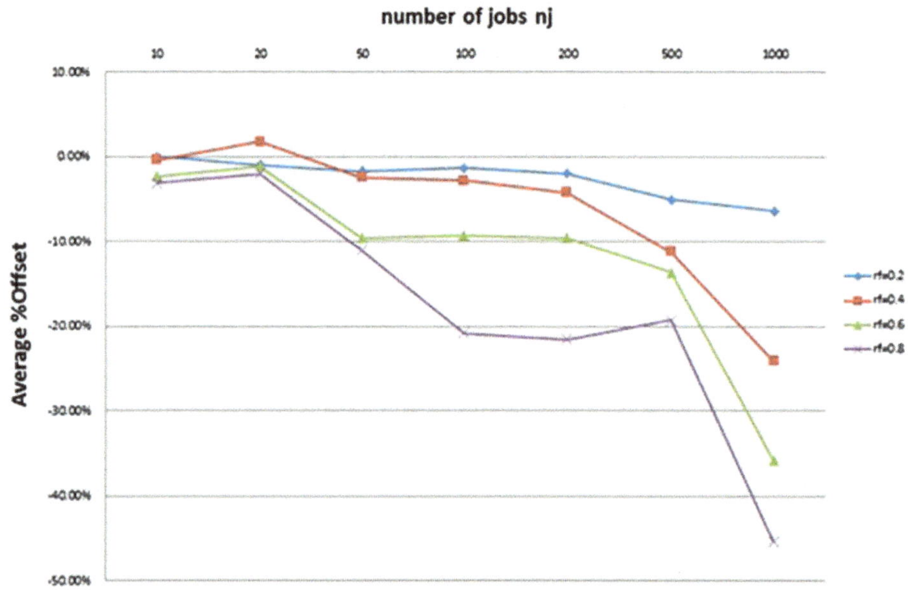

Figure 4. Average % offset values for each value of *nj* and *rf*.

rf value is higher, the average offset varies significantly with the *nj* value. For *rf* = .8, the average offset is −3.07% for *nj* = 10 and −45.33% for *nj* = 1000. Similarly, lower is the number of jobs *n* and closer are the average offsets when *rf* varies. For instance, when *nj* = 10, the average offset is .00% for *rf* = .2 and −3.07% for *rf* = .8. On the contrary, when *nj* value is higher, the average offset varies significantly with the *rf* value. For *nj* = 1000, the average offset is −6.33% for *rf* = .2 and −45.33% for *rf* = .8.

6. Conclusions

In this paper, an enhancement to the BA has been presented. The basic BA was modified to find the most promising patches, especially local optima, using SAC followed by HCA, for convergence, during the local search process. By including the improvement, both the search speed was improved and more accurate results were obtained. The enhancement to the BA was tested on 10 selected benchmark functions first and on a well-known manufacturing problem as the single machines scheduling in order to show its applicability and the performance in solving a real-world problem. Regarding the benchmark functions, according to the results of the accuracy analysis and the average evaluation, the proposed algorithm performed better on most of the benchmark functions. Based on the *T*-test results, it can be concluded that the proposed algorithm is statistically significant than the basic BA on higher dimensional and complex optimisation problems.

Regarding the single machine scheduling problem, the proposed SACHCA-BA outperformed the basic BA in almost all the considered instances, in particular when the complexity of the problem increases.

References

Ahmad, S. A. (2012). *A study of search neighbourhood in the Bees Algorithm* (PhD thesis). Cardiff University, Cardiff.

Ang, M. C., Ng, K. W., & Pham, D. T. (2013). Combining the Bees Algorithm and shape grammar to generate branded product concepts. *Proceedings of the Institution of Mechanical Engineers, Part B: Journal of Engineering Manufacture, 227*, 1860–1873.

Ang, M. C., Ng, K. W., Pham, D. T., & Soroka, A. (2013). Simulations of PCB assembly optimisation based on the Bees Algorithm with TRIZ-inspired operators. In H. B. Zaman, P. Robinson, P. Olivier, T. K. Shih, & S. Velastin (Eds.), *Advances in visual informatics* (pp. 335–346). Selangor: Springer.

Azfanizam, A. S., Pham, D. T., & Faieza, A. A. (2014). Combination of adaptive enlargement and reduction in the search neighbourhood in the Bees Algorithm. *Applied Mechanics and Materials, 564*, 614–618.

Beni, G., & Wang, J. (1989). *Swarm intelligence in cellular robotic systems.* Proceeding of NATO Advanced Workshop on Robots and Biological System, Tuscany.

Castellani, M., Pham, Q. T., & Pham, D. T. (2012). Dynamic optimisation by a modified Bees Algorithm. *Proceedings of the Institution of Mechanical Engineers, Part I: Journal of Systems and Control Engineering, 226*, 956–971.

Gould, J. L., & Gould, C. G. (1988). *The honey bee.* New York, NY: Scientific American Library.

Grosan, C., & Abraham, A. (2011). *Intelligent systems: A modern approach.* Berlin: Springer.

Hussein, W. A., Sahran, S., & Abdullah, S. N. H. S. (2014). Patch-Levy-based initialization algorithm for Bees Algorithm. *Applied Soft Computing, 23*, 104–121.

Lien, L. C., & Cheng, M. Y. (2014). Particle bee algorithm for tower crane layout with material quantity supply and demand optimization. *Automation in Construction, 45*, 25–32.

Mastrocinque, E., Yuce, B., Lambiase, A., & Packianather, M. S. (2013). A multi-objective optimisation for supply chain network using the Bees Algorithm. *International Journal of Engineering Business Management, 5*, 1–11.

OR-Library, Common due date scheduling. Retrieved April 2, 2014, from http://people.brunel.ac.uk/~mastjjb/jeb/orlib/schinfo.html

Otri, S. (2011). *Improving the Bees Algorithm for complex optimisation problems* (PhD thesis). Cardiff University, Cardiff.

Pham, D. T., Afify, A. A., & Koç, E. (2007). Manufacturing cell formation using the Bees Algorithm. In *3rd International Virtual Conference on Intelligent Production Machines and Systems (IPROMS 2007)* (pp. 523–528). Dunbeath: Whittles.

Pham, D. T., & Castellani, M. (2009). The Bees Algorithm: Modelling foraging behaviour to solve continuous optimization problems. *Proceedings of the Institution of Mechanical Engineers, Part C: Journal of Mechanical Engineering Science, 223*, 2919–2938.

Pham, D. T., Castellani, M., & Fahmy, A. A. (2008). Learning the inverse kinematics of a robot manipulator using the Bees Algorithm. In *Proceedings of INDIN* (pp. 493–498).

Pham, D. T., Castellani, M., & Ghanbarzadeh, A. (2007). Preliminary design using the Bees Algorithm. In *Proceedings Eighth International Conference on Laser Metrology, CMM and Machine Tool Performance, LAMDAMAP* (pp. 420–429). Cardiff: Euspen.

Pham, D. T., Darwish, A. H., Eldukhri, E. E., & Otri, S. (2007). *Using the Bees Algorithm to tune a fuzzy logic controller for a robot gymnast*. Proceedings of IPROMS 2007 Conference, Cardiff.

Pham, D. T., & Ghanbarzadeh, A. (2007). *Multi-objective optimisation using the Bees Algorithm*. Proceedings of IPROMS 2007 Conference, Cardiff.

Pham, D. T., Ghanbarzadeh, A., Koc, E., Otri, S., Rahim, S., & Zaidi, M. (2005). *The Bees Algorithm* (Report No. MEC 0501). Cardiff: Manufacturing Engineering Centre, Cardiff University.

Pham, D. T., Ghanbarzadeh, A., Koc, E., Otri, S., Rahim, S., & Zaidi, M. (2006). The Bees Algorithm – A novel tool for complex optimisation problems. In *Proceedings of the 2nd Virtual International Conference on Intelligent Production Machines and Systems (IPROMS 2006)* (pp. 454–459).

Pham, D. T., Koc, E., Ghanbarzadeh, A., & Otri, S. (2006). *Optimisation of the weights of multi-layered perceptrons using the Bees Algorithm*. 5th International Symposium on Intelligent Manufacturing Systems, Sakarya.

Pham, D. T., Koc, E., Lee, J. Y., & Phrueksanant, J. (2007). Using the Bees Algorithm to schedule jobs for a machine. In *LAMDAMAP, 8th International Conference on Laser Metrology, CMM and Machine Tool Performance* (pp. 430–439), Euspen, Cardiff.

Pham, D. T., Otri, S., Afify, A., Mahmuddin, M., & Al-Jabbouli, H. (2007). *Data clustering using the Bees Algorithm*. 40th CIRP International Manufacturing Systems Seminar, Liverpool.

Pham, D. T., Packianther, M. S., Imaguliyev, A., & Yuce, B. (2012, September). Early neighbourhood search and efficiency based improved the Bees Algorithm. In IMS 2012 Conference (pp. 26–29), Turkey: Sakarya University.

Pham, D. T., Soroka, A. J., Koç, E., Ghanbarzadeh, A., & Otri, S. (2007). *Some applications of the Bees Algorithm in engineering design and manufacture*. Proceedings of International Conference on Manufacturing Automation (ICMA 2007), Singapore.

Seeley, T. D. (1995). *The social physiology of honey bee colonies: The wisdom of the hive*. London: Hardvard University Press.

Sholedolu, M. O. (2009). *Nature-inspired optimisation: Improvements to the Particle Swarm Optimisation Algorithm and the Bees Algorithm* (PhD thesis). Cardiff University, Cardiff.

Talbi, E. G. (2009). *Metaheuristics*. New Jersey, NJ: Wiley.

Von Frisch, K. (1955). *The dancing bees: An account of the life and senses of the honey bee*. New York, NY: Harcourt.

Woeginger, G. J. (2003). Exact algorithms for NP-hard problems: A survey. In M. Juenger, G. Reinelt, & G. Rinaldi (Eds.), *Combinatorial Optimization: Eureka! You shrink!* (pp. 185–207). Aussois: LNCS 2570: Springer.

Wolpert, D. H., & Macready, W. G. (1997). No free lunch theorems for optimization. *IEEE Transactions on Evolutionary Computation, 1*, 67–82.

Yuce, B. (2012). *Novel computational technique for determining depth using Bees Algorithm and Blind Image Deconvolution* (PhD thesis). Cardiff: Cardiff University.

Yuce, B., Mastrocinque, E., Lambiase, A., Packianather, M. S., & Pham, D. T. (2014). A multi-objective supply chain optimisation using enhanced Bees Algorithm with adaptive neighbourhood search and site abandonment strategy. *Swarm and Evolutionary Computation, 18*, 71–82.

Yuce, B., Packianather, M. S., Mastrocinque, E., Pham, D. T., & Lambiase, A. (2013). Honey bees inspired optimization method: The Bees Algorithm. *Insects, 4*, 646–662.

5

Inventory classification based on decoupling points

Joakim Wikner*[,1] and Eva Johansson

Department of Industrial Engineering and Management, School of Engineering, Jönköping University, Jönköping, Sweden

The ideal state of continuous one-piece flow may never be achieved. Still the logistics manager can improve the flow by carefully positioning inventory to buffer against variations. Strategies such as lean, postponement, mass customization, and outsourcing all rely on strategic positioning of decoupling points to separate forecast-driven from customer-order-driven flows. Planning and scheduling of the flow are also based on classification of decoupling points as master scheduled or not. A comprehensive classification scheme for these types of decoupling points is introduced. The approach rests on identification of flows as being either demand based or supply based. The demand or supply is then combined with exogenous factors, classified as independent, or endogenous factors, classified as dependent. As a result, eight types of strategic as well as tactical decoupling points are identified resulting in a process-based framework for inventory classification that can be used for flow design.

Keywords: decoupling points; inventory classification; planning and control; materials management

1. Introduction

Increasing customer requirements for customization, cost issues, and ever shortening product life cycles result in a dynamic behaviour and structural complexity of the logistics system that are very demanding. The decisions are not only a complex matter due to the intricacy of each individual decision but also the sheer volume of decisions adds weight to the burden. In addition, the reduction of inventories in the logistics systems leads to stronger dependencies across the material flows, which increases the risk for knock-on effects of disturbances as pointed out by, among others, Goldratt and Cox (1984). Considering these characteristics in combination, it is obvious that logistics managers need tools to simplify and automate decision-making. Important concepts in this context are abstraction, in order to retain only information that is relevant for a particular decision problem, and aggregation, to focus on groups of entities where the same decision can be implemented.

From a decision-maker's perspective, the implications of applying abstraction (selecting relevant information) and aggregation (grouping of information) are significant. Abstraction is usually applied through hierarchical decision-making. At the lowest level, detailed decisions are usually made with a local and short-term focus. At higher levels, issues of more strategic character are handled. Typical examples at the strategic

*Corresponding author. Email: joakim.wikner@ju.se

[1]The paper is a revised and extended version of Wikner, Johansson, and Persson (2009).

level are to focus on constraining resources and materials critical to customer service. Aggregation can be applied by aggregating decision problems and making a decision based on a key property of the aggregate, such as a product group, and then more or less automatically applying the decision to all products of that product group. The terminology used in this paper refers to this type of aggregation as classification. By classifying an item as a specific type due to some of its properties, a group of items sharing this property is created.

These two concepts, abstraction and aggregation, are the theme of this paper. Item and inventory classifications of different types are very common in logistics management (see, e.g. Broeckelmann, 1999), and our ambition is to further emphasize the importance of a process perspective in this context. Previously, much focus has been on classifications using item properties as a point of departure, e.g. ABC classification, but with an additional process perspective, further support can be provided to logistics management. The purpose of the research presented in this paper is to develop a process-based framework for inventory classification. The application of the framework in planning and control is discussed in order to show how the results can be used.

The paper is structured as follows. Next, a general overview of the inventory management context is provided in terms of decoupling points. The concept of supply system is introduced and some key properties of decoupling points related to replenishment and withdrawal are defined. Thereafter, a generic structural model for logistics systems is introduced to identify four key delivery strategies and their respective positions of the two critical decoupling points of strategic importance. Finally, a process-based approach to inventory classification is derived from the discussion on strategic and tactical decoupling points, followed by managerial implications for planning and scheduling, and some ideas for further research.

2. Decoupling points and material flows

The context of material flows is here referred to as a supply system that can be described as a set of resources transforming raw materials into finished goods. Throughout the material flow, resources interact in the transformation process to create value in terms of form, time, and place. In an ideal state, the material flow is continuous and one piece (Ohno, 1988, p. 96) meaning that there is a steady and even flow where all the resources are synchronized and performing with a minimum of balancing loss in the system. This ideal state is, however, rarely possible to realize in the complete supply system. Usually, this type of material flow is limited to a set of resources performing a number of steps continuously but in a context of only a manufacturing cell or a line. In the context of place transformation, the only case where continuous flow can be achieved, except for in pipeline transportation, is through aggregation of several material flows to, for instance, a truck load, thus providing one-piece flow for each separate material flow. Using common lean terminology, this kind of well-defined section of one-piece flow is referred to as a process (Rother & Shook, 2003, p. 18). Each process then represents transformation and replenishes a point of inventory downstream and withdraws materials from one or more other points of inventory upstream. At a 'micro' level, the material flows can therefore be modelled as a set of stock points, i.e. inventories, each connected with processes that withdraw materials and processes that replenish materials. Each inventory can hence be seen as a point that decouples two processes (replenishment and withdrawal) and enables them to act somewhat independently. This kind of decoupling point is referred to as a DP (this definition is in line with the general perception of the concept, see, e.g. Blackstone (2008). The supply system can hence be modelled as a set of DPs that interact through replenishment and withdrawal processes. Note that a process that is a replenishment process for one DP can be a withdrawal process for another DP. The only exceptions are the replenishment processes related to supply external to the supply system (e.g. external suppliers) and withdrawal processes that are external to the supply process (e.g. external customers).

2.1. *Decoupling points categorization*

The concept of DP is a core component in the management of material flows. Considering different combinations of unexpected and expected events, related to replenishment and withdrawal, for each and every DP results in a very demanding challenge for management since the number of DPs in an enterprise or supply chain is staggering. It is very difficult in all realistic settings, at least with today's technology, to control each and every DP to any greater extent. To handle this complexity, the DPs can be differentiated into categories. In particular three categories play a key role in materials management; they are here referred to as operational, tactical, and strategic:

- *Operational DPs* connect transformation points in the physical material flows and are, for example, known as queues in front of resources.
- *Tactical DPs* are related to materials given a unique identifier, usually referred to as item number or part number. Introducing this identity code enables item-level materials management.
- *Strategic DPs* are a subset of all tactical DPs that play a role of critical importance to the interface of the supply system and its context.

2.2. *Decoupling points and planning and scheduling*

Strategic DPs relate to items that are exposed to external influence on the withdrawal side, such as customer orders, or on the replenishment side, such as purchase orders. External demand is not possible to derive from demand for other items and is hence usually unpredictable in terms of both quantity and timing. Classifying items as exposed to external influence or only as internally affected corresponds to the classification in terms of dependent and independent demand, which is the fundamental principle behind materials planning (Vollmann, Berry, Whybark, & Jacobs, 2005, p. 134). Demand for an item that can be derived from demand for other items is in this context referred to as dependent demand (DD). In its most straightforward application, demand for a component is calculated as the demand for the parent item multiplied by the explosion quantity. Demand that cannot be calculated in this way but is directly affected by external demand is thus classified as independent demand (ID). This classification of demand has a long history and is particularly associated with material requirements planning, MRP (Orlicky, 1975), of which backward scheduling is a key concept. As this scheduling approach is performed, the demand for the next level of items is calculated in the bill-of-material (BoM) explosion process. Many items with DD may also have a certain proportion of ID, if they, for example, are sold separately as spare parts besides being used in the production of another parent item. In summary, DD items are exposed to internal demand and hence of a less strategic nature. ID items, on the other hand, are exposed to external demand, e.g. customer orders, and due to their strategic nature correspond to strategic DPs. The similarities between independent and strategic DPs have so far been covered from a demand-based perspective but, as will be shown below, the corresponding similarities can also be identified from a supply-based perspective. Integrating the demand-based and the supply-based perspectives results in a more comprehensive approach to the classification of DPs.

3. The logistics system

The basic supply system outlined above corresponds to an inventory subsystem related to one DP connected with a replenishment process and a withdrawal process. In isolation, this is a relatively simple entity where item-based inventory management can be applied. An inventory subsystem is, however, usually a part of a larger logistics system consisting of a set of connected inventory subsystems. As the complete logistics system is considered, a range of dependencies is introduced, which increases the complexity of the decision problem. Instead of dealing with single-item inventory management, the decision-maker is facing a scenario with a network of inventory subsystems that interact with each other, and to some extent also interact with the context of the complete logistics system. This type of system also corresponds to the logical entity introduced by Wikner (2014) as a supplement to the traditional actor-based approach to flow design. Such a complex decision problem is very difficult to handle with a monolithic model and therefore requires some kind of hierarchical approach to system analysis providing the system analyst with an appropriate abstraction level for each decision. Above, the lower abstraction level in terms of one individual DP was introduced. Next, the higher abstraction level is introduced involving the key concepts of strategic DPs and strategic supply systems. In particular, we are interested in DPs exposed to exogenous influence. Exogenous influence refers to external influence, of which the customer order is the most important. Internal influence, on the other hand, is dependent on other internal events and is thus endogenous.

3.1. Demand and supply categories

To discriminate between different DPs with the objective of identifying the vital few, it is necessary to identify the DPs exposed to exogenous influence. Dependent and independent demands (DD and ID) were described earlier and are classified in Table 1. To the authors' knowledge, there is no corresponding general conceptual framework for bottom-up approaches where the supply is the driver in the process. More specific approaches do, however, exist, such as the forward flow scheduling approach of process flow scheduling (Taylor & Bolander, 1994). Taking supply as the point of reference would result in employing a forward scheduling approach to generate supply for the next higher level in the BoM. Note that a common denotation of levels in a BoM is that

Table 1. Summary of demand and supply categories.

	Supply	Demand
Independent	Supply that must be estimated (purchase orders)	Demand that must be estimated (forecast and customer orders)
Dependent	Supply that can be calculated (forward scheduling)	Demand that can be calculated (backward scheduling)

the top level is 0 and for each subsequent lower level the level number increases. In analogy with the demand-based concept, this is referred to as dependent supply (DS). Using the concept of forward scheduling, the supply of material on level i in the BoM would be dependent on supply on the next lower level, level $i + 1$, during the scheduling phase. In a backward scheduling scenario, the demand for items on level i would be dependent on the demand for items on the next higher level, level $i - 1$. The key property of DD and DS is that they both can be calculated based on the available endogenous information of the logistics system, such as lead times and available inventory.

The description above shows the clear analogy between DS and DD. The next step is to identify independent supply (IS) as an analogy to ID. This analogy is, however, not as clear-cut, but we define supply from a source outside the logistics system as IS since it is outside the strict control of the logistics system. Control can cover a complete manufacturing network or just a single site. The important aspect is that within the controlled supply system there is no need for traditional purchasing orders; instead internal orders (sometimes referred to as distribution orders) can be used covering both the demand process from one DP and the supply process of another DP. In summary, four types of drivers can be identified in the material flows as in Table 1, i.e. the intersections between supply/demand and independent/dependent.

Based on these two types of independent drivers (IS and ID), two types of corresponding DPs can be identified; they are discussed in the following two sections.

3.2. Demand-based decoupling point

Delivery strategies such as make-to-stock (MTS) and make-to-order (MTO) play a key role in providing an appropriate balance between efficiency and responsiveness in meeting market demand (see, e.g. Chopra & Meindl, 2007). The baseline for designing the appropriate strategy in this perspective is the relation between the delivery lead time (D), requested by the market, and the total supply lead time (S) within which the system needs to produce the output. The ratio between these two lead times is critical; it was first identified by Shingo (1981) as the D:P relation. Mather (1984) changed the relation to the inverse for practical reasons and since then it is known as the P:D ratio or P:D relation. Shingo used P as he was referring to product lead time. In a similar vein, the more general supply lead time is used and thus referred to as S (see, e.g. Wikner, 2014). S more explicitly involves, for example, also purchasing and distribution lead times. If S is shorter than D, all supply activities can be performed while the customer is waiting. But when S is longer than D some activities must be performed in advance to create semi-processed parts to be used when the customer order arrives. The supply system is thus split into two parts corresponding to two subsystems. The subsystem supplying semi-processed parts is performed before the customer order arrives and is therefore forecast-driven. The second subsystem uses the semi-processed parts in stock to deliver on the customer order. This is sometimes referred to as customer-order-driven (see, e.g. Hoekstra & Romme, 1992) but also as demand-driven (see, e.g. Christopher, 1998). The semi-processed parts are kept in stock at a decoupling point that is known as the customer order decoupling point, CODP, (see, e.g. Giesberts & van der Tang, 1992) but also as order penetration point, OPP (Sharman, 1984) or simply as decoupling point (Hoekstra & Romme, 1992). The CODP has been defined in a number of different ways but, as summarized by Wikner (2014, p. 195), '[t]he customer order decoupling point (CODP) separates decisions about initiating flow based on speculation on future customer orders from commitment against actual customer orders'. Using these

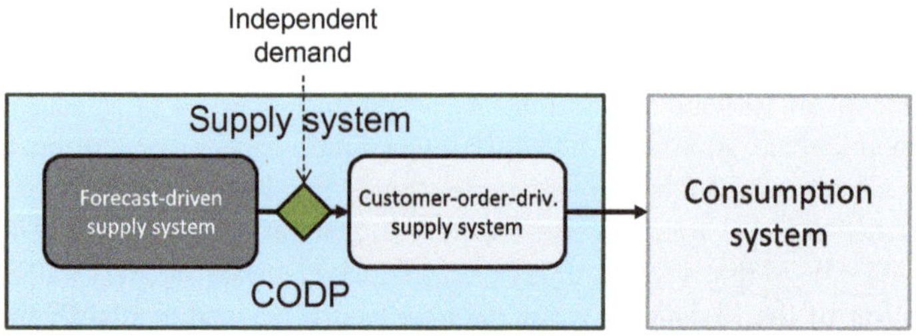

Figure 1. The CODP perspective of the supply system.

three building blocks (forecast-driven supply system, CODP, and customer-order-driven supply system), the supply system can be depicted as in Figure 1, where a consumption system has been introduced as an origin of demand and to represent the customers of the supply system. The diamond is used to represent the CODP in line with the guidelines suggested by Wikner (2014).

Considering a customer order arriving in the supply system of Figure 1, the customer must accept to wait during the time it takes to perform the customer-order-driven supply system (during the lead time D). This system lead time could involve activities outside manufacturing (see, e.g. Rudberg & Wikner, 2004). To replenish the CODP, some kind of forecast is necessary since customer orders are assumed to arrive in time to perform only the customer-order-driven supply system in Figure 1. The forecasting can be performed at the end-item level but if any parts are shared between different end items it is easier, due to aggregation effects reducing the coefficient of variation in demand, to forecast at the CODP on semi-processed parts. This corresponds to defining the CODP parts as subject to ID as shown in Figure 1. Activities downstream of the CODP may be performed under certainty of demand since a customer order is driving the activities. The parts at the CODP must, however, be produced to forecast since they are expected to be available at the moment the customer order arrives.

3.3. Supply-based decoupling point

The CODP emphasizes the relation between the delivery lead time D and the supply lead time S, and hence investigates the demand-related properties of S but does not provide any further information on what actually constitutes supply. Taking a supply perspective, an important concept is to what extent the complete supply system is integrated. In this case, the focus is on inventory and materials management and therefore also the level of integration is defined based on the level of control that can be exercised on the supply system. In an integrated supply system, the complete set of supply activities would be controlled as one entity. If, however, the supply system is not integrated, the supply system could be separated into two subsystems, here referred to as one controlled subsystem and one uncontrolled subsystem, see Figure 2. A hexagon is used to illustrate the purchase order decoupling point (PODP) in line with Wikner (2014). The uncontrolled subsystem is out of control reach for the controlled subsystem.

An example is a component supplier of a company that supplies parts per order and where there is no further integration. A purchase order is sent to replenish the inventory when required. In this case the supplier would be modelled as an uncontrolled system including not only the supplier but also all the supplier's suppliers, etc. Using this

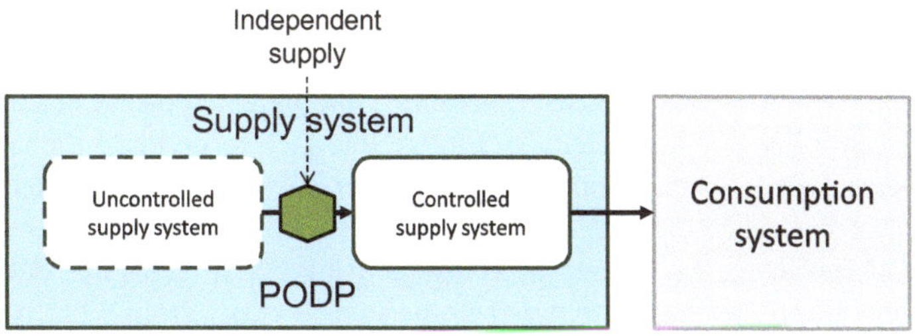

Figure 2. The PODP perspective of the supply system.

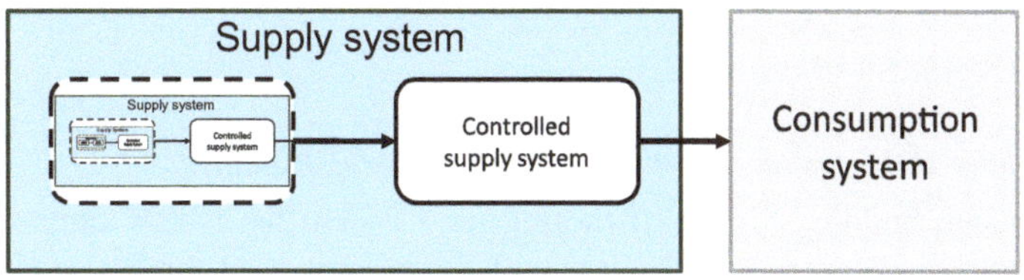

Figure 3. Recursive model of a supply system.

approach, a recursive systems definition is obtained where the controlled supply system and how it interacts with the consumption system is the focus. Remember that from the uncontrolled supply system perspective, the controlled supply system would represent the consumption system as depicted in Figure 3. This is because all upstream activities not covered by the controlled supply system are included in the uncontrolled supply system.

Within the controlled supply system, there is full transparency in terms of planning and execution which enables a cost-efficient allocation of decoupling points, i.e. buffers, from a system's perspective. The uncontrolled supply system does, however, not provide this level of transparency by definition. Since control cannot be exercised on this system by the controlled supply system, the only way to manage unexpected events occurring in this system is to introduce a buffer explicitly targeting these properties and decoupling the controlled system to protect it. As shown by Lee (2002), this relation must be carefully designed and alternative strategies for minimizing unexpected events should be considered, but if uncertainty prevails it must be handled with some type of buffer, otherwise customer service and/or cost efficiency will suffer. Since the relation with the uncontrolled supply system usually is through some kind of purchase order, we introduce the term PODP for this kind of buffer, as in Figure 2. Activities within the controlled supply system are usually coordinated by production orders and distribution orders but when it comes to coordinating with an external system, the relation is usually handled through a purchase order. This model is also related to the lead time based analysis performed by Wallin, Rungtusanatham, and Rabinovich (2006). They related the delivery lead time D to production time P in the spirit of Shingo (1981), but then made a distinction between the part of P covered by the supplier and the part covered by the focal company. Using this approach, different strategies for inventory management of purchased material were identified, but here the emphasis is on the complete supply system structure and not only purchased materials at the PODP. In Wallin et al.

(2006), the focal company is the unit of analysis. Here, however, the unit of analysis is a supply system, and to emphasize what specific supply system is considered it may be referred to as the focal supply system. In Figure 2 the focal supply system would correspond to the controlled supply system. A similar approach to dividing the supply system into subsystems based on exogenous influence has also been investigated by Wikner (2005), but then from the perspective of efficiency and responsiveness.

The PODP can hence be defined as follows: '[t]he PODP separates decisions about delimiting flow based on what is external to the logical entity from what is internal and hence controllable' (Wikner, 2014, p. 199). This indicates that the uncontrolled supply system provides input to the controlled supply system. This input contains more uncertainty compared to what is perceived from the controlled supply system and thus requires some kind of estimation (forecasting), which also is the key property of IS, see Table 1.

3.4. Modelling the logistics system based on delivery strategies

The fundamental building blocks of the logistics system have been defined above and are summarized in Figures 1 and 2. Basically, the properties of the supply systems highlighted are type of driver, related to CODP, and level of control, related to PODP. Each particular logistics system has its own unique characteristics based on its history, market conditions, available resources, etc. There are, however, some generic configurations that can provide a baseline for illustrating how the constructs introduced above can be employed to represent an enterprise from a logistics perspective. The two fundamental configurations used here are the completely customer-order-driven system (MTO) and the completely forecast-driven system (MTS). The third system (assemble-to-order, ATO) represents a compromise system with elements of both MTO and MTS (Wemmerlöv, 1984). The fourth system, finally, involves also the possibility of outsourcing not just forecast-driven activities to uncontrolled supply systems but also customer-order-driven activities; this is referred to as purchase-and-make-to-order (PMTO) (Hoekstra & Romme, 1992).

3.4.1. Make-to-stock (MTS)

When the MTS strategy is used, all activities are forecast-driven (with the possible exception of some picking and delivery activities). In this case, the CODP is positioned as far downstream in the material flow as possible. No activities are customer-order-driven and

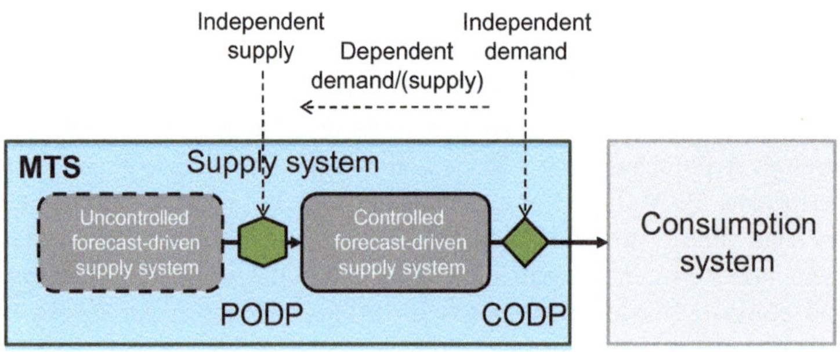

Figure 4. Modelling make-to-stock (MTS) supply systems.

usually the master schedule (see, e.g. MRPII) or pacemaker (see, e.g. rate-based master scheduling as in lean) is positioned at the end-item level, where ID is present. This allows for a materials planning approach based on DD (backward scheduling), which is the most common approach in this setting, for all items down to the PODP. The PODP hence represents the extension of the integrated planning and control approach as shown in Figure 4. In Figures 4–7, the dashed vertical arrows represent exogenous input to the supply system. The dashed horizontal lines represent endogenous propagation of demand and supply respectively, in the direction indicated by the arrow.

3.4.2. Make-to-order (MTO)

The second fundamental configuration corresponds to the MTO strategy, where all activities of the controlled system are customer-order-driven. When the customer order is received, the customer waits until the product is finished. The planning approach can be based on DD (as in MRP), where the end item is master scheduled and DD is calculated for all other items down to the CODP. A preferred scenario would, however, in some cases be to start with the CODP items availability and then to forward schedule all items (DS) up to the end item to provide, for example, information about delivery capability. This process would then preferably also involve finite loading considering any finite capacity to provide reliable delivery dates. Since the CODP here is positioned at the beginning of the controlled supply system, it coincides with the PODP as shown in Figure 5.

3.4.3. Assemble-to-order (ATO)

A preferred strategy in many cases is to establish a level in the BoM that allows for modularization. Different modules can then be used in a multitude of configurations enabling a mass customization approach. Modules are then produced to forecast using a standard MTS 'sub-strategy', whereas the final assembly is customer-order-driven and similar to the MTO approach. Setting the two models of MTS (Figure 4) and MTO (Figure 5) in tandem results in an ATO scenario as in Figure 6. The CODP is positioned between the forecast-driven and the customer-order-driven supply systems, and the PODP is also in this case positioned at the interface between the uncontrolled and the controlled supply systems.

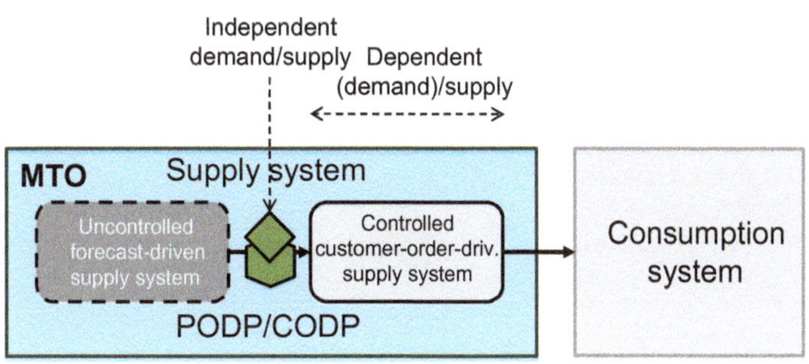

Figure 5. Modelling make-to-order (MTO) supply systems.

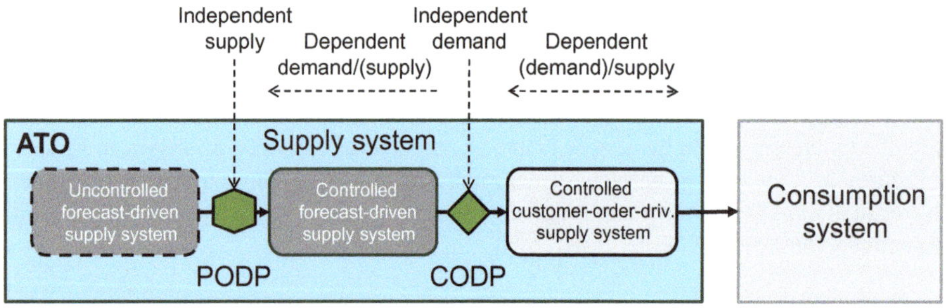

Figure 6. Modelling assemble-to-order (ATO) supply systems.

3.4.4. Purchase-and-make-to-order (PMTO)

If only the CODP is considered, the three strategies identified above, MTS, MTO, and ATO, are key scenarios. Each strategy consists of a unique combination of forecast-driven and customer-order-driven supply systems. Taking a broader perspective and also considering that the delivery lead time D may extend beyond the lead time of the controlled supply system provide leeway for a fourth scenario. This scenario involves an external supplier (here referred to as an uncontrolled supply system) that is customer-order-driven, as illustrated in Figure 7. There are now two customer-order-driven supply systems separated by the PODP, and the CODP is positioned upstream of these two supply systems. This is a possible, if not always desirable, strategy as some activities performed while the customer, i.e. the consumption system, is waiting are performed beyond the control of the supply system responsible for delivery to the customer. A possible approach in this case would be to seek to integrate, i.e. 'in-source', the uncontrolled customer-order-driven supply system and, if this is not possible, to establish a safety lead time at the PODP, unless all uncertainties between the CODP and the PODP can be eliminated.

3.4.5. Summary of the four logistics system configurations

To conclude the discussion on logistics systems configured based on the concepts of customer-order/forecast-driven and uncontrolled/controlled, the configurations are summarized in Table 2. The four concepts provide four possible configurations of supply systems as represented by the columns. In addition, it is also required, by definition, that

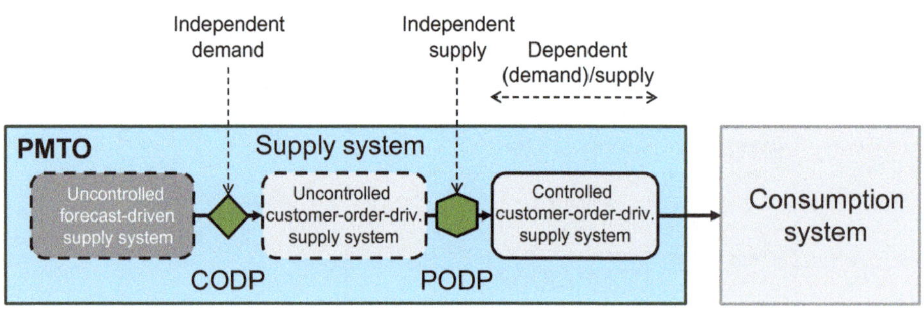

Figure 7. Modelling purchase-and-make-to-order (PMTO) supply systems.

Table 2. Summary of the four possible configurations of the logistics supply system.

	Uncontrolled forecast-driven	Controlled forecast-driven	Uncontrolled customer-order-driven	Controlled customer-order-driven
Config. MTS	X	X	–	–
Config. ATO	X	X	–	X
Config. MTO	X	–	–	X
Config. PMTO	X	–	X	X

certain precedence constraints are fulfilled. A forecast-driven supply system should not be positioned downstream of a customer-order-driven supply system. This is because the supply systems are positioned along a time axis and if it is possible to be customer-order-driven, this is assumed to always be the preferred option. Why forecast-driven if actual demand is available? The second precedence constraint is that uncontrolled supply systems are positioned upstream from controlled supply systems. One exception to this logic is when certain operations within a routing are outsourced. This precedence constraint is nevertheless assumed here.

To verify that there are no other fundamental configurations regarding these constraints, a simple analysis can be performed. To begin with, it is assumed that there is always the possibility of having uncontrolled forecast-driven activities which corresponds to purchasing materials to forecast. This means that the configurations of the leftmost column of Table 2 are valid (uncontrolled forecast-driven). Configurations MTS and ATO in Table 2 both involve a forecast-driven supply system that also is controlled. This means that there can be no uncontrolled customer-order-driven activities without invalidating any of the precedence rules. The only option available is thus to include or exclude controlled customer-order-driven activities, which results in MTS and ATO. If there are no controlled forecast-driven activities, as in configurations MTO and PMTO, that is only possible with two different configurations. Either all customer-order-driven activities are controlled (MTO) or parts of them are uncontrolled (PMTO). If all activities were uncontrolled, the investigated system would be empty and here considered as void.

These four configurations of logistics supply systems highlight the structural importance of the two types of decoupling points: CODP and PODP. Based on this insight, we return to the complete set of DPs to introduce a classification that takes this insight as a point of departure.

4. Process-based inventory classification using decoupling points

Inventory classification is usually based on some properties of the items in inventory such as value, weight, or perishability. In the case of classification based on, e.g. withdrawal frequency or supplier characteristics, more emphasis is placed on a process perspective but one of the most important properties of processes is still not considered. As processes represent all the value-adding and non-value-adding activities performed in an enterprise, a key factor is how and why these processes are initiated, i.e. what triggers a process. In the analysis of the logistics system above, two important triggers were identified. Basically, the triggers can be characterized as exogenous or endogenous. An exogenous trigger refers to external influence, of which the customer order is the most important, and an endogenous trigger refers to internal influence, corresponding to DD and DS. Finally, a process is also vulnerable when it crosses organizational boundaries, and the relation to external suppliers is of particular interest since it is a consequence of a make/buy decision and represents the reach of what is within process control of the business. In all, there are four process-based concepts that can be used in classifying the DPs and hence inventory. They are related to dependent or independent, which can be combined with a supply or demand perspective in four different ways. These four concepts (dependent, independent, demand, and supply) are combined below and then classified as strategic (related to the choice of delivery strategy and level of outsourcing) or tactical (related to other item-based DPs of importance when synchronizing the material flows).

4.1. Strategic process-based decoupling points

The customer order triggers the processes in the customer-order-driven supply system. From an inventory perspective, the DP of outstanding importance is the CODP which is subject to ID and crucial to the service level that can be achieved. The CODP can be seen as the incarnation of the delivery strategy since the position of the CODP is based on the delivery lead time. The positioning of the CODP has been identified as a key strategic decision, for example, in terms of the CODP itself (Olhager, 2003), as a consequence of balancing efficiency and responsiveness (Chopra & Meindl, 2007), as the point where lean meets agile (leagility) (Mason-Jones, Naylor, & Towill, 2000), or where the physical and innovative parts of the supply chain meet (Olhager, Selldin, & Wikner, 2006).

The positioning of the PODP is also strategic in the sense that it represents the selected level of vertical integration, i.e. the level of outsourcing. It is also crucial for buffering external disturbances and hence for providing the internal flows with the best possible conditions. From an inventory perspective, a second important property of the PODP is the amount of control that can be exercised on a process, and it is assumed that uncontrolled subsystems constitute a risk and may have a negative impact on the controlled supply system, which is downstream from the PODP.

The inventory at the CODP enables the forecast-driven supply system to focus on efficiency since the inventory at the CODP buffers against seasonal variations as well as short-term fluctuations related to demand. The dimensioning of the CODP inventory is usually quantity based, as it is the impact of the demand variations on the service level that is in focus. At the PODP, on the other hand, the situation is more concerned with the reliability of the supply lead times and hence one is more inclined to use safety lead time at this DP. In addition, the PODP also handles the lot-size inventory due to the delivery frequency from the external suppliers.

In summary, the strategic DPs are either driven by downstream exogenous factors (related to external customers through customer orders) or driven by upstream exogenous factors (related to external suppliers through purchase orders). The strategic DPs

Table 3. Strategic process-based decoupling points.

	No independent demand (NID)	Independent demand (ID)
No independent supply (NIS)	–	NIS–ID
Independent supply (IS)	IS–NID	IS–ID

are categorized in Table 3 based on how they are related to these two types of exogenous triggers. If the trigger is only related to ID with no independent supply (NIS), the decoupling point is of the type NIS–ID, better known as CODP (see Figure 1). The reverse, IS but no independent demand (NID), is indicated as IS–NID and defined above as PODP (see Figure 2). These two types of strategic DPs may be separate but still present in the same flow as shown in Figures 4, 6, and 7. However, in some particular cases, a strategic DP can be exposed to both independent supply and independent demand (IS–ID), which represents the case where a DP is both a CODP and a PODP (see Figure 5). The fourth case finally, with NIS and NID, is not included in terms of strategic DPs but will be further analysed next at the tactical level.

4.2. Tactical process-based decoupling points

Tactical decisions should follow as a consequence of the decided strategy. In the same way the positioning of DD and DS is a direct consequence of ID and/or IS. The strategic DPs create a structural framework based on the delivery strategy and the outsourcing strategy. The objective of the positioning of DD and DS DPs is to support this strategy. These DPs are endogenously triggered to support the ID of all related items using the bill-of-material/distribution as the road map. In all, three DP classes can be identified at the tactical level as shown in Table 4. DPs with no dependent supply (NDS) but with DD, i.e. type NDS–DD, are backward scheduled based on the demand downstream, and DPs with DS but no dependent demand (NDD), i.e. type DS–NDD, are forward scheduled based on the supply upstream. Since backward and forward scheduling in practice are difficult to combine at the same DP, the combination of dependent supply and dependent demand (DS–DD) is rarely established. An example could however be when one item is part of two different products one of which is made to customer order and the other is sold from finished goods, i.e. a hybrid of MTS (see Figure 4) and MTO (see Figure 5) for that particular item. The combination involving no dependent supply and no dependent demand (NDS–NDD) is not included at the tactical level since a tactical driver is not present. DPs classified as tactical are generally used during the planning phase to establish the appropriate level of inventory over time at these decoupling points.

4.3. Summary of process-based inventory classifications

Three types of DPs were identified above as strategic and related to ID (CODP-based) and IS (PODP-based). If NID and NIS were present (upper left cell of Table 3), the DPs would be classified as tactical and could still be differentiated as exposed to DD or DS as in Table 4. In total, these two tables highlight six different types of DPs. By combining these two tables, two of the strategic DPs, IS–NID and NIS–ID, may be further differentiated. Compared to Table 3, NID is differentiated into no demand (ND) and DD. If there is NID there could still be DD. NIS is differentiated into no supply

Table 4. Tactical process-based decoupling points.

	No dependent demand (NDD)	Dependent demand (DD)
No dependent supply (NDS)	–	NDS–DD
Dependent supply (DS)	DS–NDD	(DS–DD)

Table 5. Process-based inventory classifications.

	No demand (ND)	Dependent demand (DD) (backward scheduled)	Independent demand (ID) (CODP based)
No supply (NS)	–	NS–DD	NS–ID
Dependent supply (DS) (forward scheduled)	DS–ND	(DS–DD)	DS–ID
Independent supply (IS) (PODP based)	IS–ND	IS–DD	IS–ID

(NS) and DS in a corresponding way as in Table 5. This results in five different types of strategic DPs, as highlighted in Table 5. In combination with the three types of DPs in Table 4, there are in total eight types of DPs (inventory) that have been identified using the process-based approach suggested here, with a foundation in planning and control. Five of them are of a strategic character involving either IS or ID. Three classes of DPs are of a tactical character involving NIS and NID as summarized in Table 5. Table 5 is a combination of Table 3 (mainly represented by the white area) and Table 4 (represented by the shaded area). 'No Demand (ND)' in Table 5 represents that the use of inventory in the DP is not directly driven by demand. Correspondingly, 'No Supply (NS)' refers to the fact that the replenishment of the DP is not driven by supply pushing items into the DP. A DP with a combination of ND and NS would be completely disconnected from the system and is therefore not classified in Table 5.

In essence, Table 5 represents eight (nine) classes of decoupling points, i.e. inventory points. Hence, the table constitutes a classification of inventory based on the intersection between the demand perspective (the columns) and the supply perspective (the rows). Note that these DPs are tactical or strategic meaning that all DPs are related to items. From this perspective, Table 5 can be seen as providing eight different boxes where all items in an enterprise can be allocated to at least one box. In some cases, one individual item can be related to two boxes if it, for example, is used in two different end products where one is MTS and the other is MTO. It is important to note that Table 5 does not provide any information on if an item is forecast-driven or customer-order-driven, except for the CODP items which are forecast-driven since they are expected to be in stock when the customer order arrives. The table does not provide a set of possible systems but rather the building blocks that can be used for designing any type of system. The NS–ND category should be empty since all items are either driven by demand or supply.

5. Managerial implications

The eight different classes of DPs in Table 5 represent an abstraction in the sense that these are the DPs that should be handled at a strategic and at a tactical level. They also represent an aggregation in the sense that each cell in Table 5 represents a type, or group, of DPs that have certain prerequisites concerning, for example, planning and control. Using the terminology of the ABC classification (see, e.g. Vollmann et al., 2005), the process-based classification in Table 5 can be summarized as generating three classes of inventory (DPs) to focus on. Class A is considered as the most important class and is fundamental for the order fulfilment process. Class B is the second in rank and relates to exogenous uncertainty that is difficult to control, and Class C, finally, is of least importance since it is within the controlled system:

- Class A: CODP inventory (column 3 in Table 5), since it is key to the delivery strategy chosen and a foundation for the delivery reliability to the customer.
- Class B: PODP inventory (row 3 in Table 5), since it is key to the sourcing strategy chosen and based on the delivery reliability of the suppliers.
- Class C: Tactical DPs (grey cells in Table 5), as they represent the link between PODPs, CODPs, and the customers.

Interpreting the different types of DPs in terms of planning and control methods provides some further insights into the significance of the eight types of DPs. The details of this mapping are not the subject of this paper but some guidelines have already been outlined above and some further insights are summarized below.

The framework suggested above and summarized in Table 5 provides a baseline for materials management but does not in itself provide any methods for planning and control. It does, however, highlight some key properties of when a particular planning and control method is appropriate. A key aspect of the framework is process triggering; in materials management this aspect has received considerable interest and in general a process is assumed to be triggered in basically two ways (Hopp & Spearman, 2000, p. 340):

- A customer to the process demands that the process should be performed (frequently referred to as pull). The customer can either be an external customer or an internal customer represented by another process or a point of inventory.
- A schedule demands that the process should be performed (frequently referred to as push). In some instances it is assumed that the schedule in push is based on forecasted demand (see, e.g. Marchwinski, Shook, & Schroeder, 2008), but this is a confusing and misleading assumption in this context. How should, for example, customer-order-driven manufacturing using scheduling software be classified under this assumption? The key issue here is rather that the mode is proactive and the schedule can be based on a wide range of information, whereas the pull approach operates in a reactive mode.

The terminology used in these two cases is somewhat different but still from a functional view they are perfectly in line with the framework introduced here, to which process triggering is central. To provide further support for this conclusion, three examples are presented of what kind of scenarios three different approaches are designed to handle. First, the standard MRPII approach is discussed, followed by takt-based planning and control, and finally an example from the process industry is presented.

The DPs NS–DD and NS–ID correspond to Orlicky's (1975) framework for MRP, where NS–ID is referred to as ID and should be master scheduled, whereas NS–DD is referred to as DD and should be handled by the MRP logic. In case there is NDD (only single-level bill-of-materials), the NS–ID could be managed using traditional inventory control methods such as an order point system. Another example is the classical ATO scenario where the order promising process considers stock availability of modules (NS–ID) at the CODP and the order promising logic checks availability downstream using forward scheduling (DS–ND) to determine possible delivery dates.

In the lean environment, the interpretation of Table 5 requires a distinction between how the system is dimensioned during planning (takt time analysis, line balancing, estimation of number of kanbans, and size of FIFO buffers, etc. (see, e.g. Duggan, 2013) and how it actually works within those boundaries during execution. The pacemaker is frequently positioned at the CODP (see, e.g. Rother & Shook, 2003). Downstream from the pacemaker, the DPs would then be considered as DS–ND and upstream as NS–DD during the planning phase. During the execution phase, all DPs would act as NS–ID responding to pull signals from downstream processes considered basically as ID.

In some process industries, the incoming raw material is not controllable but rather a consequence of long-term contracts, such as in farming. The livestock may be supplied to the abattoir at a certain age and must be processed without delay and therefore represents IS–ND and requires a planning and control approach based on divergent product profiles. Referring to process flow scheduling (Taylor & Bolander, 1994), the DPs covered by reverse flow scheduling (referred to as backward scheduled here) would be NS–DD except for the final stage, which would be NS–ID. Forward flow scheduling (forward scheduled) would be DS–ND except for the first stage, which would be IS–ND or NS–ID depending on whether demand or supply drives the flow. Mixed flow scheduling finally would then be a combination with some central point in the flow being NS–ID and then upstream NS–DD and downstream DS–ND.

These examples highlight how inventory can be classified and related to different scheduling and control techniques. As can be seen from these examples, the approach for the operational DPs here follows from the approach selected for the corresponding item DP (which is not generally necessary if control can be differentiated on an operations level). This means that if the item DPs are of the type NS–DD, then also all the operations to produce that item are assumed to be DD (backward scheduled).

As shown above, the classification framework introduced can be used in a contextual way for a wide range of environments to classify different types of DPs. The approach is hence generic and not only applicable in certain environments. The approach rests on a set of fundamental system properties and can therefore be applied in most environments.

6. Conclusions and further research

Materials management is a comprehensive topic usually involving several thousands of items and a multitude of aspects to analyse such as risk, investment, durability, etc. Using abstraction and aggregation as outlined in this paper, the management of inventory can be more explicitly targeted at focusing on the inventory of strategic importance, i.e. the strategic DPs. Process-based classification enables evaluation of inventory management across different delivery strategies (different positions of the CODP), which are of particular interest to companies with hybrid strategies where, for example, both MTO and MTS products are present. From a customer service perspective the management of the CODP inventory is critical to all kinds of delivery strategies. In combination with evaluation of how the inventory at the PODP is managed, a comprehensive view of management of strategic inventory can be obtained. The CODP is also the key starting point for most planning and control methods for manufacturing and logistics. The framework does not in itself provide a new approach to materials management but it provides a structured way to better understand the role of different DPs, i.e. inventory points, and their role in materials management.

With this classification as a point of reference, there are many possible lines of research to pursue. One of the most important is probably that this classification may guide researchers on inventory and materials management in terms of which DPs to focus on. Also, in case of evaluation of manufacturing strategies across different delivery strategies, the strategic DPs may serve as a point of reference for delineation between lean and agility (see, e.g. Mason-Jones et al., 2000). Giesberts and van der Tang (1992) pointed out that in many cases companies have 'hybrid production situations' and they showed how information systems may take this into account. The impact of this classification scheme on hybrid strategies remains an area of interesting opportunities for further research that could also take the framework suggested by Wikner (2014) into consideration. In the same spirit, the practitioners can use this classification to focus their investment on inventory and materials management where it gives the best leverage. The classification may also work as a tool for categorizing different concepts such as planning and control methods to provide guidelines on how to implement these methods based on strategic as well as tactical decoupling points.

Acknowledgements

This research has been performed in collaboration with six companies in the projects KOPeration and KOPtimera.

Disclosure statement

No potential conflict of interest was reported by the authors.

Funding

The research has been funded by the Swedish Knowledge Foundation (KKS), Jönköping University and the participating companies.

ORCID

Joakim Wikner http://orcid.org/0000-0003-2252-5337
Eva Johansson http://orcid.org/0000-0003-3783-0633

References

Blackstone, J. H. (Ed.). (2008). *APICS dictionary.* Chicago, IL: APICS The Association for Operations Management.

Broeckelmann, R. G. (1999). *Inventory classification innovation: Paving the way for electronic commerce and vendor managed inventory.* Boca Raton, FL: St. Lucie Press.

Chopra, S., & Meindl, P. (2007). *Supply chain management: Strategy, planning, and operation.* Upper Saddle River, NJ: Pearson Prentice Hall.

Christopher, M. (1998). *Logistics and supply chain management: Strategies for reducing cost and improving service* (2nd ed.). London: Financial Times/Prentice Hall.

Duggan, K. J. (2013). *Creating mixed model value streams: Practical lean techniques for building to demand* (2nd ed.). Boca Raton, FL: CRC Press.

Giesberts, P. M. J., & van der Tang, L. (1992). Dynamics of the customer order decoupling point: Impact on information systems for production control. *Production Planning & Control, 3,* 300–313.

Goldratt, E. M., & Cox, J. (1984). *The goal: A process of ongoing improvement.* Croton-on-Hudson, NY: North River Press.

Hoekstra, S., & Romme, J. (1992). *Integrated logistics structures: Developing customer oriented goods flow.* London: McGraw-Hill.

Hopp, W. J., & Spearman, M. L. (2000). *Factory physics* (2nd ed.). New York, NY: McGraw-Hill.

Lee, H. L. (2002). Aligning supply chain strategies with product uncertainties. *California Management Review, 44*, 105–119.

Marchwinski, C., Shook, J., & Schroeder, A. (Eds.). (2008). *Lean lexicon – A graphical glossary for lean thinkers* (4th ed.). Cambridge, MA: Lean Enterprise Institute.

Mason-Jones, R., Naylor, B., & Towill, D. R. (2000). Engineering the leagile supply chain. *International Journal of Agile Management Systems, 2*, 54–61.

Mather, H. (1984). *Attack your P:D ratio*. APICS Conference Proceedings, Falls Church, VA, 84–88.

Ohno, T. (1988). *Toyota production system: Beyond large-scale production*. Cambridge, MA: Productivity Press.

Olhager, J. (2003). Strategic positioning of the order penetration point. *International Journal of Production Economics, 85*, 319–329.

Olhager, J., Selldin, E., & Wikner, J. (2006). Decoupling the value chain. *International Journal of Value Chain Management, 1*, 19–32.

Orlicky, J. (1975). *Material requirements planning: The new way of life in production and inventory management.* New York, NY: McGraw Hill.

Rother, M., & Shook, J. (2003). *Learning to see: Value stream mapping to create value and eliminate muda*. Cambridge, MA: Lean Enterprise Institute.

Rudberg, M., & Wikner, J. (2004). Mass customization in terms of the customer order decoupling point. *Production Planning & Control, 15*, 445–458.

Sharman, G. (1984). The rediscovery of logistics. *Harvard Business Review, 62*, 71–79.

Shingo, S. (1981). *Study of 'Toyota' production system from industrial engineering viewpoint.* Tokyo: Japan Management Association.

Taylor, S. G., & Bolander, S. F. (1994). *Process flow scheduling: A scheduling framework for flow manufacturing*. Falls Church, VA: APICS.

Vollmann, T. E., Berry, W. L., Whybark, D. C., & Jacobs, F. R. (2005). *Manufacturing planning and control systems for supply chain management*. New York, NY: McGraw-Hill.

Wallin, C., Rungtusanatham, M. J., & Rabinovich, E. (2006). What is the 'right' inventory management approach for a purchased item? *International Journal of Operations & Production Management, 26*, 50–68.

Wemmerlöv, U. (1984). Assemble-to-order manufacturing: Implications for materials management. *Journal of Operations Management, 4*, 347–368.

Wikner, J. (2005). *Hantering av osäkerhet i försörjningskedjor med utgångspunkt i frikopplingspunkter* [Management of uncertainty in supply chains from a decoupling point perspective]. Proceedings of Produktionslogistik 2005 – Plans forsknings-och tillämpningskonferens, Borås, 221–236.

Wikner, J. (2014). On decoupling points and decoupling zones. *Production & Manufacturing Research, 2*, 167–215.

Wikner, J., Johansson, E., & Persson, T. (2009). *Process based inventory classification*. The 21st Annual Nofoma Conference, Jönköping Sweden.

6

Inclusive design drivers and barriers – a manufacturing perspective from Pakistan

Amjad Hussain[a], Aftab Ahmad[a] and Keith Case[b]*

[a]Department of Industrial and Manufacturing Engineering, University of Engineering and Technology, Lahore 54890, Pakistan; [b]Mechanical and Manufacturing Engineering, Loughborough University, Loughborough, Leicestershire LE11 3TU, UK

The demographics of older people and people with disabilities in developing countries are discussed in the context of inclusive design and the drivers and barriers to inclusive design have been identified. Data were collected from 50 individuals from various industrial sectors in Pakistan. Corporate social responsibility (CSR) relates to inclusive aspects of products, but most respondents either did not know about CSR or did not have a CSR post in their organizations, but 64% had awareness of inclusive design terminology. The study concluded that motivation through social responsibility; innovation and differentiation; demographics and consumer trends; brand enhancement; customer satisfaction; new market opportunity; and legislation were the perceived drivers for manufacturers in Pakistan. Most respondents felt that lack of resources and guidance, lack of awareness about inclusive design, difficulty in changing the business culture, lack of government regulations, and the perception that inclusive design is expensive were the most significant barriers.

Keywords: inclusive design; drivers and barriers; manufacturing; corporate social responsibility (CSR); developing countries

1. Introduction

Inclusive design aims to address the design needs of the largest proportion of the population where special consideration is given to accommodating older people and people with disabilities along with the rest of the population in a single design solution. The percentage of the older population is increasing throughout the world; however, this trend is quite prominent in developed countries such as the USA, UK, Australia, Japan, Canada, and Germany because of better living facilities, medical treatment, and healthier working environments (U.N.O., 2009). On the other hand, in developing countries, while the average age of the population is lower than that of developed countries, these countries constitute a significant proportion of the overall world population. Attention has previously been given to assessing the level of awareness and exploring drivers and barriers for the promotion of the inclusive design approach in product, process, environment, and service design in developed countries like the UK, USA, and Japan. In spite of the challenging demographics of developing countries considered as the main contributors to the world population, no effort has been made to increase the level of

*Corresponding author. Email: k.case@lboro.ac.uk

awareness of the inclusive design approach, or to explore and highlight the drivers and barriers to the implementation of this design approach. This research is an effort to contribute in this respect by assessing the perspectives of manufacturers on the implementation of the inclusive design method.

2. Literature review

2.1. Demographics

The total population of many countries is increasing. However, this trend is quite significant in developing countries like China, India, Pakistan, and Bangladesh, where the population of these four countries is about 42% of the world population. In the same way, the populations of these countries contain a significant proportion of older people (37% of the world population,) and people with disabilities (6% of the world population) (CIA World Fact Book, 2014). These statistics clearly highlight the need for the implementation of an inclusive design approach in these countries as their contribution to the world population is quite significant.

In general, it is clear from the global demographics that older people and people with disabilities are a considerable proportion of the world population, and this justifies financial as well as legislative incentives for including these groups in the design of products, environments, and services (Coleman, 2001). However, the trends in this demographic change are different in different parts of the world. For example, the overall percentage of older workers is increasing in majority of the developed countries and this justifies their accommodation in the design of products, processes, environments, and services. However, in the developing countries, the same increase is prominent as the overall population of these countries is growing very quickly, and the proportion of older people and people with disabilities is increasing accordingly. These facts draw the attention of designers, ergonomists, engineers, psychologists, planners, and entrepreneurs to seriously consider the provision of healthy living conditions for this significant proportion of the world population. The following sections describe how these challenges can be met by design scenarios that are equally acceptable for the broadest range of the population and also identify the drivers and barriers in the promotion of these ideas.

2.2. Inclusive design – drivers and barriers

'Design is the process of converting an idea or market need into the detailed information from which a product or system can be made' (Royal Academy of Engineering, 2005).

The British Standards Institute (2005) defined inclusive design as 'The design of *mainstream* products and/or services that are accessible to, and *usable* by, *as many people* as reasonably possible … without the need for special adaptation or specialized design'. Subsequently, the inclusive design term has also been related to providing quality of life and independent living for the aging population (Waller & Clarkson, 2009). It can be said that approaches like 'Inclusive Design,' 'Universal Design,' and 'Design for All' have been developed to help designers in developing design strategies that can promote design scenarios that are equally acceptable for all, including older people and people with disabilities (European Institute for Design & Disability, 2006; Keates & Clarkson, 2004; Preiser & Ostroff, 2001).

Published literature shows that there have been previous efforts in different parts of the world to explore the drivers and barriers for inclusive design. Vanderheiden and Tobias (2000) conducted telephone interviews of 26 manufacturers of consumer products in the USA, and identified a range of barriers and motives including government regulation, market data, training, consumer demands, technical complexity and unavailability of highly relevant knowledge, data and techniques. A similar survey was conducted in Japan where 307 companies from different industrial categories were surveyed. Interestingly, Japanese companies provided results that were similar to the USA companies (Unpublished report, 2000).

In the UK, Keates & Clarkson (2004) found that few industries knew about inclusive design and that there were misconceptions in the fundamental understanding of this design method. Companies believed that inclusive design meant designing only for older people and people with disabilities. In another survey conducted by Sims (2003), 32 design professionals working with different types and sizes of companies were surveyed, and it was concluded that 'design for all' is widely known but unfortunately not practiced within the design community. The majority of designers were aware of the philosophy of 'design for all' but rarely considered the approach because of the perceived time and financial costs. Underwood and Metz (2003) and Bellerby and Davis (2003) also discussed how inclusive design methods can be promoted and design-related issues could be addressed. They suggested that the provision of guidelines and standards could be important drivers, as currently these are not presented appropriately. Moreover, legislation and brand imaging can also play an effective role as generic business drivers.

Dong, Keates, and Clarkson (2004) and Dong, Clarkson, Ahmed, and Keates (2004) conducted a more comprehensive study with SMEs, where a survey was carried out with 38 manufacturing and retailing companies, along with 35 design consultancies. It was concluded that different companies perceive different factors as major barriers. However, drivers within these groups were found to be the same. For example, manufacturers and retailers mentioned key barriers because of the assumptions that inclusive design is more expensive, difficult to practice and learn and time consuming. In 2006 Goodman, Dong, Langdon, and Clarkson (2006) unlike Dong, Keates, et al. (2004) and Dong, Clarkson, et al. (2004), targeted large organizations along with SMEs and used a survey method for obtaining a more detailed insight into the drivers and barriers for inclusive design, and used the same questionnaire for comparison purposes. Complete responses were collected from 101 UK companies and organizations, and a detailed analysis was carried out. Barriers most frequently identified were a lack of time and budget for supporting inclusive design, lack of knowledge and tools to practice it, and it not being perceived as a need of the end users. Moreover, the perception that there was no justifiable business case for inclusive design was considered extremely important by most of the respondents.

As mentioned earlier, the population of developing countries is growing rapidly and the proportion of older and disabled people is also increasing accordingly. There are important questions as to how the living standards and quality of life of these people can be maintained and improved. Currently, in these countries the needs of older people and people with disabilities are not considered sufficiently by designers, planners, architects, and ergonomists. It is critically important to create awareness for policy-makers, planners, manufacturers, and designers about the importance of the inclusive design approach for creating environments, products, and services that meet the diverse and changing needs of the whole population.

3. Research focus

The main focus of the research was to assess the level of awareness of inclusive design in developing countries like China, India, Pakistan, and Bangladesh as a significant proportion of the world population lives in these countries. Until now, no research has been conducted in these countries to explore the main drivers that can motivate manufacturers and barriers that are potentially the reasons for lack of interest and resistance to its promotion. This is potentially a huge market where manufacturers can gain benefits by implementing design solutions that are equally acceptable for a wide range of the population. An underlying purpose of the research was to provide awareness to the manufacturers about the inclusive design approach, so that manufacturing industries in this part of the world can create business opportunities by attracting older customers and customers with special needs (disabilities) along with rest of the population.

4. Research method

The main objective of the survey was to identify barriers and drivers for inclusive design from the perspective of manufacturers in Pakistan. Initially, a list of manufacturing companies was prepared from which some were recruited after telephone inquiries about their willingness to participate in the survey. Manufacturing companies from textile, beverage, sports, automotive, design consultancy, and automotive sectors were included in this research. The survey questionnaire was designed with reference to the method adopted by the Engineering Design Centre's Inclusive Design survey at the University of Cambridge (Coleman, 2001) and is shown in abbreviated form as Appendix A. However, additional drivers and barriers found in research studies from the USA and Japan were also included in the study. A pilot survey was carried out to assess the reliability of the method. During the pilot study, it was found that some people had no understanding of any of the terms, 'inclusive design,' 'universal design,' and 'design for all.' So, prior to final distribution, a revised questionnaire was developed including a two-page summary for basic understanding of these terms, elaborated with the help of practical examples, so that the underlying objective of increasing the level of awareness about inclusive design could be achieved. (the illustrations of the practical examples have been removed from Appendix A, but they covered a wide range of activities of daily living that people with disabilities might have difficulty with such as accessing cash machines and dealing with escalators). It was further concluded that questionnaires should be completed in the presence of a research project team member, so that more reliable results could be collected. Representatives from 8 manufacturing companies (linked with the design process within their organizations) were selected, and a total of 50 responses were collected. Pakistan has 6417 manufacturing companies, 3590 of them in Punjab state where this study was conducted (Pakistan Bureau of Statistics, 2006).

5. Results and discussion

Responses to the distributed questionnaires were analyzed using descriptive statistics, where frequencies, average scores (means), and standard deviations (SD) were calculated, keeping in view the type of question and which output parameters could be used as indicators. The overall analysis was divided into three main areas: awareness, drivers, and barriers.

5.1. Awareness

The first portion of the questionnaire was about the assessment of the level of awareness and current understanding of the terms 'Inclusive Design,' 'Universal Design,' and 'Design for All'. As mentioned, the data were collected from 50 respondents, from companies whose profiles with respect to the number of employees are shown in Table 1. Table 2 shows the categories of manufacturing organizations included in the study and their proportional relevance in the sample population. Table 1 shows that about 70% of the sample population belonged to organizations having between 251 and 5000 employees. Moreover, 26% were from large organizations having over 5000 employees. The respondents were asked if their organization had a post for corporate social responsibility (CSR), and all claimed to understand the term CSR. However, 42% admitted that they did not know whether or not their organization had such a post. This clearly indicates that many of those involved in design and manufacturing activities have some preliminary understanding of CSR, but that there is no clear focus on this in their organizations. On the other hand, 30% responded that their organizations had a CSR post and 28% clearly claimed that their organizations had no CSR post. Overall, it can be concluded that in nearly 70% of organizations those with responsibility for design and manufacturing either did not know of or had no clear focus on CSR (shown in Figure 1).

Current understanding or awareness of different terms like 'Inclusive Design,' 'Universal Design,' and 'Design for All' were explored by asking respondents if they had heard of the terms and if they understood their meaning. By their responses, 36% showed no understanding of these terms. However, 64% did show understanding and the most popular term was 'Design for All' with 30% having previously heard of the term. Awareness or understanding of 'Inclusive Design,' and 'Universal Design' were recorded as 14 and 20%, respectively (Figure 2).

The above statistics clearly indicate a lack of awareness about the terminology and consequently of design inclusivity. Similar results were found in the pilot study and on the basis of this a two-page comprehensive summary of these terms along with useful and relevant examples were prepared, and made a necessary part of the questionnaire so that those who had no knowledge of the concepts could at least understand and give appropriate feedback about drivers and barriers to inclusive design. During final data collection, respondents were not allowed to see this information until they had recorded their responses about current understanding and awareness. As mentioned earlier, all the responses were collected in the presence of a member of the project team, so that the respondents were able to ask questions on the scope of drivers and barriers. A large majority of the respondents actively participated in discussion sessions by asking interesting questions and gained a clear understanding about these initiatives. Interestingly, 100% of the respondents requested our team members to send them final conclusions and findings of the research study, and this clearly shows their interest in this area. The following sections discuss findings related to drivers and barriers.

Table 1. Company profiles: number of employees.

No of employees	Frequency	Percent
Up to 250	3	6.0
251–1000	14	28.0
1001–5000	20	40.0
Over 5000	13	26.0
Total	50	100.0

Table 2. Types of manufacturing industries and number of responses.

Type	Responses
Textile	7
Beverages	10
Electronics	10
Design consultancies	12
Sports	3
Automotive	8
Total	50

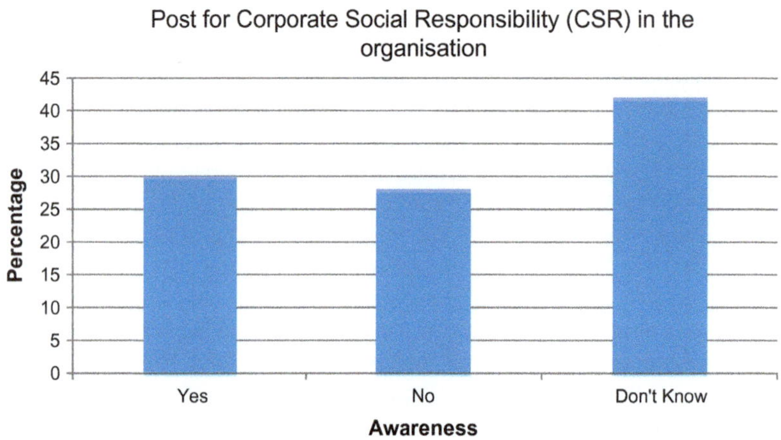

Figure 1. Corporate social responsibility (CSR) awareness.

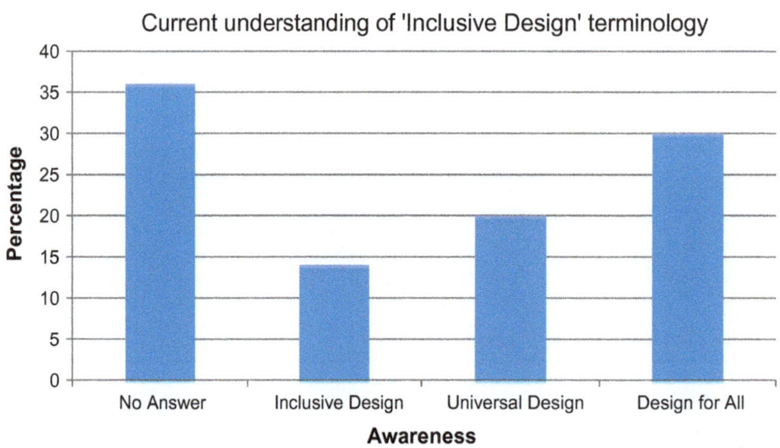

Figure 2. Current understanding of 'Inclusive Design' terminology.

5.2. Drivers

The perceptions of drivers of inclusive design from manufacturers have been computed as average scores (mean values) and SD. Responses were collected on a 1–4 Likert-type scale, where 22 drivers were listed and divided into two categories. Drivers, such as legislation, The Employment and Rehabilitation Act, social responsibility, demographic and consumer trends and brand enhancement were included in category one, where responses were collected in the form of agreement or disagreement ('strongly disagree' to 'strongly agree'). In the second category, all other drivers mentioned in Table 2 were included and responses were again collected on a four-point Likert-type scale, where effectiveness in terms of commercial benefits was measured ('not effective' to 'very effective'). The responses indicated that the vast majority of individuals in manufacturing felt that motivation through social responsibility awareness would be the most significant driver for the promotion of inclusive design in their organizations. As discussed in the preceding section, there is a need to promote CSR initiatives in organizations so that the level of awareness about inclusive aspects of products, environment, or service design might be increased. Motivation through social responsibility awareness was perceived as the most effective driver (mean 3.28 and SD .970) in the implementation of design inclusivity. Similarly, other drivers that were perceived as significant were: source of innovation and differentiation (mean 3.28); demographic and consumer trends (3.26); brand enhancement (3.22); increase customer satisfaction (3.16); and new market opportunity (3.12). However, the least significant drivers were: consumer dissatisfaction with current products (mean 2.40); assessment of how many people are excluded (2.58); champion for inclusive design on company board (2.66); major competitors adoption of inclusive design (2.68); and increase the large share of current market (2.70). Full details of means and SD values for all 22 drivers are shown in Table 3.

5.3. Barriers

This section discusses the results of data analysis concerning perceived barriers to inclusive design. The perspectives of respondents on 27 potential barriers to inclusive design promotion among manufacturers in Pakistan were sought. Table 4 shows the results of the data analysis, and highlights the percentages of respondents in agreement with the barrier statements. It was concluded that lack of resources/guidance and awareness of inclusive design are the most prominent perceived barriers as about 84% (42 respondents out of 50) and 80% (40 out of 50) respondents showed agreement, respectively. Moreover, change in culture of business (76%), lack of government regulations (76%), perception that inclusive design is more expensive (68%), and that it increased time to market (68%) were perceived as significant barriers to inclusive design. Interestingly, about 80% of the respondents perceived that implementation of inclusive design was not an unachievable goal, and a significant number of people felt that inclusive design was not a passing trend and did not compromise the esthetics of design. Similarly, more than 50% of respondents from manufacturing organizations did not feel that inclusive design was a too difficult thing to practice. These findings highlight that the majority of manufacturing-related people were of the opinion that inclusive design was not too difficult to implement. However, inclusive design can be promoted by launching an organized campaign about the associated usefulness and benefits, by providing proper resources and guidance, along with changes in organizational culture and enforcement of focused regulations regarding inclusive design. The percentage agreements of the respondents with other barriers are shown in Table 4.

Table 3. Perceptions of manufacturers regarding 'drivers' for inclusive design.

Drivers	Mean	SD
Motivation through social responsibility	3.28	.970
Source of innovation and differentiation	3.28	.858
Demographic and consumer trends	3.26	1.121
Brand enhancement	3.22	1.166
Increase customer satisfaction	3.16	.934
Employment and Rehabilitation Act 1981 will help in practicing inclusive design	3.16	1.131
New market opportunity	3.12	.872
Chances of innovation by practicing inclusive design	3.10	.931
Entrance to new market	3.08	.944
Legislation	3.08	1.209
Increase revenue through increased usage	3.06	.913
Increase customer loyalty	3.04	.903
Increase the size of potential market	2.94	.793
Availability of tools/methods to help practicing	2.88	.872
Public awareness of inclusive design	2.88	.982
Availability of expert consultation on inclusive design	2.82	.941
Availability of training opportunities on inclusive design	2.80	.948
Increase the large share of current market	2.70	.863
Major competitors adoption of inclusive design	2.68	.935
Champion for inclusive design on company board	2.66	.823
Assessment of how many people are excluded	2.58	.950
Consumer dissatisfaction with current products	2.40	1.030

From the analysis of the data presented above, many of the drivers and barriers found important in this study were the same as those highlighted by other studies conducted in the UK, USA, and Japan. For example, drivers such as new market opportunity, brand enhancement, and source of innovation were commonly perceived drivers in all these studies. Similarly, barriers like awareness of inclusive design, lack of resources and guidance, and lack of time and budget were common barriers presented in all of these studies. Moreover, barriers such as inclusive design was too difficult and an unachievable goal that compromised the esthetics of design were perceived as relatively less significant barriers. This shows that almost all studies have concluded that people who have seen the relevance of the promotion and implementation of inclusive design felt that there was a great need to increase levels of awareness within companies especially amongst designers. At the same time, they felt that design for inclusivity is not an unachievable goal; it can be achieved by increasing awareness about its importance, usefulness, and relevance with our daily life and by providing technical and financial assistance at the organizational and country level.

Tables 5 and 6 present a comparison of results on perceived drivers and barriers for the different manufacturing sectors (textile, beverages, electronics, sports, automotive and design consultancy). It is quite clear that there were variations in the perceptions of inclusive design drivers and barriers depending upon the type of manufacturing sector. For example, in the electronics manufacturing sector, legislation was perceived as the most prominent driver with a mean score of 3.9, not only highest in this sector but overall as well (Table 5). Similarly, in the beverage sector, it was commonly perceived that putting effort into an increase in customer satisfaction by applying inclusive design methodologies was the most important driver. In the automotive sector, it was felt that new market opportunity and increase in customer loyalty were the most prominent drivers, whereas the same had relatively less importance (in the middle of the list) in

Table 4. Perceptions of manufacturers regarding 'barriers' for inclusive design.

Barriers	No. of responses	Percentage agreed
Lack of resources/guidance on inclusive design	42	84
Lack of awareness of inclusive design	40	80
Difficulty in changing the culture of business	38	76
Lack of government regulations	38	76
Implementing inclusive design could require significant cultural change	35	70
Perception that inclusive design is more expensive	34	68
Perceived longer development time to market	34	68
Perception – more complex to design inclusively	34	68
Lack of interest in inclusive design	34	68
Stigma associated with inclusive design	33	66
Lack of time and budget	33	66
Lack of company policy on inclusive design	32	64
Inclusive design is a perceived need of our end user	32	64
Lack of methods/tools for practicing inclusive design	32	64
Lack of motivation for tackling inclusive design	32	64
Perception of brand association with disabled/older people	32	64
Lack of availability of good design examples	31	62
Perception – there is no need to practice inclusive design	31	62
Working for short term financial objectives	30	60
Perception that inclusive design represents a niche market	29	58
Unavailability of internal support for inclusive design	29	58
No justifiable business case to support inclusive design	27	54
Lack of knowledge and tools to practice inclusive design	26	52
Perception – inclusive design is too difficult	23	46
Perception – inclusive design is a passing trend	20	40
Inclusive design compromises the aesthetics of design	18	36
Perceived as an unachievable goal	10	20

the overall priority list. Generally, it can be said that the textile, beverages, design consultancy, and electronics manufacturing sectors provided similar kinds of results to those of the overall analysis. As shown, only three respondents from sports industries recorded their responses, however, those were not of high significance because of the smaller sample size. The sports sector is one of the leading export-based manufacturing sectors in Pakistan, and so there is a need to explore this sector in a detailed way.

A comparison of different perceived barriers among a variety of manufacturing sectors is shown in Table 6, where differences in perceptions can be observed. For example, in the automotive sector, the vast majority of respondents agreed that there was no need to practice inclusive design, that it reduces short-term financial advantage and was associated with stigma. In comparison with the overall perception, these factors showed relatively less significance. A detailed comparison is shown in Table 6.

6. Future work

Future work will focus initially on capturing the perceptions about inclusive design drivers and barriers from designers working in different areas like architecture, civil engineering, town planning and design, transportation design and management in Pakistan, so as to widen the scope of the investigations. Subsequently, the same study will be carried out in China, India, and Bangladesh, and efforts have been started to

Table 5. Comparison of different manufacturing sectors on perceived inclusive design drivers.

Drivers	Overall		Textile		Sports		Beverages		Design consultancies		Electronics		Automotive	
	Mean	SD	Mean	SD	Mean	SD	Mean	SD	Mean	SD	Mean	SD	Mean	SD
Motivation through social responsibility	3.28	.970	3.57	.787	3.00	1.000	2.90	.876	3.67	.651	3.40	1.265	2.88	1.126
Source of innovation and differentiation	3.28	.858	3.71	.488	3.33	1.155	3.20	.919	3.33	.985	2.90	.994	3.38	.518
Demographic and consumer trends	3.26	1.121	3.00	.816	2.33	1.528	3.10	1.287	3.67	.985	3.50	1.179	3.13	1.126
Brand enhancement	3.22	1.166	3.00	1.155	2.33	1.528	2.90	1.287	3.75	1.215	3.60	.966	2.88	.835
Increase customer satisfaction	3.16	.934	3.57	.787	3.33	1.155	3.30	1.059	3.17	.937	2.60	.843	3.25	.886
Employment and Rehabilitation Act 1981will help in practicing inclusive design	3.16	1.131	3.29	1.113	4.00	1.732	3.10	.994	3.42	1.240	3.00	1.054	2.63	1.061
New market opportunity	3.12	.872	3.71	.488	2.67	1.528	2.80	.919	2.92	1.084	3.20	.632	3.38	.518
Chances of innovation by practicing inclusive design	3.10	.931	3.86	.378	3.00	1.732	2.70	.949	3.25	.965	2.70	.823	3.25	.707
Entrance to new market	3.08	.944	3.29	1.113	3.00	1.732	3.10	.876	3.17	1.030	3.10	.738	2.75	.886
Legislation	3.08	1.209	2.86	1.069	4.33	.577	3.20	1.229	2.58	1.311	3.90	.876	2.38	.916
Increase revenue through increased usage	3.06	.913	3.43	.787	2.67	1.155	3.20	.919	2.83	1.030	3.10	.876	3.00	.926
Increase customer loyalty	3.04	.903	3.00	1.000	3.00	1.000	3.20	1.033	2.58	.996	3.20	.632	3.38	.744
Increase the size of potential market	2.94	.793	3.29	.951	3.00	1.000	2.90	.876	2.83	.718	2.80	.632	3.00	.926
Availability of tools/methods to help practicing	2.88	.872	2.86	.900	2.33	1.528	3.00	1.054	2.67	.888	3.00	.471	3.13	.835
Public awareness of inclusive design	2.88	.982	3.29	.756	3.00	1.732	2.80	1.033	3.00	1.044	2.70	.823	2.63	1.061
Availability of expert consultation on inclusive design	2.82	.941	3.14	.900	2.67	1.528	2.90	.994	3.00	.953	2.50	.707	2.63	1.061
Availability of training opportunities on inclusive design	2.80	.948	3.00	1.000	2.33	1.528	2.40	1.174	2.92	.793	2.70	.823	3.25	.707
Increase the large share of current market	2.70	.863	3.00	1.000	2.67	1.528	2.70	.949	2.50	.798	2.60	.516	2.88	.991
Major competitors adoption of inclusive design	2.68	.935	2.71	1.113	2.33	1.155	2.80	.919	2.83	1.030	2.20	.632	3.00	.926
Champion for inclusive design on company board	2.66	.823	2.71	.951	2.33	.577	2.50	1.080	2.75	.622	2.60	.699	2.88	.991
Assessment of how many people are excluded	2.58	.950	3.00	.816	2.33	1.528	2.40	.843	2.75	1.215	2.50	.850	2.38	.744
Consumer dissatisfaction with current products	2.40	1.030	2.43	1.272	1.33	.577	2.10	.994	2.25	.965	2.80	.789	2.88	1.126

Table 6. Comparison of different manufacturing sectors on perceived barriers to inclusive design.

Barriers	Overall	Textile	Sports	Beverages	Design consultancies	Electronics	Automotive
Lack of resources/guidance on inclusive design	84	85.7	100.0	90	91.7	80	62.5
Lack of awareness of inclusive design	80	85.7	66.7	80	91.7	70	75.0
Difficulty in changing the culture of business	76	71.4	66.7	70	83.3	70	87.5
Lack of government regulations	76	71.4	100.0	90	91.7	60	50.0
Implementing inclusive design could require significant cultural change	70	85.7	66.7	80	66.7	50	75.0
Perception that inclusive design is more expensive	68	71.4	33.3	70	91.7	40	75.0
Perceived longer development time to market	68	42.9	66.7	50	75.0	90	75.0
Perception – more complex to design inclusively	68	57.1	66.7	70	91.7	40	75.0
Lack of interest in inclusive design	68	57.1	66.7	80	83.3	50	62.5
Stigma associated with inclusive design	66	85.7	33.3	30	83.3	60	87.5
Lack of time and budget	66	57.1	66.7	50	75.0	70	75.0
Lack of company policy on inclusive design	64	71.4	66.7	60	75.0	40	75.0
Inclusive design is a perceived need of our end user	64	85.7	66.7	40	66.7	80	50.0
Lack of methods/tools for practicing inclusive design	64	85.7	66.7	70	75.0	40	50.0
Lack of motivation for tackling inclusive design	64	85.7	100.0	70	83.3	20	50.0
Lack of availability of good design examples	62	71.4	66.7	50	75.0	50	62.5
Perception – there is no need to practice inclusive design	62	71.4	33.3	60	58.3	50	87.5
Working for short term financial objectives	60	42.9	100.0	40	66.7	50	87.5
Perception that inclusive design represents a niche market	58	42.9	66.7	40	75.0	50	75.0
Unavailability of internal support for inclusive design	58	57.1	66.7	60	58.3	40	75.0
Perception of brand association with disabled/older people	54	57.1	100.0	70	66.7	60	62.5
No justifiable business case to support inclusive design	54	57.1	33.3	40	50.0	70	62.5
Lack of knowledge and tools to practice inclusive design	52	57.1	66.7	40	58.3	40	62.5

engage relevant people in these countries. Moreover, during this research exercise, efforts will be made to educate the relevant people about the importance of inclusive design and its impact on our daily life in terms of work productivity and better living. In addition to the social aspects of inclusive design, another important associated factor is the potential business opportunities, and efforts will be made to develop useful business cases for companies to motivate them in the promotion of inclusive design.

7. Conclusion

Changing demographics demand the promotion of the inclusive design method throughout the world. Developing countries like China, India, Pakistan, and Bangladesh make a major contribution to the world population, and the overall population in these countries is increasing and the proportion of older people and people with disabilities is increasing accordingly. This article reveals the need to explore the main barriers and drivers for inclusive design from the perspective of manufacturers, so that drivers can be promoted and barriers can be removed. It has been concluded that people from manufacturing organizations feel that motivation through social responsibility, source of innovation, brand enhancement, and new market opportunities are the main drivers for inclusive design in Pakistan. On the other hand, lack of awareness about inclusive design, lack of resources, guidance, change in culture, government regulations, time and budget are perceived as barriers toward the implementation of design inclusivity in manufacturing organizations. The conclusions on drivers are in close agreement with those found by similar studies in the UK, USA, and Japan as are those on barriers except that this study found that inclusive design was not considered too difficult, less achievable, or likely to compromise esthetics.

However, a difference in the perceptions on drivers and barriers has been observed in different manufacturing sectors that shows that inclusive design promotion can be carried out by focusing on the prioritized relevant factors in each sector. Hence, for example, the electronics industry showed considerable concern for legislation, whereas for the automotive industry customer loyalty and market opportunities were prominent. This supports the idea that efforts are needed to make people aware of the usefulness of inclusive design, not only in the social responsibility context, but also as a potential business opportunity.

The promotion of inclusive design practices will not only be useful in improving living conditions of the people of the developing world, but also will open greater business opportunities for companies. It is not only the responsibility of manufacturing companies, but also of governments to provide adequate resources/guidance that can help in increasing awareness about the usefulness of inclusive design. Development and implementation of proper legislation is also an important area on which to focus where regulatory agencies would have to make sure that organizations are properly implementing inclusive design strategies in their true spirit. These investigations can help in shaping different strategies for the promotion of inclusive design awareness in manufacturing industries in general; and textile, automotive, electronics, beverages, and design consultancies in particular.

This research has contributed in terms of promoting a better understanding of the inclusive design approach and this will ultimately help manufacturers and designers in designing products, services, processes, and environments that are equally safe, healthy, productive, and acceptable for a wide range of population. The aim is to achieve an improvement in living conditions of older people and people with disabilities along with the rest of the population, while developing a business case for manufacturers as these countries constitute a huge potential market of product consumers.

Acknowledgements

This research was carried out in collaboration with the Department of Industrial and Manufacturing Engineering, University of Engineering and Technology, Lahore, Pakistan and the Wolfson School of Mechanical and Manufacturing Engineering, Loughborough University, UK.

Disclosure statement

No potential conflict of interest was reported by the authors.

References

Bellerby, F., & Davis, G. (2003). Defining the limits of inclusive design. *Include 2003* (pp. 1:00–1:17). London: Royal College of Art.

British Standards Institute. (2005). British Standards 7000-6:2005. Design Management Systems – Managing Inclusive Design – Guide.

CIA World Fact Book. (2014). www.cia.gov/library/publications/the-world-factbook/

Coleman, R. (2001). Designing for our future selves. In W. F. E. Preiser & E. Ostroff (Eds.), *Universal design handbook* (pp. 4.1–4.25). New York, NY: McGraw-Hill.

Dong, H., Clarkson, P. J., Ahmed, S., & Keates, S. (2004). Investigating perceptions of manufacturers and retailers to inclusive design. *The Design Journal, 7*, 3–15.

Dong, H., Keates, S., & Clarkson, P. J. (2004). Inclusive design in industry: Barriers, drivers and the business case. In 8th ERCIM Workshop "User Interface for All", Vienna, Austria.

European Institute for Design and Disability. (2006). www.design-for-all.org.

Goodman, J., Dong, H., Langdon, P., & Clarkson, P. J. (2006). Increasing the uptake of inclusive design. In industry. *Gerontechnology, 5*, 140–149.

Keates, S., & Clarkson, P. J. (2004). *Countering design exclusion*. London: Springer.

Pakistan Bureau of Statistics. (2006). *Census of Manufacturing Industries 2005–06, District-wise Report*. Islamabad: Government of Pakistan.

Preiser, W., & Ostroff, E. (2001). *Universal design handbook*. New York, NY: McGraw-Hill.

Royal Academy of Engineering. (2005). Educating engineers in design. Retrieved from www.raeng.org.uk

Sims, R. E. (2003). *'Design for All': Methods and data to support designers* (PhD thesis). Loughborough University, Loughborough, UK.

Underwood, M., & Metz, D. (2003). Seven business drivers of inclusive design. In *Include 2003*. 1, 1:39–1:44.

U.N.O. (2009). *Ageing*. Retrieved from http://social.un.org/index/Ageing.aspx

Unpublished report. (2000). Kyoyo-hin (Universal Design) in Japan. Available from the i~ design collection, Helen Hamlyn Research Centre, Royal College of Art, UK.

Vanderheiden G., & Tobias, J. (2000). Universal design of consumer products: Current industry practice and perceptions; 2000. Retrieved from http://trace.wisc.edu/docs/ud_consumer_products_hfes2000/index.htm

Waller, S., & Clarkson, P. J. (2009). Tools for inclusive design. In C. Stephanidis (Ed.), *The universal access handbook*. Boca Raton, FL: Taylor and Francis.

Appendix A. SURVEY Inclusive Design Survey for assessing the level of awareness, drivers and barriers among Designers

This survey consists of six parts and will only take a few minutes to complete.
Please answer all Questions:

Contact Name
Company name
Position in Company
Telephone Number
E-mail Address
Company Web Site:

Please Note: We will not share this information with any third party or use these detail to contact you unless you ask us to share the results from this study where these results will be presented in an anonymized form.

- Please provide an email address and tick this box if you would like us to send you the anonymized result from this study.
- Please tick this box if you would like us to contact you to discuss how inclusive Design may assist your organization.

1.Please describe what your organization does:
2.Please list down the names of the products/Services your organization provides

Part-1: Company Profile

1. Number of Employees approximately	☐ Up to 250 ☐ 251-1000 ☐ 1001-5000 ☐ Over 5000
2. How many people use your products/services?	☐ Up to 1000 ☐ 1001-10,000 ☐ 10,001-100,000 ☐ 100,001-1 Million ☐ Over 1 Million
3. How many companies represent most of your sales?	☐ Up to 5 ☐ 6-50 ☐ 51-500 ☐ Over 500 ☐ Direct to customer
4. Global Profile (tick all if apply)	☐ Pakistan ☐ Asia ☐ Europe ☐ USA ☐ Global
5. Does your organization have a post for corporate social responsibility (CSR)?	☐ Yes ☐ No ☐ Don't know ☐ Don't understand

Part-2: Current Understanding

1. Have you heard of the following terms? (tick all if apply)	☐ Inclusive Design ☐ Universal Design ☐ Design for all
2. What do you think they mean?	

Part-3: Current Company Position

In answering the following. Please think of the definition of inclusive design."A process whereby designers, manufacturers, and service providers ensure that their products and services address the needs of the widest possible audience."	
(Please select the appropriate number)	
1. What level of awareness of Inclusive Design is there in your organization?	☐ 1 (No awareness) ☐ 2 ☐ 3 ☐ 4 (Extremely aware)
2. How inclusive do you think your products/services are?	☐ (Not inclusive) ☐ 2 ☐ 3 ☐ 4 (very inclusive))
3. What level of effort is utilized to make your products/ services inclusive?	☐ 1 (Little or no effort) ☐ 2 ☐ 3 ☐ 4 (Large amount of effort)
4. Is your company interested in making your products/ services more inclusive?	☐ 1 (Not interested) ☐ 2 ☐ 3 ☐ 4 (Extremely interested)
5. How aware is your company or organization?	**A.** The (Employment and Rehabilitation) Act,1981 of Pakistan ☐ 1 (No awareness) ☐ 2 ☐ 3 ☐ 4 (Extremely aware) **B.** Other legislation, Policies and, or codes of practice? ☐ 1 (No awareness) ☐ 2 ☐ 3 ☐ 4 (Extremely aware)
6. Have you ever heard about the (Employment and Rehabilitation) Act,1981	☐ Yes ☐ No
7. Please specify and particular products/services that are designed to be inclusive.	
8. What level of effort is currently utilized to make your products/services inclusive?	☐ 1 (Little or no effort) ☐ 2 ☐ 3 ☐ 4 (Large amount of effort)
Please elaborate on any specific initiatives that you are able to discuss and comment on their success.	

Part-4: Drivers for inclusive Design

1. How much do you agree with the following five statements?	**1**. Legislation is a major driver for inclusive design within our organization. ☐ 1 (Strongly disagree) ☐ 2 ☐ 3 ☐ 4 (Strongly agree) ☐ Don't know **2**. The (Employment and Rehabilitation) Act, 1981 will help us to practice and manage inclusive design. ☐ 1 (Strongly disagree) ☐ 2 ☐ 3 ☐ 4 (Strongly agree) ☐ Don't know **3**. Social responsibility motivates our organization to consider inclusive design. ☐ 1 (Strongly disagree) ☐ 2 ☐ 3 ☐ 4 (Strongly agree) ☐ Don't know **4**. Demographic and consumer trends are driving our organization commitment to inclusive design. ☐ 1 (Strongly disagree) ☐ 2 ☐ 3 ☐ 4 (Strongly agree) ☐ Don't know **5**. Brand enhancement is a key driver for inclusive design within our organization. ☐ 1 (Strongly disagree) ☐ 2 ☐ 3 ☐ 4 (Strongly agree) ☐ Don't know
2. How effective do you think inclusive design could be in helping you to achieve the following commercial benefits?	Entrance to a new market. ☐ 1 (Not effective) ☐ 2 ☐ 3 ☐ 4 (very effective) Increase the size of your potential market. ☐ 1 (Not effective) ☐ 2 ☐ 3 ☐ 4 (very effective) Increase the size of your potential market. ☐ 1 (Not effective) ☐ 2 ☐ 3 ☐ 4 (very effective)

(Continued)

Achieve a large share of your current market.
- ☐ 1 (Not effective)
- ☐ 2
- ☐ 3
- ☐ 4 (very effective)

Increase customer loyalty
- ☐ 1 (Not effective)
- ☐ 2
- ☐ 3
- ☐ 4 (very effective)

Increase revenue through increased usage
- ☐ 1 (Not effective)
- ☐ 2
- ☐ 3
- ☐ 4 (very effective)

Increase customer satisfaction
- ☐ 1 (Not effective)
- ☐ 2
- ☐ 3
- ☐ 4 (very effective)

A source of innovation and differentiation
- ☐ 1 (Not effective)
- ☐ 2
- ☐ 3
- ☐ 4 (very effective)

New market opportunities by practicing inclusive design
- ☐ 1 (Not effective)
- ☐ 2
- ☐ 3
- ☐ 4 (very effective)

Assessment of how many people are excluded
- ☐ 1 (Not effective)
- ☐ 2
- ☐ 3
- ☐ 4 (very effective)

Consumer dissatisfaction with current products
- ☐ 1 (Not effective)
- ☐ 2
- ☐ 3
- ☐ 4 (very effective)

Chances of innovation by practicing inclusive design
- ☐ 1 (Not effective)
- ☐ 2
- ☐ 3
- ☐ 4 (very effective)

Public/consumer awareness of inclusive design
- ☐ 1 (Not effective)
- ☐ 2
- ☐ 3
- ☐ 4 (very effective)

Availability of expert consultation on inclusive design
- ☐ 1 (Not effective)
- ☐ 2
- ☐ 3
- ☐ 4 (very effective)

(Continued)

	Availability of tools/methods to help the practice of inclusive design ☐ 1 (Not effective) ☐ 2 ☐ 3 ☐ 4 (very effective) Availability of training opportunities on inclusive design for staff/designers ☐ 1 (Not effective) ☐ 2 ☐ 3 ☐ 4 (very effective) Champion for inclusive design on company boards ☐ 1 (Not effective) ☐ 2 ☐ 3 ☐ 4 (very effective) Major competitor's adoption of inclusive design ☐ 1 (Not effective) ☐ 2 ☐ 3 ☐ 4 (very effective)
If there are other key Drivers for your organization, please specify?	

Part-5: Barriers to inclusive Design

1. Please respond to the following statements on barriers to inclusive design with respect to your organization.	There is little or no internal support for inclusive design. ☐ Yes ☐ No Implementing inclusive design could require significant cultural change. ☐ Yes ☐ No We lack the knowledge and tools to practice inclusive design. ☐ Yes ☐ No There is no justifiable business case to support inclusive design. ☐ Yes ☐ No There is a lack of time and budget to support inclusive design. ☐ Yes ☐ No There is a perception that inclusive design is too difficult, ☐ Yes ☐ No Inclusive design compromises the aesthetics of the design. ☐ Yes ☐ No Inclusive design is a perceived need of our end users. ☐ Yes ☐ No There is a stigma associated with inclusive design. ☐ Yes ☐ No Inclusive Design is seen as an unachievable goal. ☐ Yes ☐ No Lack of awareness of inclusive design ☐ Yes ☐ No Lack of interest in inclusive design ☐ Yes ☐ No Lack of motivation for tackling inclusive design ☐ Yes ☐ No Perception that inclusive design is more expensive ☐ Yes ☐ No Perception that it can be more complex to design inclusively ☐ Yes ☐ No

(*Continued*)

Perception that inclusive design represents a niche market
- ☐ Yes
- ☐ No

Perceived problems of brand association with disabled/order people
- ☐ Yes
- ☐ No

Perceived longer development time to market
- ☐ Yes
- ☐ No

Perception that there is no need to practice inclusive design
- ☐ Yes
- ☐ No

Perception that inclusive design is a passing trend
- ☐ Yes
- ☐ No

Lack of resources/guidance on inclusive design
- ☐ Yes
- ☐ No

Lack of availability of good design examples
- ☐ Yes
- ☐ No

Lack of government regulations
- ☐ Yes
- ☐ No

Lack of methods/tools for practicing inclusive design
- ☐ Yes
- ☐ No

Difficulty in changing the culture of business
- ☐ Yes
- ☐ No

Lack of company policy on inclusive design
- ☐ Yes
- ☐ No

Working for short-term financial objectives
- ☐ Yes
- ☐ No

2. Which of the following barriers do you perceive as most important.
- ☐ Most important.
- ☐ Second most important.
- ☐ Third most important.

3. Which of the above barriers do you perceive as least important and.
- ☐ Least important.
- ☐ Second least important.
- ☐ Third least important.

4. Are there any other significant barriers to the adoption of inclusive Design within your organization?

Part-6: Tools to increase the Usage of Inclusive Design

1. How important do you perceive the following tools to encourage inclusive Design within your organization?	(A) Convincing arguments (business case) for top level management. ☐ 1 (not important) ☐ 2 ☐ 3 ☐ 4 (extremely important) (B) An increased understanding of inclusive design. ☐ 1 (not important) ☐ 2 ☐ 3 ☐ 4 (extremely important) (C) Skills or tools to assist with designing inclusively ☐ 1 (not important) ☐ 2 ☐ 3 ☐ 4 (extremely important) (D) Effective tools to market inclusive Design ☐ 1 (not important) ☐ 2 ☐ 3 ☐ 4 (extremely important)
Which of these (A-D) is the most important?	Most important Second Most important Third most important
2. Which of the following would be useful to your business (tick at that apply)	**A.** A national awareness campaign. **B.** A national centre of excellence for inclusive Design. **C.** Attending an Inclusive Design event. **D.** Receiving an exclusion assessment on your product/service. **E.** Consultancy and/or training on inclusive design for your organization. **F.** Other, please specify:
Which of these (A-F) is most important, and why?	Most important Second Most important Third most important

7

End mill tools integration in CNC machining for rapid manufacturing processes: simulation studies

Muhammed Nafis Osman Zahid, Keith Case* and Darren Watts

Mechanical and Manufacturing Engineering, Loughborough University, Loughborough, Leicestershire LE11 3TU, UK

Computer numerical controlled (CNC) machining has been recognized as a manufacturing process that is capable of producing metal parts with high precision and reliable quality, whereas many additive manufacturing methods are less capable in these respects. The introduction of a new layer-removal methodology that utilizes an indexing device to clamp the workpiece can be used to extend CNC applications into the realm of rapid manufacturing (CNC-RM) processes. This study aims to improve the implementation of CNC machining for RM by formulating a distinct approach to integrate end mill tools during finishing processes. A main objective is to enhance process efficiency by minimizing the staircasing effect of layer removal so as to improve the quality of machined parts. In order to achieve this, different types of end mill tools are introduced to cater for specific part surfaces during finishing operations. Virtual machining simulations are executed to verify the method and the implications. The findings indicate the advantages of the approach in terms of cutting time and excess volume left on the parts. It is shown that using different tools for finishing operations will improve the capabilities of CNC machining for rapid manufacturing applications.

Keywords: end mill tools; CNC machining; rapid manufacturing

1. Introduction

In machining processes, it is crucial to ensure good surface finish quality as this will influence the performance of machined parts and production costs (Davim, 2001). Improving surface finish has become a major concern particularly in rapid manufacturing processes where the objective is to achieve final geometry directly. A major consideration in rapid manufacturing processes is that planning time needs to be substantially reduced so that overall time from design to finished part is also reduced. In additive processes (AM), relatively long fabrication times are compensated by the limited planning time required. In traditional computer numerical controlled (CNC) the opposite situation exists, and whilst machining can be very rapid, the planning process may be complex and time-consuming. In trying to use CNC for rapid applications, the main effort is in maintaining a simple approach to planning whilst possibly sacrificing some of the available machining efficiency.

An approach that has some of the advantages of additive manufacturing involves the use of an indexable fourth axis device to hold and clamp the workpiece onto the table

*Corresponding author. Email: K.Case@lboro.ac.uk

of a three-axis CNC milling machine (Figure 1). This approach allows a layer-removal process to take place at various rotations about one axis and complex shapes and features can be created by layer-based removal of material (Frank, Wysk, & Joshi, 2006). The indexable device permits rotation about one axis while also clamping the workpiece. Layers are removed from several orientations to reveal all surfaces of the part without refixturing (Frank et al., 2006). Determination of the necessary orientations is central to the process and this is achieved by visibility algorithms (a minimum set of orientations are sought where the surfaces are 'visible' from the direction of the tool).

Within each orientation, roughing and finishing operations are performed sequentially using different tool sizes and machining parameters. Machining starts with the first roughing operation that cuts until the furthest possible surface is reached or the workpiece is fully cut (Frank, 2007). Then, finishing operations are executed no further than the central axis of the cylindrical workpiece. In order to reduce effort in the planning tasks, feature recognition on machined parts is avoided. Therefore, the selection of a cutting tool becomes straightforward and is based on the smallest diameter and a flat end mill tool is most likely to be selected.

Despite the effectiveness of reducing complexities in the planning phase, the use of a single end mill tool in finishing processes leads to several drawbacks. First, due to the geometry of flat end mill cutters, a staircasing effect is visible on part surfaces particularly on contours and non-flat regions. This effect still can be seen even if very small cutting layer depths are adopted (Frank, Joshi, & Wysk, 2002). In addition, using small cutting depths will obviously increase cutting times considerably. Another limitation can be seen in the assessment of tool accessibility of part features. Analysis that focuses on machinability has resulted in some uncut material being left on the part due to inaccessibility of the cutting tool (Li & Frank, 2006) and inevitably this will affect dimensional accuracy. Figure 2 illustrates potential non-machined regions generated on a toy jack part. To date, no specific method has been developed to cater for the integration of different end mills in rapid machining processes. Therefore, this study attempts to show that using different end mill tools in finishing operations can improve surface finish by reducing the excess volume left on the parts. The next section of the paper will review the previous approach in cutting tools selection for CNC-RM processes. A methodology to integrate end mill tools is then described considering the implications for process planning and cutting operations. A framework for the simulation studies is established that is later used to generate the results to validate the proposed method. The findings

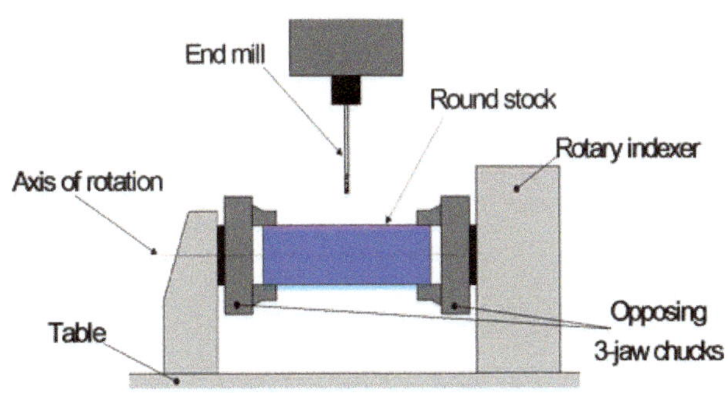

Figure 1. Setup for CNC-RP (Frank, Wysk, & Joshi, 2004).

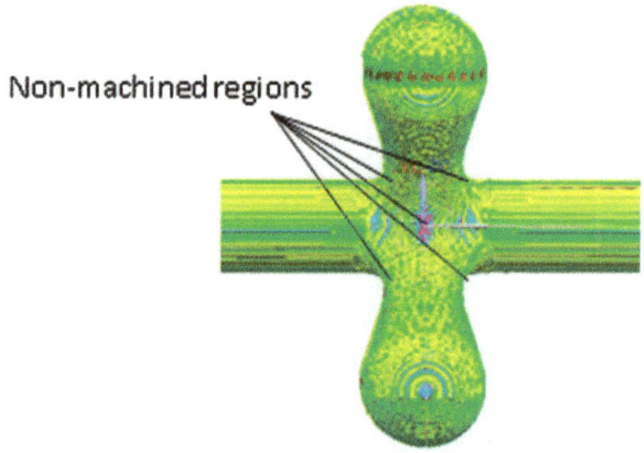

Figure 2. Possible non-machined regions using a flat end mill tool (Li & Frank, 2006).

are described further to highlight the achievements and limitations of the approach before drawing conclusions.

2. Cutting tools and machining operations

Initially, due to the objective of simplifying process planning, only a single cutting tool with the smallest possible diameter is used in CNC machining for rapid prototyping processes (Frank, 2003). Since the visibility analysis is based on 2D cross-sectional layers of part geometries, a flat end mill tool is usually selected to finish cut the part. Consequently, the machined parts exhibit the same staircase effect as presented in most of the AM processes. Figure 3 shows the staircase effect formed on contoured surfaces of a part. However, CNC machines are capable of reducing the layer thickness down to 0.0127 mm. Therefore, the step appearance is minimized within an acceptable range for making prototypes. From a planning perspective, the use of a single cutting tool has generalized the cutting areas to cover all surfaces in each orientation. This avoids any feature recognition tasks during machining process. Only one cutting operation is performed in each orientation that considerably reduces the number of operations and planning complexities. However, a small cutting tool requires more passes, which will result in longer machining times. Besides, the removal of thin layers is inefficient especially when the bulk of the material needs to be removed during early phases of machining.

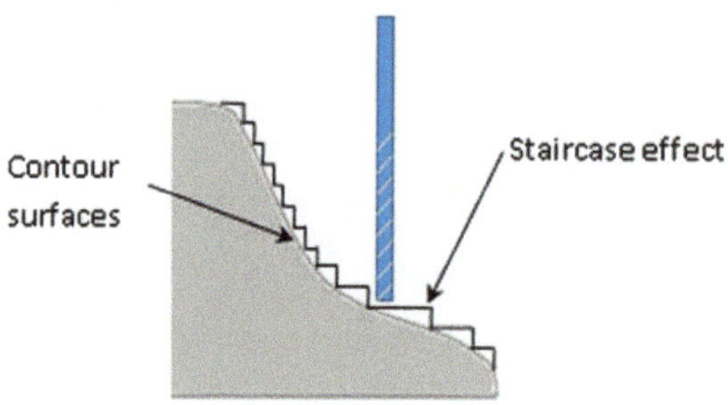

Figure 3. Staircase effect presented on contoured surfaces.

These considerations tend to restrict the capabilities of CNC machining even if some level of automation is embedded in the process planning.

Due to the inefficiency caused by a single cutting tool, later developments suggest the use of different tool sizes to handle roughing and finishing operations (Frank, 2007). As a result, roughing operations utilize larger tool diameters selected based on the size of the workpiece. Hence, high volumes of material can be removed during roughing operations whilst leaving sufficient material for finishing processes. Meanwhile, the approach of selecting the smallest tool diameter remains the same for executing finishing operations. This allows the cutting tool to reach all features presented on the part. Both roughing and finishing operations employ the same cutting areas. Thus, within one cutting orientation, two different toolpaths are developed and this directly increases the number of machining operations. Additionally, tool changes are required in each orientation and can be carried out through automated systems built into the CNC machine. This combination of different tool sizes based on roughing and finishing operations succeeds in enhancing the application of CNC machines in producing parts rapidly.

The introduction of different cutting tool sizes has resulted in a further study that attempts to assist in the selection of a proper combination of end mill tool sizes for roughing and finishing operations. Different tool sizes possess different abilities in removing the material and shaping the part. Large tools are capable of removing more material in shorter time, but are restricted in accessing small cutting regions. To the contrary, small tools have much improved accessibility but very inferior rate of material removal. Therefore, an appropriate combination of tool sizes is necessary to machine the workpiece effectively. The region that is accessible to different tool sizes can be calculated through the Tool Access Volume (TAV) method. Then, the analysis is extended to identify appropriate tool selection and sequencing by using a method known as Relative Delta Volume Clearance (RDVC) (Lim, Corney, Ritchie, & Clark, 2001). Based on information gained from these analyses, tool sizes and a set of cutting orientations can be identified by applying a genetic algorithm technique (Renner, 2008). Accordingly, the machining times are substantially reduced by using carefully selected combinations of tool sizes.

A recent study has proposed another approach to further improve the machining operations in CNC-RM (Osman Zahid, Case, & Watts, 2014). It was successfully carried out by manipulating the orientations for roughing operations. Instead of performing roughing and finishing operations using the same set of orientations, rough cuts were extracted and performed through another set of orientations. This set of orientations was based on a combination of four cutting orientations, 0–90–190–270°. A customized programme was developed to simulate the cutting process and propose efficient cutting directions. This method reduced the required cutting tool depth during roughing operations for avoidance of tool breakage. Since roughing and finishing operations are based on different sets of orientations, tool changes occur only once throughout the process. The time spent on roughing operations frequently increases, but it was found that this resulted in a reduction of total machining time.

Over the years, much research has been aimed at reducing machining time by manipulating cutting parameters as a part of process planning (Bouzid, 2005; Lavernhe, Tournier, & Lartigue, 2008; Palanisamy, Rajendran, & Shanmugasundaram, 2007). A good surface finish with low machining time can be obtained through the proper selection of machining parameters (Bharathi Raja & Baskar, 2012). The prominent parameters to be controlled are cutting speed, feed rate and depth of cut. Based on a multiple regression model, a previous study indicated that the depth of cut was the most influential parameter in determining the level of surface roughness achieved (Agarwal, 2012). Therefore, it is undeniable that cutting parameters are among the major factors that influence surface quality and machining time. However, in CNC-RM applications, the aim is to rapidly generate machining operations that reflect the planning and execution stages. Optimizing cutting parameters would require the handling of a large number of variables, and this tends to complicate the planning tasks and limits the level of automation that can be adopted in the planning stage. Therefore, the study conducted here integrates different end mill tools and observes the implications for machining time and surface quality. However, efficient cutting parameters suggested from previous studies could still be incorporated during the planning stage. Based on specific materials and tool sizes, a set of efficient parameters could be incorporated using decision-making tools in the development of machining plans.

Generally, the developments in selection of cutting tools and machining operations have enhanced the ability of CNC machining for rapid processes. The time required to machine parts has been substantially reduced with the introduction of roughing and finishing operations, appropriate tool size combinations and different rough cut sets of orientations. All these have enhanced the application of end mill tools in CNC machining, particularly for rapid manufacturing. However, far too little attention has been paid to the geometries of the end mill tool while developing the machining operations. Hence, there are no clear guidelines on how to combine the different tool geometries in the processing steps of CNC-RM. Nevertheless, several studies have strongly addressed the significance of cutting tool geometries on the level of quality achieved in the machining process (Patel, 2010; Reddy & Rao, 2005). Even if cutting can proceed at a very small depths of cut, neglecting the tool geometries tends to constrain the abilities of CNC machining in producing high quality parts. Therefore, considering the rapid manufacturing application, integrating different types of end mill tools is thought to be an approach to fulfil the quality requirements of end-use products.

3. Methodology

In attempts to implement CNC machining for rapid manufacturing processes, finishing tool selection is strictly defined based on the smallest diameter single end mill tool. This approach minimizes the tasks required in the planning phase and realizes the rapid generation of machining operations. Nevertheless, there is an implication for part quality as a single end mill tool is not capable of catering for a wide range of part surfaces. Therefore, integrating different end mill tools is a possible way of resolving the problem, but the guidelines must be clearly defined so that process planning can be handled effectively. Several important tasks are identified including separation of cutting regions and the types of end mill tool selected.

3.1. Cutting tools and surface classification

In the CNC-RM approach, finishing operations are executed through predetermined orientations and aim to shape the material completely. Hence, this process cuts the material with very small depths of cut to obtain a fine surface finish. Previous developments have implemented different cutting tools to finish cut parts built by welding process through a layer deposition method (Akula & Karunakaran, 2006). However, this approach utilized various type of tools and several surface categories, and seems impractical as it tends to complicate the planning tasks. However, some modifications to minimize and eliminate irrelevant setups will help to adopt this method in CNC-RM processes.

From one particular cutting direction, surfaces of the part can be classified into flat and non-flat surfaces. A flat surface is defined as a surface that is perpendicular to the cutting tool direction whereas the rest of the surfaces are categorized as non-flat surfaces. These classifications allow specific cutting tools to be assigned to machine the part effectively. Therefore, one orientation will perform up to two cutting operations if both types of surface are present. The first areas that will be cut are the flat surfaces using a flat end mill cutter. The process continues by machining non-flat surfaces using a ball nose end mill. Based on Figure 4, assuming cutting starts at 0° cutting orientation, the tool is engaged with the workpiece from ZC direction. In this particular direction, the dark areas represent flat surfaces that are perpendicular to the cutting tool direction. The first finishing operation is performed on this surface using a flat end mill tool. Then, the CNC machine replaces the current tool with a similar-sized ball nose end mill. This tool is used to machine the other surfaces shown as light grey areas on the part.

Generally, there are three common end mill tools (Figure 5) that differ based on the shapes of the tool tip. These are flat, bull nose and ball nose end mills. In this study, only two tools are selected to be utilized in finishing operations. This selection is expected to greatly reduce the complications for process planning as only two variables are present instead of three. A flat end mill is capable of machining plane surfaces with a minimum scallop size compared to other tools (Elber, 1995; Ryu, Choi, & Chu, 2006). Therefore, this is exactly the right tool to cut the flat surfaces classified on parts. The second selection utilizes a ball nose tool to cater for non-flat surfaces. Based on previous studies, the capabilities of this tool in handling sculptured surfaces are proven and high-quality surface finish can be produced (Chen, Huang, & Chen, 2005; Chen & Shi, 2008; Vijayaraghavan, Hoover, Hartnett, & Dornfeld, 2008). Moreover, this tool can be easily guided to engage with part surfaces and produces simpler NC code for machining (Chen & Shi, 2008). Meanwhile, a bull nose end mill is capable of removing high volumes of material with efficient machining times (Patel, 2010). But, due to the tool tip geometry, there is limited access to certain part features and this can result in

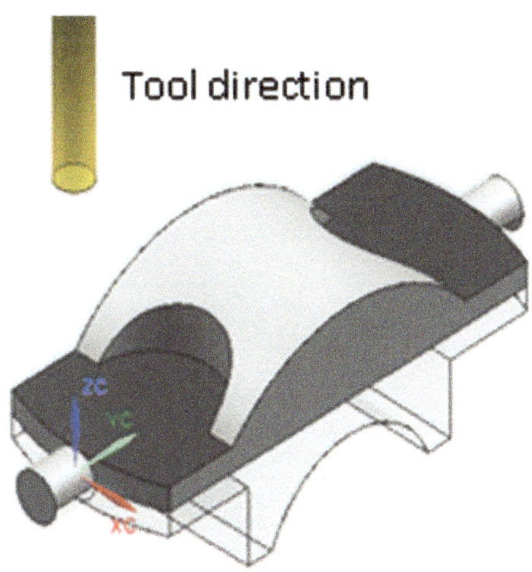

Figure 4. Classification of flat and non-flat surfaces in one cutting direction.

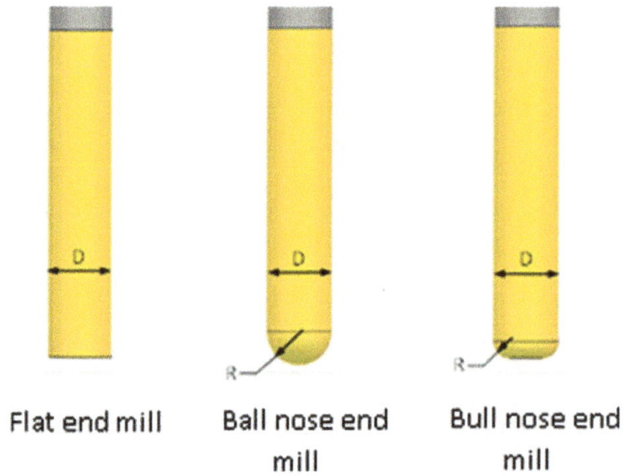

Figure 5. Three common shapes of end mill tool (Engin & Altintas, 2001).

high volumes of excess material. This limitation is shown in Figure 6. Therefore, in this study, only flat and ball nose end mills are selected for cutting tool integration in the CNC-RM processes.

3.2. Machining and verification processes

The introduction of different end mill tools requires a revision of the machining parameters, especially the cutting levels. Similar to roughing operations, finishing operations continue until the central axis of the cylindrical workpiece is reached. However, this is only applicable if a flat end mill tool is used in the operation. In the case where both tools are used, the ball nose cutter requires extended cutting levels to completely remove the remaining material. This is critically important for a part that contains closed regions that only permit tool access in limited directions. Figure 7 illustrates the situation where there is excess material due to there being insufficient cutting levels. Therefore, the cutting levels for a ball nose tool are extended by at least half the cutting tool diameter from the centre of the workpiece. Other parameters such as cutting feed and spindle speed are dependent on the diameter of the end mill tool. The finishing

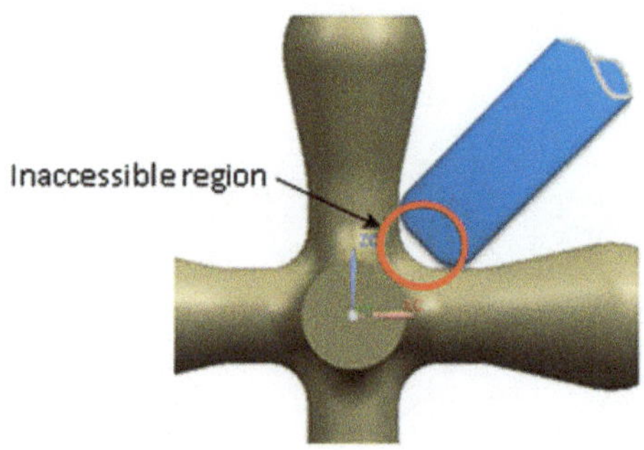

Figure 6. Restricted accessible regions for bull nose end mill tool.

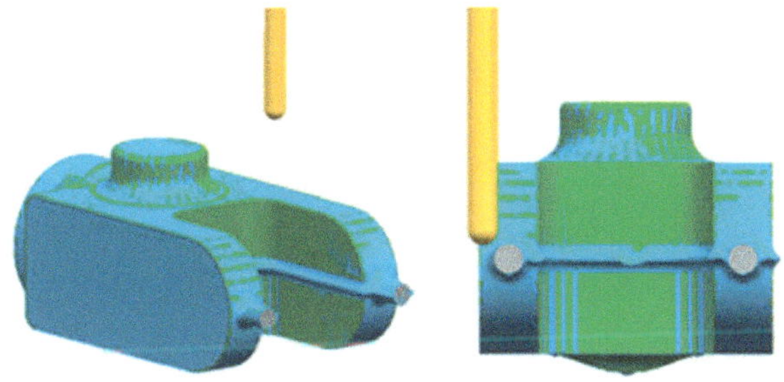

Figure 7. Remaining uncut materials left by ball nose end mill tool.

operations in this study employ the same tool diameter regardless of the type of tool used. Hence, both share the same value of spindle speed and feed rate. In this study, spindle speed and feed rates were determined based on the size of the cutting tools. The parameters are proposed by the CAD/CAM software itself and are not necessarily optimum cutting parameters. Depending on the models, depths of cut ranged between 1 and 1.5 mm for rough machining and 0.1 and 0.3 mm for finishing operations.

The simulation analysis was carried out with NX 7.5 software using the manufacturing application. A series of machining operations were created and the software used to estimate machining time to fabricate the parts. In order to calculate the excess volume left, the analysis was extended by using CGTech VERICUT ® 7.2.3 (CGTech, 2012). Recently, this software has been interfaced to NX which provides convenient access to run the simulation once operations have been created. There are several pre-processing steps while calculating the excess volume. First, the machining operations are translated into a cutter location source file (.clsf). Then, the file is run on VERICUT® software to simulate the cutting operations. Once completed, the 'X-caliper' tool is activated to

Table 1. Results for drive shaft and knob model.

Approach Criteria		Drive shaft		Knob	
		Approach 1	Approach 2	Approach 1	Approach2
Machining time	h:min:s	04:13:20	04:02:35	03:50:09	03:52:40
Finishing time		02:38:30	02:28:39	02:44:49	02:27:16
Non-cutting time		00:16:58	00:16:08	00:09:11	00:10:20
Roughing time		01:17:52	01:17:48	01:15:04	01:15:04
Cutting orientations		90–180–280–0–32–180–0°	90–180–280–0–32–180–0°	90–180–280–0–180–45–315°	90–180–280–0–180–45–315°
Machined volume	mm^3	150,273.23	150,432.83	112,946.97	113,203.60
Current part volume		53,244.05	53,084.45	21,660.63	21,403.91
Excess volume		514.82	355.22	527.99	271.27

obtain volumetric information. This includes stock volume, machined volume and current part volume. Prior to that, the machined part volume is calculated by a 'measure bodies' tool available in the NX system. All this information is gathered and the excess volume is calculated by subtracting the current part volume from the machined part volume. Finally, the results obtained are used to evaluate the effectiveness of integrating end mill tools in the machining processes.

4. Results and discussion

The simulation studies were conducted using different models including drive shaft, knob, salt bottle and toy jack models. Figure 8 illustrates the models adopted for the simulation studies. To visualize the implications of the suggested approach, similar cutting operations relying totally on the use of a flat end mill tool were developed. Therefore, Approach 1 was based on a process that utilized a single flat end mill whereas Approach 2 represented the method developed in this study that integrates different cutting tools. In this study, machining parameters were determined based on the size of the cutting tools. Since there are two types of operations (roughing and finishing), two sets of cutting parameters were involved. Meanwhile in the finishing operations, two different tools are introduced (flat and ball nose end mill). However, the machining parameters are similar since only one size of cutting tool is used. It is important to note that all the machining parameters are constant between Approach 1 and 2. Thus, the improvements gained from this study totally relied on the implication of different end mills during finishing operations. The study is not meant to validate the optimum cutting parameters, but the values could be adopted in the planning stage.

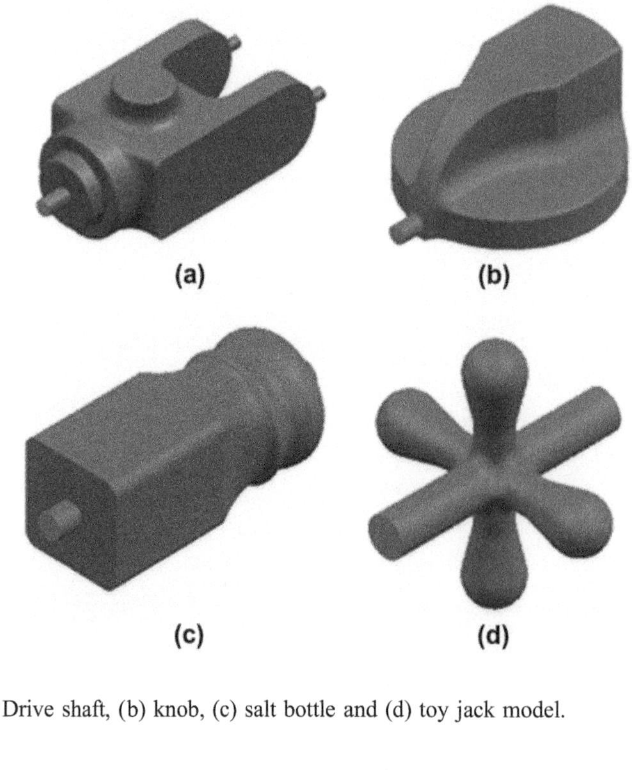

Figure 8. (a) Drive shaft, (b) knob, (c) salt bottle and (d) toy jack model.

There are two main evaluation criteria that are based on machining time and excess volume. Tables 1 and 2 summarize the results produced from simulation studies. Machining time is recorded in (h:min:s) format and volume information is in mm^3. The amount of material removed from the workpiece is presented as machined volume in the table. Part and stock volumes are calculated earlier based on the virtual models. It is apparent from these tables that there are considerable differences between Approaches 1 and 2. Out of the four models, three indicate a reduction of cutting time by implementing the approach developed in this study. Interestingly, all models achieved low excess volumes once different cutting tools were used in finishing operations. Assigning specific tools to particular surfaces, the ranges of excess volume reduce from 0.9–3.3% down to 0.7–1.6% of the total part volume.

4.1. *Implications for machining time*

Data gathered in the tables clearly indicate the benefits of implementing different cutting tools in CNC-RM processes. The majority of the models studied show a decreasing trend of machining time compared to a single tool approach. Finishing cutting times contribute most of the savings. This signals the effectiveness of cutting tools removing material on assigned surfaces. Depending on part geometries, cutting times differ from 2 min up to 11 min. For example, results for the drive shaft model in Table 1 indicate that the current method in Approach 2 manages to reduce machining time by about 11 min compared to Approach 1. Nonetheless, some models show only a small reduction of machining time such as the toy jack model. In this model, only ball nose tools were used as only non-flat surfaces were present on the part. For a medium production run, this reduction might still be important in minimizing cost and time. Oppositely, the knob model indicated a slightly longer machining time compared to Approach 1. The variety of flat and non-flat surfaces present on the part are possibly the main factors influencing cutting time.

Table 2. Results for salt bottle and toy jack model.

Approach Criteria		Salt bottle		Toy jack	
		Approach 1	Approach 2	Approach 1	Approach 2
Machining time	h:min:s	04:32:02	04:23:26	03:25:22	03:21:42
Finishing time		02:38:44	02:29:33	02:16:04	02:12:44
Non-cutting time		00:11:57	00:12:28	00:14:06	00:13:44
Roughing time		01:41:21	01:41:26	00:55:12	00:55:12
Cutting orientations		0–90–190–270–45–180–270°	0–90–190–270–45–180–270°	50–140–240–320–49–140–228–320°	50–140–240–320–49–140–228–320°
Machined volume	mm^3	76,676.80	76,845.85	109,662.15	109,792.30
Current part volume		34,499.39	34,330.33	7769.86	7639.71
Excess volume		417.56	248.5	252.74	122.59

Table 3. Excess materials distribution based on studied models.

Drive shaft		Knob	
Approach 1	Approach 2	Approach 1	Approach 2
514.82 mm^3	355.22 mm^3	527.99 mm^3	271.27 mm^3

Salt bottle		Toy jack	
Approach 1	Approach 2	Approach 1	Approach 2
417.56 mm^3	248.5 mm^3	252.74 mm^3	122.59 mm^3
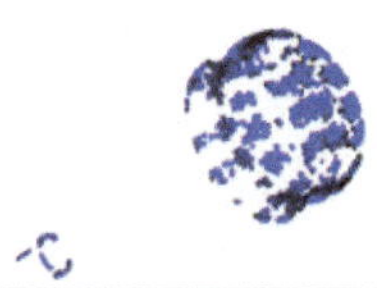			

The simulation studies adopted different sets of orientations for roughing and finishing operations as proposed from previous research (Osman Zahid et al., 2014). Therefore, integrating different end mill tools did not add to the number of orientations, but it increased the number of operations and tool changes. The presence of flat and non-flat surfaces in one cutting orientation will require both end mill tools and one tool change operation. Taking the salt bottle as example, there are two orientations consisting of flat and non-flat surfaces. Hence, four finishing operations are required within the two cutting directions. Meanwhile, in a single tool approach (Approach 1), only one finishing operation is needed in each orientation and this is quite straightforward compared to Approach 2. However, considering the effectiveness of material removal, it is worthwhile integrating the cutting tools.

4.2. Volume of excess materials

Previous studies have validated the capabilities of ball nose cutters on sculptured surfaces. Hence, assigning the tools based on flat and non-flat surfaces manages to minimize the excess volume on parts. Surprisingly, simulation results consistently indicate that excess volume was reduced for all models studied. Some models show a considerable reduction. For instance, excess volume for the knob model decreases dramatically from 527 mm^3 down to 271 mm^3. In CNC machining, uncut materials can be reduced by modifying cutting parameters in finishing operations including depth of cut, number of passes, feed rates and speed. However, in the CNC-RM approach, all these parameters remain constant except for the type of cutting tools used. Therefore, the achievements of this study show that the cutting tool combination is the main factor that influences the results.

Furthermore, the implications can be visualized by analysing the material distribution diagram as shown in Table 3. Comparing Approach 1 and 2, it can be seen that most of the excess material on non-flat surfaces of the part have been reduced. This indicates that the ball nose cutter effectively minimizes the staircasing effect. Keeping the flat end mill cutter to machine flat surfaces is useful in overcoming the weakness of ball nose tools.

5. Limitations and future work

Using multiple end mill tools in finishing operations manages to greatly reduce the excess volume left on the machined parts. The results of simulation studies indirectly portray the benefits of the approach in terms of surface quality. But, observation of the cutting simulations has revealed a limitation in shaping the sacrificial support of the part. If the support structure is connected to a flat vertical surface, a ball nose end mill faces a difficulty in completely shaping the structure. An excess volume (fillet shape) is created on the edge of the support and part. This problem is shown in Figure 9. On the other hand, a flat end mill is capable of forming the sharp edge, but is limited to areas that are perpendicular to the cutter direction. Excess volume will still be present in the rest of the connection area. Currently, the only way to minimize the problem is by using the smallest possible end mill tool during finishing processes. Larger tools will increase the excess volume and cause more work at the post-processing stage.

In real applications, several variations can be expected on the method proposed in this study. First, the rate of material removal is highly dependent on the size of cutting tools adopted in roughing and finishing processes. The implications can be seen based on accessibility and machining time. Selection of the smallest cutting tool in finishing operations will guarantee the accessibility, but machining time will probably be extended. Using optimization tools like TAV and RDVC that have been discussed earlier could be helpful in reducing the variations. Another factor that will cause variation between simulation and real machining is process interventions. In the simulation study, the estimated cutting time represents the complete operation including cutting and non-cutting time, but times for tool changes and workpiece rotations were not considered. If both are fully automated, it would have less impact on the cutting time. However, if the operations are carried out manually, which requires the machine to stop, then the estimated time is likely to be extended. For this reason, it is highly recommended that CNC-RM operations are executed in an automated manner.

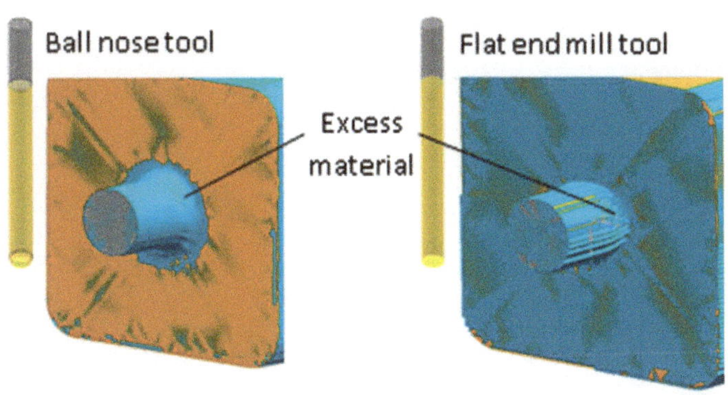

Figure 9. Excess material formed at the edge of sacrificial support.

The integration of cutting tools has benefited the production stage of the CNC-RM, but, it affects the planning stage as more variables need to be handled. Previously, process planning in CNC machining restricted implementation in RM as it involves many manual tasks and is highly dependent on human skill (Anderberg, Beno, & Pejryd, 2009; Relvas & Simoes, 2004). However, a recent development in CAD/CAM has introduced several features that can be used to effectively handle the process planning tasks (Agrawal, Soni, & Dwivedi, 2013). Therefore, an assisting tool can be developed to integrate the cutters while constructing the machining operation in the system and make it adaptable to any kind of part features. On the other hand, the identification of flat and non-flat surfaces within the cutting orientations needs to be performed correctly to ensure the right end mill tools are selected. Flat surfaces are not only the horizontal and vertical surfaces of the CAD model, but, depending on cutting direction, inclined surfaces can also be considered as flat surfaces if they are perpendicular to the direction of tool movement. Considering these issues, the medium of communication between the user and the system must be preserved to assist decision-making during process planning. To realize this, customized programming tools and computerized systems can be implemented in the CNC-RM process plan.

6. Conclusion

This simulation study has revealed the advantages of using different cutting tools in finishing operations particularly for CNC-RM processes. The present study was designed to improve part quality by reducing uncut material left after completion of machining. Generally, using different tools to cater for flat and non-flat surfaces is an effective way to reduce excess volume. Furthermore, machining times also can be reduced depending on part geometries. In order to visualize the implications, the results were compared to a previous method that adopted a single flat end mill tool throughout the cutting operations. Distinctly, the comparisons show the method developed in this study is capable of enhancing part quality and appearance. Further studies could focus on integrating this approach into CNC-RM process planning. The surface classification task needs to be handled carefully so that it will not complicate the planning task. Also, further verification is still required through machining experiments so that real parts can be fabricated and analysed completely. In the simulation studies, the excess volume was a key indicator to determine the surface quality of the part. However, in real machining operations, the surface quality needs to be examined through roughness analysis.

Disclosure statement

No potential conflict of interest was reported by the authors.

References

Agarwal, N. (2012). Surface roughness modelling with machining parameters (speed, feed and depth of cut) in CNC milling. *MIT International Journal of Mechanical Engineering, 2*, 55–61.

Agrawal, A., Soni, R. K., & Dwivedi, N. (2013). Development of integrated CNC-RP system through CAD/CAM environment. *International Journal of Mechanical and Production Engineering Research and Development, 3*, 1–10.

Akula, S., & Karunakaran, K. (2006). Hybrid adaptive layer manufacturing: An Intelligent art of direct metal rapid tooling process. *Robotics and Computer-Integrated Manufacturing, 22*, 113–123.

Anderberg, S., Beno, T., & Pejryd, L. (2009). CNC machining process planning productivity – A qualitative survey. *Proceedings of the International 3rd Swedish Production Symposium*, Göteborg, Sweden, 228–235.

Bharathi Raja, S., & Baskar, N. (2012). Application of particle swarm optimization technique for achieving desired milled surface roughness in minimum machining time. *Expert Systems with Applications, 39*, 5982–5989.

Bouzid, W. (2005). Cutting parameter optimization to minimize production time in high speed turning. *Journal of Materials Processing Technology, 161*, 388–395.

CGTech. (2012). *VERICUT 7.2: Release notes.*

Chen, J., Huang, Y., & Chen, M. (2005). A study of the surface scallop generating mechanism in the ball-end milling process. *International Journal of Machine Tools and Manufacture, 45*, 1077–1084.

Chen, T., & Shi, Z. (2008). A tool path generation strategy for three-axis ball-end milling of free-form surfaces. *Journal of Materials Processing Technology, 208*, 259–263.

Davim, J. P. (2001). A note on the determination of optimal cutting conditions for surface finish obtained in turning using design of experiments. *Journal of Materials Processing Technology, 116*, 305–308.

Elber, G. (1995). Freeform surface region optimization for 3-axis and 5-axis milling. *Computer-Aided Design, 27*, 465–470.

Engin, S., & Altintas, Y. (2001). Mechanics and dynamics of general milling cutters: Part I: Helical end mills. *International Journal of Machine Tools and Manufacture, 41*, 2195–2212.

Frank, M. C. (2003). *The development of a rapid prototyping process using computer numerical controlled machining* (PhD thesis). The Pennsylvania State University.

Frank, M. C. (2007). Implementing rapid prototyping using CNC machining (CNC-RP) through a CAD/CAM interface. *Proceedings of the Solid Freeform Fabrication Symposium*, Autin, TX, 112–123.

Frank, M., Joshi, S. B., & Wysk, R. A. (2002). CNC-RP: A technique for using CNC machining as a rapid prototyping tool in product/process development. *Proceedings of the 11th Annual Industrial Engineering Research Conference*, Orlando, FL, 19–22.

Frank, M. C., Wysk, R. A., & Joshi, S. B. (2004). Rapid planning for CNC milling - A new approach for rapid prototyping. *Journal of Manufacturing Systems, 23*, 242–255.

Frank, M. C., Wysk, R. A., & Joshi, S. B. (2006). Determining setup orientations from the visibility of slice geometry for rapid computer numerically controlled machining. *Journal of Manufacturing Science and Engineering, 128*, 228–238.

Lavernhe, S., Tournier, C., & Lartigue, C. (2008). Optimization of 5-axis high-speed machining using a surface based approach. *Computer-Aided Design, 40*, 1015–1023.

Li, Y., & Frank, M. C. (2006). Machinability analysis for 3-axis flat end milling. *Journal of Manufacturing Science and Engineering, 128*, 454–464.

Lim, T., Corney, J., Ritchie, J., & Clark, D. (2001). Optimizing tool selection. *International Journal of Production Research, 39*, 1239–1256.

Osman Zahid, M. N., Case, K., & Watts, D. (2014). Optimization of roughing operations in CNC machining for rapid manufacturing processes. *Production & Manufacturing Research, 2*, 519–529.

Palanisamy, P., Rajendran, I., & Shanmugasundaram, S. (2007). Optimization of machining parameters using genetic algorithm and experimental validation for end-milling operations. *The International Journal of Advanced Manufacturing Technology, 32*, 644–655.

Patel, K. (2010). *Web based automatic tool path planning strategy for complex sculptured surfaces* (Master thesis). University of Waterloo.

Reddy, N. S. K., & Rao, P. V. (2005). Selection of optimum tool geometry and cutting conditions using a surface roughness prediction model for end milling. *The International Journal of Advanced Manufacturing Technology, 26*, 1202–1210.

Relvas, C., & Simoes, J. (2004). Optimization of computer numerical control set-up parameters to manufacture rapid prototypes. *Proceedings of the Institution of Mechanical Engineers, Part B: Journal of Engineering Manufacture, 218*, 867–874.

Renner, A. (2008). *Computer aided process planning for rapid prototyping using a genetic algorithm* (Master thesis). Iowa State University.

Ryu, S. H., Choi, D. K., & Chu, C. N. (2006). Roughness and texture generation on end milled surfaces. *International Journal of Machine Tools and Manufacture, 46*, 404–412.

Vijayaraghavan, A., Hoover, A., Hartnett, J., & Dornfeld, D. (2008). Improving end milling surface finish by workpiece rotation and adaptive toolpath spacing. *International Journal of Machine Tools & Manufacture, 49*, 89–98.

8

Development of a predictive model for determining mechanical properties of AA 6061 using regression analysis

Joseph Achebo*

Department of Production Engineering, University of Benin, Benin City, Nigeria

Aluminum alloys (AA) such as AA 6061 are difficult to weld, and their application to high-strength demands tends to be limited because of their inherently low-strength threshold levels when compared to other alloys. To effectively use this alloy to its full potential, its tensile strength, is investigated. In actual manufacturing settings, it has proven beneficial to attempt a prediction of the tensile properties of potential weld joints. To achieve this, a model was developed using the multiple linear regression analysis to predict these tensile properties such as the ultimate tensile strength, the yield strength (YS), and percentage elongation (% Elong). It was found from the scattered diagrams that the measured and predicted values were almost a perfect fit with a coefficient of determination of between 0.99 and 1.0. The analysis of variance further validated the adequacy of the model. The analysis showed that the claims observed in the study match with those of other investigators. The predictive model obtained is expected to help the welding community to pre-determine the tensile properties of AA 6061 weldment using selected values for each of the process parameters applied in this study.

Keywords: aluminum alloy 6061; mechanical properties; process parameters; regression analysis

1. Introduction

Materials can either be regarded as being ductile or brittle depending on their ability to undergo plastic deformation under the application of load without fracturing. Ductile materials are often used for engineering projects because they usually exhibit homogeneous load distribution. Brittleness is regarded as the opposite of ductility. In a brittle material, at critical conditions, fractures may likely occur suddenly when loads are applied because the YS equals the ultimate tensile strength (UTS). Therefore, because it is expected that most structural materials are bound to be subjected to repeat loading stretching over a long period, and at the same time, are expected to absorb impact energies without fracturing, the elastic property of these materials is of great importance in engineering applications, and ductility is therefore a highly prized quality.

Ductility could be measured by determining the % Elong and reduction in area of the structural material. Percent elongation as considered in this study shows the extent to which a material can endure the application of stress before a fracture can occur. The ability of an engineering material to withstand stresses cannot be overemphasized, in terms of equipment maintainability, coupled with the fact that replacement and

*Email: josephachebo@yahoo.co.uk

breakdown maintenance due to material failure could be very costly. Such unwarranted failures invariably lead to the resultant increase in idle time or downtime. Thus, in today's engineering landscape, structural materials are designed to carry even heavier loads, so that their service lives could exceed expectations with optimum efficiency. For instance, in bridge construction, structural materials are designed to serve for a very long period, as it would be quite uneconomical to sustain continued major routine overhauling processes.

Since ductility is a major factor to consider in selecting materials for structural engineering purposes, it has become imperative to develop a model to predict the tensile properties that define ductility. In this study, the ductile material considered is the AA 6061 aluminum alloy. Kumar, Kolhe, Morey, and Datta (2011) wrote that aluminum alloy 6061 (Al-Mg-Si) is widely used for the fabrication of food processing equipment, chemical containers, passenger cars, road tankers, and railway transport systems due to its high specific strength, excellent weldability, and resistance to corrosion. These aluminum alloy 6061 plates were welded with the gas metal arc welding (GMAW) equipment used to make weld deposits. Singla, Singh, and Deepak (2010) defined the GMAW as a process in which the heat source produces an arc between the consumable electrode and the workpiece.

Prediction of mechanical properties to optimize material production performance has been attempted by many investigators, such as Chiteka (2014). They aimed at using artificial neural networks in predicting UTS of a weldment, given rotational speed as well as the welding speed in a friction stir welding environment. Kim, Son, and Jeung (2001) applied novel algorithms to develop mathematical models for optimizing bead width for multi-pass welding using neural network as well as multiple regression methods. Xue et al. (2005) used the fuzzy regression method for predicting and controlling bead width in a robotic arc welding process. Mostafa and Khajavi (2006) developed a model for predicting the value of weld penetration using regression analysis. Karthikeyan and Mahadevan (2014) wrote on the modeling of the friction stir welded Al 6351 plates reinforced with SiC particles in the butt-joint region. Taguchi's experimental design was used to design the experiment matrix and a mathematical model was developed to predict the mechanical properties of the welded Al 6351 plates. The predicted results were compared with the actual experimental results Khorram, Ghoreishi, Yazdi, and Moradi (2011) employed the response surface methodology to model and predict weld bead geometry parameters. Palanivel, Mathews, and Murugan (2011) used mathematical model to predict the mechanical properties of AA 6351 aluminum alloy. Cavaliere, Campanile, Panella, and Squillace (2006) studied the effect of welding parameters on mechanical and microstructural properties of AA 6056 joints produced by friction stir welding. Sreeraj, Kannan, and Maji (2013) used a five-level five-factor full factorial design matrix based on central composite rotatable design technique for the mathematical development of the model used for predicting clad bead geometry of austenitic stainless steel deposited by GMAW. They also used a simulated annealing algorithm tool available in MATLAB 7 software which was efficiently employed for predicting this same clad bead geometry. Heidarzadeh, Mahmoudi, Khodaverdizadeh, and Nazari (2012) worked on tensile behavior of friction stir welded AA 6061-T4 aluminum alloy joints. Elangovan, Balasubramanian, and Babu (2009) predicted the tensile strength of friction stir welded AA 6061 aluminum alloy joints using mathematical model. Sen, Mukherjee, and Pal (2014) developed a mathematical model using a multiple linear regression analysis in MINITAB 13.1 to predict the weld bead geometry and the authors also checked for the adequacy of the model by conducting analysis of variance (ANOVA). It was found from their study that the predicted optimum bead geometry correlates very well with the confirmation test result.

In this study, a model was developed to predict tensile properties and to optimize the extent of ductile content in an aluminum alloy 6061 using the multiple linear regression analysis. Dhas and Satheesh (2012) acknowledged the claim made by other investigators that weld parameter optimization through experimental methods and mathematical models has grown substantially over time to achieve a common goal of improving higher weld efficiency. The model which was developed in this study, successfully predicted the tensile properties of AA 6061. The results obtained were almost perfect match with the corresponding experimental results obtained from the weldments produced using GMAW.

2. Materials and methods

2.1. Materials

Five tensile specimens each were prepared for 16 different clusters or combinations of process parameters, giving a total of 80 weld joints. These joints were prepared according to ASTM E8M-04 procedures (see Figure 1) using a CNC Lathe machine as shown in Figure 2. Tensile tests were carried out in a 100 kN computer-controlled Universal Testing Machine (see Figure 3) as used by Prasad, Rao, and Rao (2011). The specimens were loaded at a rate of 1.5 kN/min as per ASTM specifications, so that these tensile specimens could undergo the deformation process. From the stress strain curve, UTS, YS, and % Elong of the weld joints were evaluated and the average of the five test results is presented in Table 3.

2.2. Methods

The following steps were conducted:

(1) The 16 sets of welding process parameters (see Table 3) were applied, using a GMAW operation, to make weld deposits at five joints each on 60 mm × 140 mm × 10 mm AA 6061 aluminum plates (see Figure 4).
(2) Following the above GMAW welding operations, each of the 80 weld deposits was sawn off to form individual weldment.
(3) Each weldment was machined using the CNC Lathe machine resulting in individual tensile specimens according to ASTM E8M-04 procedures.

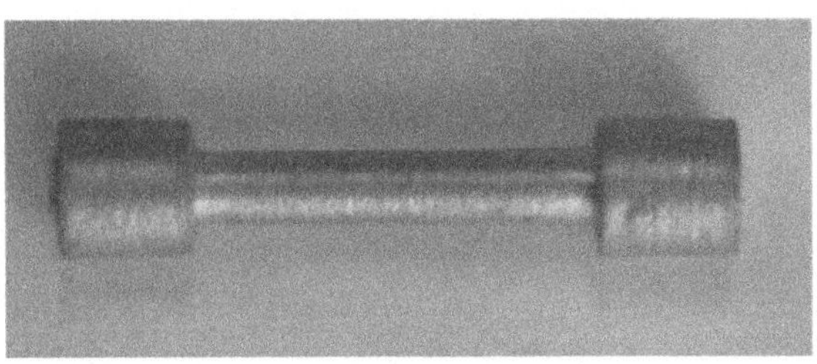

Figure 1. Tensile specimen.

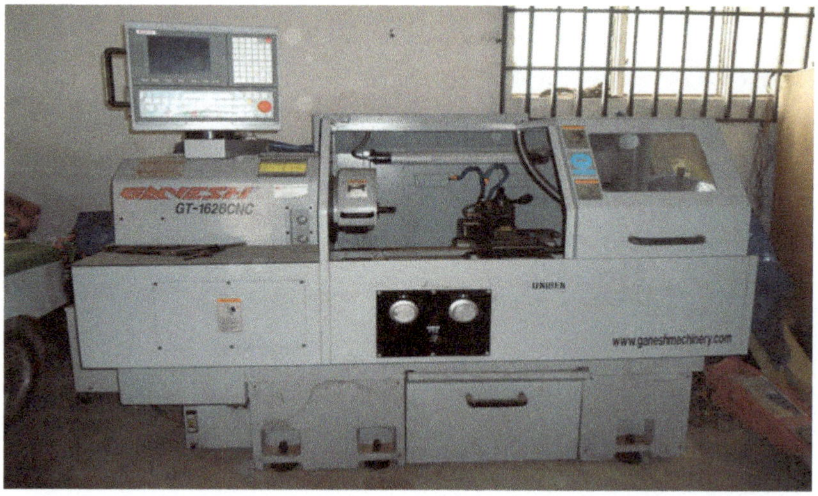

Figure 2. CNC Lathe machine.

Figure 3. 100 kN computer-controlled Universal Testing Machine.

(4) These machined tensile specimens were loaded into a 100 kN computer-controlled Universal Testing Machine.
(5) Each of the 80 loaded tensile specimens was measured for their UTS, YS, and % Elong.
(6) The results of these tensile tests were displayed in a digital screen.
(7) The average of the five tests conducted within each set for the 16 clusters of process parameters made was recorded.

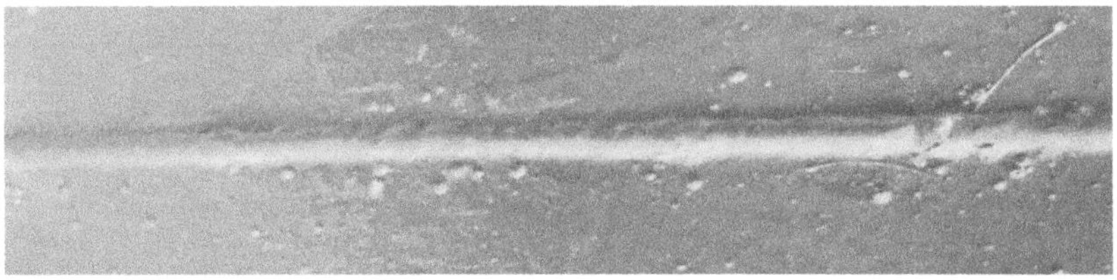

Figure 4. AA 6061 weldment sample.

(8) A statistical model was developed to predict the tensile properties of AA 6061 aluminum weldments.
(9) The effect of each of the 16 process parameters on the tensile properties of the aluminum weldments was examined.

3. Results and discussions

3.1. Results

Table 1 shows the design matrix containing both the process parameters (i.e. A, B, C, and D) and their interactions.

Table 2 shows the process parameters and their levels that were used for this study.

In Table 1, only the main parameters were considered, which comprises A, B, C, and D because linear multiple regression analysis would be used here to predict the tensile properties of aluminum plates treated in this study. The process parameters and their levels in Table 2 were substituted into the main parameters of the matrix design and used for welding the aluminum plates at their joints. Each weldment was tested for its UTS, YS, and % Elong and the measured values are recorded in Table 3.

Table 1. Design matrix.

	A	B	C	D	–AB	–BC	–CD	ABD	–AC	–BD	ABC	BCD	–ABCD	ACD	–AD
0	1	2	3	4	5	6	7	8	9	10	11	12	13	14	15
+	+	−	−	−	+	−	−	+	+	−	+	−	+	+	+
+	+	+	−	−	−	+	−	−	+	+	−	+	−	+	+
+	+	+	+	−	−	−	+	−	−	+	+	−	+	−	+
+	+	+	+	+	−	−	−	+	−	−	+	+	−	+	−
+	−	+	+	+	+	−	−	−	+	−	−	+	+	−	+
+	+	−	+	+	+	+	−	−	−	+	−	−	+	+	−
+	−	+	−	+	+	+	+	−	−	−	+	−	−	+	+
+	+	−	+	−	+	+	+	+	−	−	−	+	−	−	+
+	+	+	−	+	−	+	+	+	+	−	−	−	+	−	−
+	−	+	+	−	+	−	+	+	+	+	−	−	−	+	−
+	−	−	+	+	−	+	−	+	+	+	+	−	−	−	+
+	+	−	−	+	+	−	+	−	+	+	+	+	−	−	−
+	−	+	−	−	+	+	−	+	−	+	+	+	+	−	−
+	−	−	+	−	−	+	+	−	+	−	+	+	+	+	−
+	−	−	−	+	−	−	+	+	−	+	−	+	+	+	+
+	−	−	−	−	−	−	−	−	−	−	−	−	−	−	−

Table 2. The process parameters and their levels.

Process parameters	Unit	Notations	Levels	
			Low	High
Current	A	A (x_1)	320	390
Voltage	V	B (x_2)	18	20
Electrode diameter	mm	C (x_3)	1.6	3.2
Welding speed	mm/s	D (x_4)	80	120

Table 3. Process parameters with the corresponding measured and predicted (calculated) tensile properties.

x_1	x_2	x_3	x_4	Measured	Predicted	Measured	Predicted	Measured	Predicted
(*A*)	(*B*)	(*C*)	(*D*)	UTS (MPa)	UTS (MPa)	YS (MPa)	YS (MPa)	Elong (%)	Elong (%)
390	18	1.6	80	298	318.96	265	301.23	4.74	5.10
390	20	1.6	80	342	311.61	316	288.38	5.8	5.27
390	20	3.2	80	260	270.77	238	246.56	3.7	4.75
390	20	3.2	120	160	261.14	143	238.77	5.2	4.83
320	20	3.2	120	216	181.24	198	153.51	4.7	4.12
390	18	3.2	120	155	268.49	136	251.61	4.9	4.66
320	20	1.6	120	248	222.07	226	195.35	5.2	4.64
390	18	3.2	80	320	278.12	308	259.39	3.8	4.66
390	20	1.6	120	175	301.98	168	280.60	3.2	5.35
320	20	3.2	80	158	190.87	132	161.30	4.6	4.03
320	18	3.2	120	324	188.58	297	166.37	3.6	3.95
390	18	1.6	120	310	309.33	279	293.44	4.7	5.18
320	20	1.6	80	269	231.71	240	203.13	4.8	4.55
320	18	3.2	80	210	198.21	188	174.14	4.3	3.87
320	18	1.6	120	336	229.42	314	208.19	5.8	4.47
320	18	1.6	80	220	239.05	202	215.97	5.1	4.39

3.1.1. Developing the regression analysis model

The regression analysis model as proposed by Achebo (2008) is expressed as follows:

$$\mathrm{SSE} = \sum_{i=1}^{n}(y_i - \bar{y})^2 = \sum_{i=1}^{n}(y_i - b_o - b_1x_1 - b_2x_2 - b_3x_3 - b_4x_4) = 0 = \text{minimum} \quad (1)$$

where SSE is the sum of square error

$$\text{but}\, y_i = b_o + b_1x_1 + b_2x_2 + b_3x_3 + b_4x_4 \quad (2)$$

$$\text{therefore, } b_0 = y - b_1x_1 - b_2x_2 - b_3x_3 - b_4x_4 \quad (3)$$

Differentiating (1) with respect to b_0

$$\frac{\partial \mathrm{SSE}}{\partial b_0} = -2x_1\sum_{i=1}^{n}(y - b_0 - b_1x_1 - b_2x_2 - b_3x_3 - b_4x_4) = 0 \quad (4)$$

From (4), it is deducted that

$$\sum_{i=1}^{n} y - b_0 - b_1 \sum_{i=1}^{n} x_1 - b_2 \sum_{i=1}^{n} x_2 - b_3 \sum_{i=1}^{n} x_3 - b_4 \sum_{i=1}^{n} x_4 = 0 \quad (5)$$

Making b_0 the subject of the equation

$$b_0 = \sum_{i=1}^{n} y - b_1 \sum_{i=1}^{n} x_1 - b_2 \sum_{i=1}^{n} x_2 - b_3 \sum_{i=1}^{n} x_3 - b_4 \sum_{i=1}^{n} x_4 \quad (6)$$

Differentiating (6) with respect to b_1

$$\frac{\partial \mathrm{SSE}}{\partial b_1} = -2x_1 \sum_{i=1}^{n} (y - b_0 - b_1 x_1 - b_2 x_2 - b_3 x_3 - b_4 x_4) = 0 \quad (7)$$

Rearranging (7), we have:

$$\sum_{i=1}^{n} (yx_1 - b_o x_1 - b_1 x_1^2 - b_2 x_1 x_2 - b_3 x_1 x_3 - b_4 x_1 x_4) = 0 \quad (8)$$

Rearranging (8), we have

$$\sum_{i=1}^{n} (yx_1) - b_o \sum_{i=1}^{n} x_1 - b_1 \sum_{i=1}^{n} x_1^2 - b_2 \sum_{i=1}^{n} x_1 x_2 - b_3 \sum_{i=1}^{n} x_1 x_3 - b_4 \sum_{i=1}^{n} x_1 x_4 = 0 \quad (9)$$

Differentiating (1) with respect to b_2

$$\frac{\partial \mathrm{SSE}}{\partial b_2} = -2x_2 \sum_{i=1}^{n} (y - b_0 - b_1 x_1 - b_2 x_2 - b_3 x_3 - b_4 x_4) = 0 \quad (10)$$

Rearranging (10), we have:

$$\sum_{i=1}^{n} (yx_2) - b_o \sum_{i=1}^{n} x_2 - b_1 \sum_{i=1}^{n} x_1 x_2 - b_2 \sum_{i=1}^{n} x_2^2 - b_3 \sum_{i=1}^{n} x_2 x_3 - b_4 \sum_{i=1}^{n} x_2 x_4 = 0 \quad (11)$$

Differentiating (1) with respect to b_3

$$\frac{\partial \mathrm{SSE}}{\partial b_3} = -2x_3 \sum_{i=1}^{n} (y - b_0 - b_1 x_1 - b_2 x_2 - b_3 x_3 - b_4 x_4) = 0 \quad (12)$$

Rearranging (12), we have:

$$\sum_{i=1}^{n} (yx_3) - b_o \sum_{i=1}^{n} x_3 - b_1 \sum_{i=1}^{n} x_1 x_3 - b_2 \sum_{i=1}^{n} x_2 x_3 - b_3 \sum_{i=1}^{n} x_3^2 - b_4 \sum_{i=1}^{n} x_3 x_4 = 0 \quad (13)$$

Differentiating (1) with respect to b_4

$$\frac{\partial \mathrm{SSE}}{\partial b_4} = -2x_4 \sum_{i=1}^{n} (y - b_0 - b_1 x_1 - b_2 x_2 - b_3 x_3 - b_4 x_4) = 0 \quad (14)$$

Rearranging (14), we have:

$$\sum_{i=1}^{n} (yx_4) - b_o \sum_{i=1}^{n} x_4 - b_1 \sum_{i=1}^{n} x_1 x_4 - b_2 \sum_{i=1}^{n} x_2 x_4 - b_3 \sum_{i=1}^{n} x_3 x_4 - b_4 \sum_{i=1}^{n} x_4^2 = 0 \quad (15)$$

From (9), we have that

$$b_1 = \frac{\sum_{i=1}^{n}(yx_1) - b_o\sum_{i=1}^{n}x_1 - b_2\sum_{i=1}^{n}x_1x_2 - b_3\sum_{i=1}^{n}x_1x_3 - b_4\sum_{i=1}^{n}x_1x_4}{\sum_{i=1}^{n}x_1^2} \quad (16)$$

From (11), we have that

$$b_2 = \frac{\sum_{i=1}^{n}(yx_2) - b_o\sum_{i=1}^{n}x_2 - b_1\sum_{i=1}^{n}x_1x_2 - b_3\sum_{i=1}^{n}x_2x_3 - b_4\sum_{i=1}^{n}x_2x_4}{\sum_{i=1}^{n}x_2^2} \quad (17)$$

From (13), we have that

$$b_3 = \frac{\sum_{i=1}^{n}(yx_3) - b_o\sum_{i=1}^{n}x_3 - b_1\sum_{i=1}^{n}x_1x_3 - b_2\sum_{i=1}^{n}x_2x_3 - b_4\sum_{i=1}^{n}x_3x_4}{\sum_{i=1}^{n}x_3^2} \quad (18)$$

From (15), we have that

$$b_4 = \frac{\sum_{i=1}^{n}(yx_4) - b_o\sum_{i=1}^{n}x_4 - b_1\sum_{i=1}^{n}x_1x_4 - b_2\sum_{i=1}^{n}x_2x_4 - b_3\sum_{i=1}^{n}x_3x_4}{\sum_{i=1}^{n}x_4^2} \quad (19)$$

Process parameters values in Table 3 were substituted into Equations (6), (16)–(19) and the model coefficients in Table 4 were obtained.

The developed predictive models as contained in Table 4 are expressed below:

(1) UTS, is
MPa $= -0.0099 + 1.1415x_1 - 3.6731x_2 - 25.5238x_3 - 0.2408x_4$

(2) YS, is
MPa $= -0.7846 + 1.2179x_1 - 6.4208x_2 - 26.1426x_3 - 0.1946x_4$

(3) % Elongation, is
% Elong. $= -0.0089 + 0.0101x_1 + 0.0844x_2 - 0.3251x_3 + 0.0021x_4$

The above predictive models were used to predict the various ductile properties in Table 3.

3.1.2. *Confirmation tests*

Confirmation tests were done to ascertain the adequacy of the developed model. Five weldments were tested for their tensile properties using the computerized tensile test machine. For each weldment, five tensile test results were obtained. The average of the five results was recorded for each of the tensile properties and this process was repeated for the five weldments. The developed model was used to predict the values of these tensile properties to confirm the adequacy of the model by comparing its values with those of the corresponding measured ones, as presented in Table 5.

Table 4. Regression model coefficients.

Regression coefficients	UTS (MPa)	YS (MPa)	Elong (%)
b_0	–0.0099	–0.7846	–0.0089
b_1	1.1415	1.2179	0.0101
b_2	–3.6731	–6.4208	0.0844
b_3	–25.5238	–26.1426	–0.3251
b_4	–0.2408	–0.1946	0.0021

3.1.3. *Effect of process parameters on tensile properties*

The effects of the process parameters on the tensile properties of aluminum alloy plates were studied as shown in Figures 5–8.

Figure 5 depicts the observed effects of welding current on tensile properties of each specimen. It can be seen from the graph that the curves representing UTS, YS, and % Elong increase following graduated increments of the current applied. This indicates that a small increase in current can alter the tensile property of the aluminum alloy. The atoms that make up this aluminum alloy were still within their elastic limits until the current reached 350 A. At this current, the atoms become more agitated and the atomic collision increases. When this molten aluminum alloy cools under controlled conditions, the material experiences some strain hardening effects. This means that as it cools down, the atoms experience less agitation and dislocation. This resistance to dislocation formation makes the aluminum alloy more plastic, thereby establishing the strength properties of the tensile specimen.

As the current is increased from 360 to 380 A, the parameters reduced. This is because as more current is applied the molten aluminum drops in a process known as necking down. This process negatively affects UTS, YS, and % Elong properties of the aluminum alloy leading to a reduction in these properties. Achebo (2012) was of the opinion that welds at certain high temperatures tend to absorb moisture rapidly. Moisture lowers the cooling rate. At low cooling rate, more time is available for the molten aluminum alloy to mix with moisture. This action leads to the coarsening of the weld grains by making the grains angular in shape. This angular shape of the grains creates voids. The presence of voids in welds lowers the quality of such welds. Prasad et al. (2011) observed that the formation of coarser grains in the fusion zone is responsible for lower UTS of the welded joints.

As the current reaches 380 A, the strain hardening effect of the aluminum alloy becomes plastic. This is because at this current, the temperature also increases, further altering the atomic nature of the molten aluminum alloy, making it less susceptible to moisture interference while cooling. As a result of the reduction in the effect of moisture, the grain formation of the aluminum alloy weldment becomes less coarse or finer, thereby increasing its plastic nature. This plastic nature increases the ability of the alloy to absorb energy and this eventually increases the alloy's UTS and YS. The increase in percentage elongation indicates that the material is still ductile and at the same time slightly malleable but has almost exceeded its proportionality limits. This observation compares well with the claim made by Prasad et al. (2011), who emphasized that rapid cooling due to the decrease in current leading to decrease in heat input could lead to the formation of finer grains which eventually increases the UTS.

Table 5. Measured and predicted results of mechanical properties.

x_1	x_2	x_3	x_4	Measured UTS (MPa)	Predicted UTS (MPa)	Measured YS (MPa)	Predicted YS (MPa)	Measured Elong (%)	Predicted Elong (%)
340	19	1.6	110	251	250.99	227	228.07	4.74	4.72
360	18	1.6	80	286	284.71	266	264.69	4.79	4.81
330	20	3.2	100	201	197.47	168	169.59	4.18	4.23
380	18	3.2	90	265	264.30	242	245.27	4.50	4.54
390	18	1.6	85	319	317.75	301	300.25	5.11	5.09

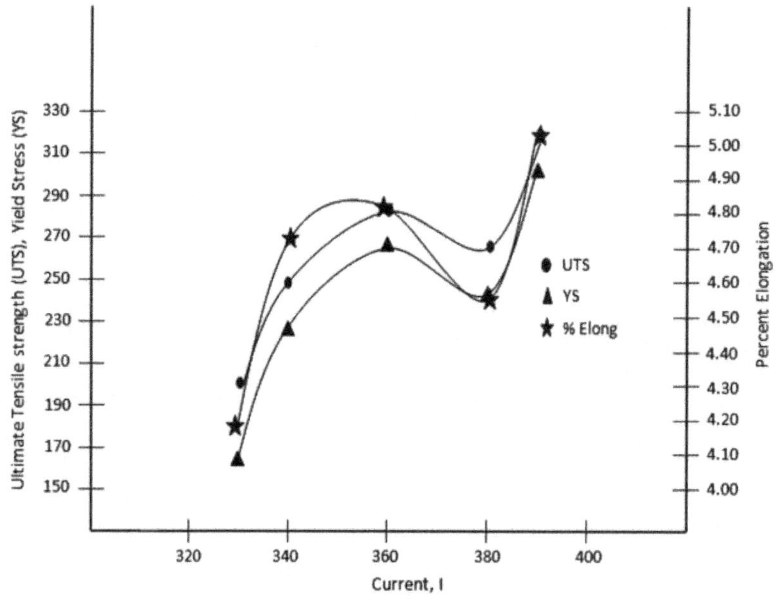

Figure 5. Effect of current on UTS, YS, and % elongation.

Figure 6 shows the relationship between tensile properties and the welding voltage. It is observed that at a constant welding voltage of 18 V, UTS, YS, and % Elong increased steeply. However, as the voltage increases from 18 V, the UTS, YS, and % Elong reduced drastically. This indicates that at 18 V, the aluminum weld was still at its solid–plastic transition state where the grains are assumed to have disintegrated repeatedly and eventually become finer and denser, but as the voltage increases, the weld transits from plastic to liquid. The hot liquid weld metal absorbs moisture and becomes coarser. This action reduces the quality of the weld which negatively affects the weld's UTS, YS, and eventually its % Elong. Its reduction in percentage elongation indicates that the weldment is not as ductile. Bahman and Alialhosseini (2010) showed that as the arc voltage increases, hardness, UTS, and YS decrease. Yang, Chandel, and Bibby (1992) said that at high voltage, bead widths are large and large bead widths are associated with a poor-quality welding process, which invariably negatively affects the UTS, YS, and % Elong.

Figure 7 shows the relationship between tensile properties and the electrode diameter. It is observed that when 1.6-mm electrode diameter was used for the welding process, UTS, YS, and % Elong increased steeply and as the electrode diameter increases, the UTS, YS, and % Elong decrease. This indicates that the 1.6-mm electrode diameter gave a more localized welding process where spatter was eliminated. This theoretically shows that at this electrode diameter, the weld dilution and deposition rate would have been high with adequate and controlled weld penetration, which is assumed to have occurred due to the Marangoni flow process. However, as the diameter of the electrode exceeds 1.6 mm, the increase in the size of the electrode produces beads with large widths forming spatter, which oxidizes rapidly and forms coarse grains, eventually reducing the UTS, YS and % Elong of the aluminum weld. Too much of an oxidization process of the spatter causes the weld to contract.

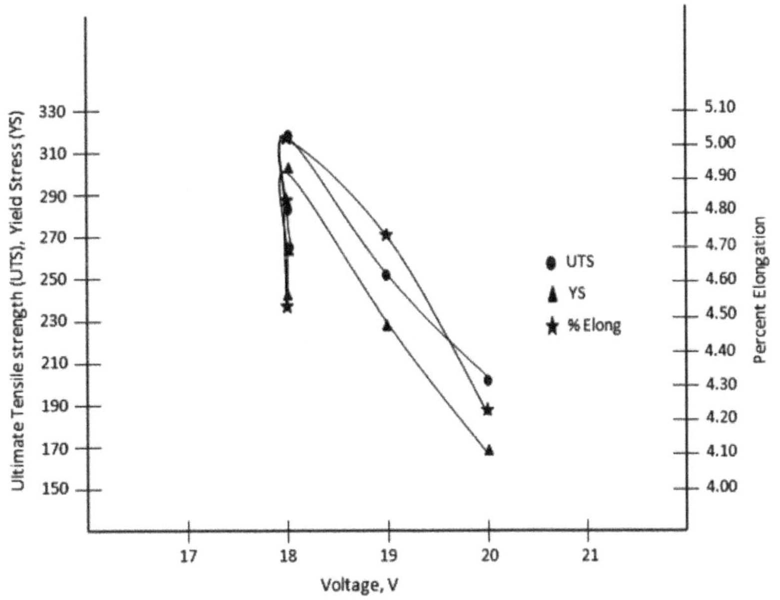

Figure 6. Effect of voltage on UTS, YS, and % elongation.

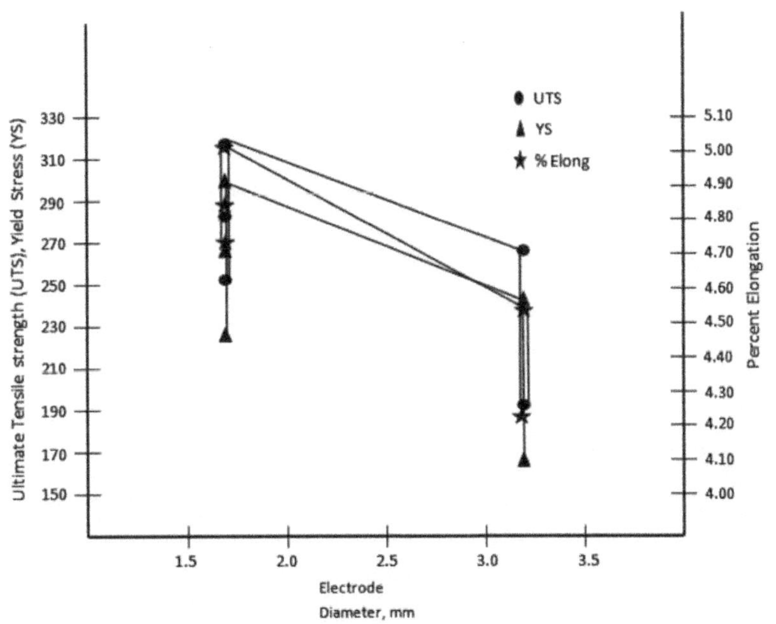

Figure 7. Effect of electrode diameter on UTS, YS, and % elongation.

Figure 8 shows the relationship between tensile properties and the welding speed. It is observed that as the welding speed increases from 80 to 85 mm/s, the UTS, YS, and % Elong also increase. However, when the welding speed increased above 85 mm/s, the UTS, YS, and % Elong decreased. These tensile properties also increased when the

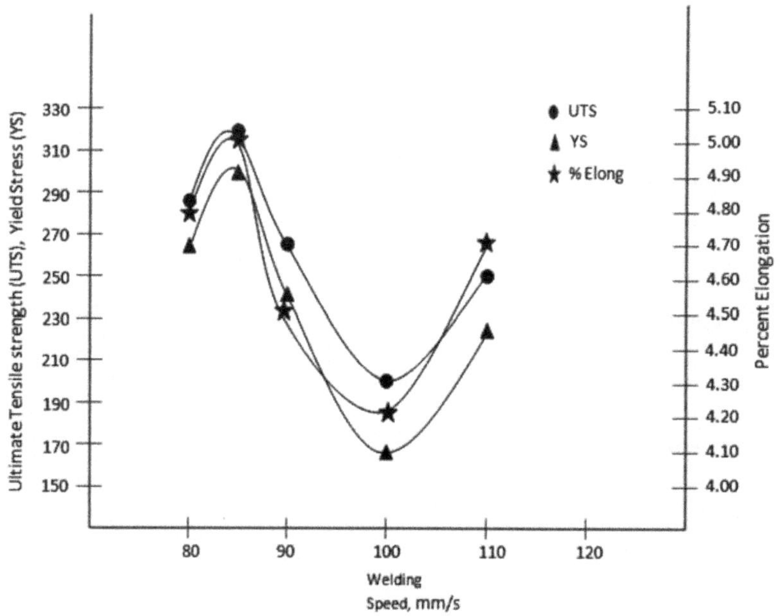

Figure 8. Effect of welding speed on UTS, YS, and % elongation.

welding speed increased from 100 to 110 mm/s. The tensile behavior of the weld indicates that the increase in welding speed, reduces the time spent in welding, this leads to a smaller heat affected zone (HAZ) and a small bead width, and as a result, less heat is utilized for the welding process. This process is used to achieve increased UTS and YS of weldment. These observations confirm the claim made by Bahman and Alialhosseini (2010) that increasing the welding speed, increase hardness, YS, and UTS of weld metal and the situation occurs when a low heat input is achieved. Conversely, prolonged heating at incremental welding speeds produces welds with a large HAZ, this eventually reduces the tensile properties of the weld.

3.1.4. *Validation of the developed model*

The adequacy of the developed model tested by making a scattered diagram of the results in Table 5, to show the correlation between the measured and the predicted (calculated) UTS of the aluminum alloy studied, as shown in Figure 9.

The adequacy of the developed model tested by making a scattered diagram of the results in Table 5, to show the correlation between the measured and the predicted (calculated) YS of the aluminum alloy studied, as shown in Figure 10.

The adequacy of the developed model tested by making a scattered diagram of the results in Table 5, to show the correlation between the measured and the predicted (calculated) % Elong of the aluminum alloy studied, as shown in Figure 11.

3.1.5. *Analysis of variance*

Table 6 presents the ANOVA at 95% confidence level. In addition, the coefficient of determination was also determined for each of the measured and predicted (calculated) tensile property.

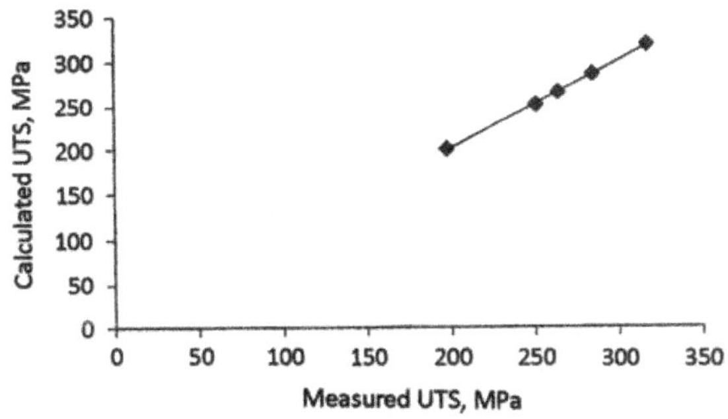

Figure 9. Scattered diagram of the UTS.

3.2. Discussion

A predictive model is developed for determining the tensile properties of AA 6061. These tensile properties include UTS, YS, and % Elong. Process parameters as shown in Table 2, and explicitly distributed in Table 3, were used to conduct 16 welding experiments. Each welding experiment was used to produce five weld samples, the five weld samples were subjected to tensile tests and the average of the responses from the tensile test results obtained such as UTS, YS, and % Elong were recorded for the 16 welding experiments conducted.

These test results were used to develop a model for predicting the tensile properties that define ductility. These mechanical properties are UTS, YS, and % Elong. In this study, the measured UTS is in the range of 155–342 MPa, the predicted value UTS is within the range of 180.24–317.96 MPa. The measured YS is in the range of 132–316 MPa, the predicted YS is found to be in the range of 153.51–301.23 MPa whereas, the measured % Elong is in the range of 3.2–5.8%, the predicted % Elong is within the range of 3.87–5.35%. Lakshminarayanan and Balasubramanian (2009) in their study measured AA 7039 weld joint whose tensile strength was obtained to be within the range of 170–317 MPa using friction stir welding process. Venkateswarlu, Mandal, Mahapatra, and Harsh (2013) determined UTS and % Elong of AA 7039 alloy, the measured UTS is in the range of 125.67–267.05 MPa, the predicted UTS is 116.49–261.97 MPa. Also these researchers determined the measured % Elong of AA 7039 to

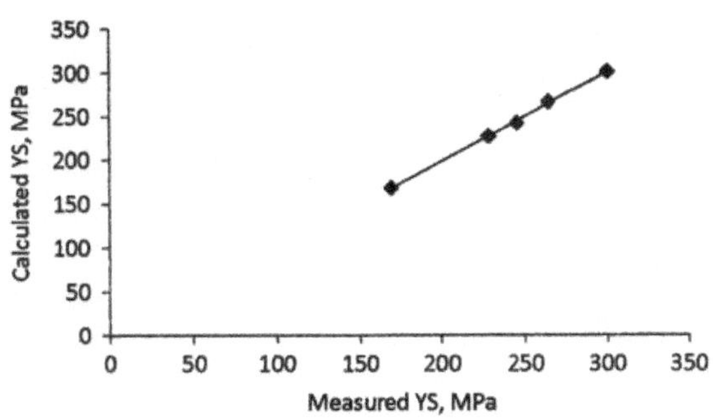

Figure 10. Scattered diagram of the yield strength.

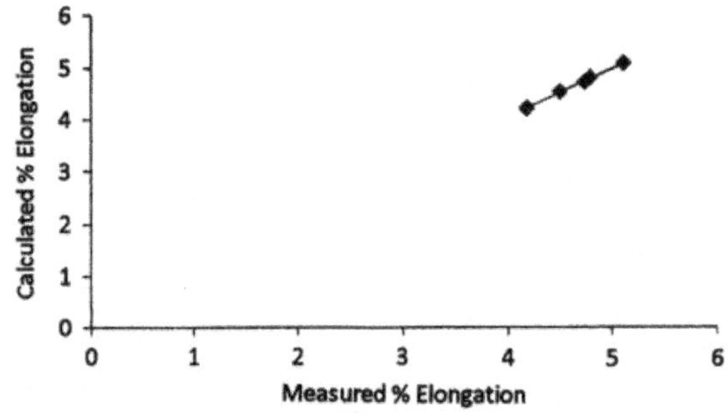

Figure 11. Scattered diagram of the % elongation.

Table 6. ANOVA and coefficient of determination.

Mechanical properties		SSE	F_{test}	F_{table}	r^2
UTS (MPa)	Predicted	0.000049	1.1767×10^{10}		1
	Measured	16.198869	35594.88209		0.99
YS (MPa)	Predicted	0.000016	3.039×10^{10}	230	1
	Measured	16.664116	29183.65826		0.99
Elong (%)	Predicted	0.000046	3940066.116		0.99
	Measured	0.005146	35220.17903		0.99

be between 1.84 and 5.9%, whereas the predicted value of the % Elong is within the range of 1.72–6.32%. Kumar et al. (2011) applied the modified Taguchi method to optimize process parameters. In their research work, they determined UTS and % Elong of AA 6061, the experimental results obtained for UTS are within the range of 256.40–386.07 MPa, with a corresponding predicted UTS of 389.31 MPa. The measured % Elong results are within the range of 7.84–12.94%, whereas the predicted % Elong is 12.98%. Hsieh and Chen (2008) determined the tensile strength of AA 6061 T-6 alloy to be within the range of 225.4–247.9 MPa and the % Elong of 5.5–6.8%. Palanivel et al. (2011) used regression model to predict UTS, YS, and % Elong of AA 6351 aluminum alloy. They obtained for measured UTS, a range of 193–219 MPa, YS within the range of 166–190 MPa and % Elong of between 3.61 and 5.24%, whereas the predicted values for UTS are within the range of 199.70–216.54 MPa, predicted YS is within the range of 172.92–187.29 MPa, and predicted % Elong is within the range of 3.52–5.08%. The mechanical properties obtained in this study falls within the range of values obtained in other reported literature. From the correlation analysis, it is shown that the measured and predicted (calculated) values are nearly a perfect fit with coefficient of determination of between 0.99 and 1.0. The effects of the process parameters used for the welding process on the tensile properties of the AA 6061 were examined. It was proposed that a prolonged increase in current could have a negative effect on the tensile properties of the weld. The increased current also increases in the HAZ profile of the weld. It was found that the increase in voltage could lead to a tremendous reduction in the UTS, YS, and % Elong. The 1.6-mm electrode diameter appears to produce welds with better mechanical properties than the 3.2 mm ones. In Figure 4, it is shown that the increase in welding speed also increases tensile properties.

An ANOVA was done to determine the significance of the level of tensile properties to the quality of the weld at 95% confidence. In each case, the Fisher test was conducted for both measured and predicted (calculated) weld properties. The predicted (calculated) and measured Fishers test values were in each case greater than the table value. This shows that the predictive model has great potential when compared with the measured values.

4. Conclusion

A predictive model using linear regression analysis was painstakingly developed along with a detailed methodology, showing a step-by-step approach. The developed model has proved to be very potent and quite beneficial. It has been used to predict the mechanical properties of AA 6061 with a focus on its UTS, YS, and Fisher Test . These tensile properties are known to be major determinant factors regarding the ductility level of the said alloy considered in this study. The proposed predictive models for determining these tensile properties are expected to ensure that engineers are equipped with a reliable means of pre-determining these tensile properties to ensure that weld operations are properly planned and concomitantly efficient. The correlation analysis and ANOVA conducted confirmed the adequacy of the model. The confirmation tests showed that the variances between the measured and calculated tensile properties were not significant. Also the measured and predicted values of these tensile properties were compared with those obtained in related literature. It was found that values obtained in this study compare well with those from other investigators. This also further confirms the potency of the derived predictive models.

Disclosure statement

No potential conflict of interest was reported by the author.

References

Achebo J. I. (2008). *Development of aluminium welding flux from local raw materials* (PhD Thesis). Department of Production Engineering, University of Benin, Nigeria.

Achebo J. I. (2012, July 4–6). *Optimization of fluence energy in relation to weld properties based on vogel approximation method*. World Congress on Engineering: International Conference on Mechanical Engineering, 3. London.

Bahman, A. R., & Alialhosseini, E. (2010). Change in hardness, yield strength and ultimate tensile strength of welded joints produced in st 37 grade steel. *Indian Journal of Science and Technology, 3*, 1162–1164.

Cavaliere, P., Campanile, G., Panella, G., & Squillace, E. (2006). Effect of welding parameters on mechanical and microstructural properties of AA6056 joints produced by friction stir welding. *Journal of Materials Processing Technology, 180*, 263–270. doi:10.1016/j.jmatprotec.2006.06.015

Chiteka, K. (2014). Artificial neural networks in tensile strength and input parameter prediction in friction stir welding. *International Journal of Mechanical Engineering and Robotics Research, 3*, 145–150.

Dhas, R. E. J., & Satheesh, M. (2012). Multiple objective optimization of submerged arc welding process parameters using grey based fuzzy logic. *Advances in Production Engineering & Management, 7*, 5–16.

Elangovan, K., Balasubramanian, V., & Babu, S. (2009). Predicting tensile strength of friction stir welded AA6061 aluminium alloy joints by a mathematical model. *Materials & Design, 30*, 188–193. doi:10.1016/j.matdes.2008.04.037

Heidarzadeh, A., Mahmoudi, A., Khodaverdizadeh, H., & Nazari, E. (2012). Tensile behavior of friction stir welded AA 6061-T4 aluminum alloy joints. *Materials & Design, 37*, 166–173. doi:10.1016/j.matdes.2011.12.022

Hsieh, H., & Chen, J. L. (2008). Influence of welding parameters on mechanical properties of friction stir welded 6061-T6 launch box. *Materials Transactions, 49*, 2179–2184.

Karthikeyan, P., & Mahadevan P. K. (2014, August 17). *A study of modeling on the mechanical properties of friction stir welded al 6351 plates added with* sic *particles in the butt-joint region*. Proceedings of Eleventh IRF International Conference, 62–65, Chennai. ISBN: 978-93-84209-47-6

Khorram, A., Ghoreishi, M., Yazdi, M. R. S., & Moradi, M. (2011). Optimization of bead geometry in CO_2 laser welding of Ti 6Al 4 V using response surface methodology. *Engineering, 3*, 708–712. doi:10.4236/eng.2011.37084

Kim, I., Son, S., & Jeung, Y. (2001). Control and optimization of bead width for multipass welding in robotic arc welding process. *Australian Welding Journal, 46*, 43–47.

Kumar, P., Kolhe, K. P., Morey, S. J., & Datta, C. K. (2011). Process parameters optimization of an aluminium alloy with pulsed gas tungsten arc welding (GTAW) using gas mixtures. *Materials Sciences and Applications, 2*, 251–257. doi:10.4236/msa.2011.24032

Lakshminarayanan, A. K., & Balasubramanian, V. (2009). Comparison of RSM with ANN in predicting tensile strength of friction stir welded AA7039 aluminium alloy joints. *Transactions of the Nonferrous Metals Society of China, 19*, 9–18. doi:10.1016/S1003-6326(08)60221-6

Mostafa, N. B., & Khajavi, M. N. (2006). Optimization of welding parameters for weld penetration in FCAW. *Journal of Achievements in Materials and Manufacturing Engineering, 16*, 132–138.

Palanivel, R., Mathews, K. P., & Murugan, N. (2011). Development of mathematical model to predict the mechanical properties of friction stir welded AA6351 aluminium alloy. *Journal of Engineering Science and Technology Review, 4*, 25–31.

Prasad K. S., Rao C. S., & Rao D. N. (2011). Optimizing pulsed current micro plasma arc welding parameters to maximize ultimate tensile strength of inconel 625 nickel alloy using response surface method. *International Journal of Engineering Science and Technology, 3*, 226–436. doi: http://dx.doi.org/10.4314/ijest.v3i6.18

Sen M., Mukherjee, M., & Pal, T. K. (2014, December 12–14). *Prediction of weld bead geometry for double pulse gas metal arc welding process by regression analysis*. 5th International & 26th All India Manufacturing Technology, Design and Research Conference, 814-6, IIT Guwahati, Guwahat.

Singla, M., Singh, D., & Deepak, D. (2010). Parametric optimization of gas metal arc welding processes by using factorial design approach. *Journal of Minerals & Materials Characterization & Engineering, 9*, 353–363.

Sreeraj, P., Kannan, T., & Maji, S. (2013). Prediction and control of weld bead geometry in gas metal arc welding process using simulated annealing algorithm. *International Journal of Computational Engineering Research, 3*, 213–222.

Venkateswarlu, D., Mandal, N. R., Mahapatra, M. M., & Harsh, S. P. (2013). Tool design effects for FSW of AA7039. *Welding Journal, 92*, 41s–47s.

Xue, Y., Kim, I. S., Son, J. S., Park, C. E., Kim, H. H., Sung, B. S., … Kang, B. Y. (2005). Fuzzy regression method for prediction and control the bead width in the robotic arc-welding process. *Journal of Materials Processing Technology, 164–165*, 1134–1139. doi:10.1016/j.jmatprotec.2005.02.174

Yang, L. J., Chandel, R. S., & Bibby, M. J. (1992). The effects of process variables on the bead width of submerged-arc weld deposits. *Journal of Materials Processing Technology, 29*, 133–144. doi:10.1016/0924-0136(92)90430-Z

9

Key success factors for implementing Taiwan TrainQuali System (TTQS) in Taiwanese enterprises

Tzong-Ru (Jiun-Shen) Lee[a], Szu-Hai Lin[a], Yang Liu[b]* and Shpend Imeri[c]

[a]Department of Marketing, National Chung Hsing University, Taichung, Taiwan; [b]Department of Production, University of Vaasa, Vaasa, Finland; [c]Department of Production, LEORON Professional Development Institute, Dubai, United Arab Emirates

Facing the change of economic environment and the need of enhancing competitive advantages, small and medium-sized enterprises in Taiwan utilize 'Taiwan TrainQuali System (TTQS)' on training talents. In order to maximize the benefits of TTQS implementation and increase the training efficiency, this paper aims to use 'grey relational analysis (GRA)' to extract key success factors (KSFs) from 46 factors, which influence the implementation of TTQS. Moreover, this study also verifies the KSFs with nine case companies which have won gold medals in TTQS evaluation and sum up 12 concrete practices as references for other enterprises in order to show how to implement the TTQS training quality system better in the future, and in turn enhance the quality and performance of human resource training.

Keywords: ISO 10015; Investors in People (IIP); Taiwan TrainQuali System (TTQS); key success factors (KSFs); grey relational analysis (GRA)

1. Introduction

Taiwan TrainQuali System (TTQS) is originated based on 'Guidelines and Action Plans for Service Industry Development (2004–2008)' of the Taiwanese Government (Taiwan TrainQuali System, 2012). The objective is to construct an integrated certification system of training quality focused on the development of talent training industry, by importing and combining ISO 10015, European vocational training policies, British Investors in People (IIP) and Australian active vocational training policies, by taking into account of the challenges for Taiwan to face globalized knowledge economy.

Due to the lack of human and capital resources in small and medium-sized enterprises (SMEs), there is less willingness to do vocational training, and there seldom exists professional training organizations (Small and Medium Enterprise Administration, Ministry of Economic Affairs, 2007). Although the Taiwanese Government provides free counselling, with limited resource they can't set it into real action. If the enterprises can effectively introduce TTQS system covering five aspects (including Plan, Design, Do, Review and Outcome), we can evaluate the training quality performance and find the key success factors (KSFs) to pass the evaluation, which can greatly improve the training performance and lead to a higher competitiveness. If we can extract the KSFs of TTQS, we believe that it would help a lot for those SMEs to implement training system better. This paper aims to study from the committee members and consultants of

*Corresponding author. Emails: yli@uva.fi, yli@puv.fi

TTQS, and the enterprises that passed the assessment, in order to: (1) use grey relational analysis (GRA) to find out the key factors and their weight of TTQS, and by sorting their weights to find the most important key factors; (2) provide the practical methods for enterprises to implement the key factors of TTQS.

This paper is organized as follows. Section 2 reviews the related studies. Section 3 introduces the data collection. Section 4 analyses the results. Section 5 draws the conclusions.

2. Literature review

2.1. Criteria for international training quality and ISO 10015

International Organization for Standardization (ISO) was founded in Geneva – Switzerland, 1947, and its primary job is to establish various international standards on behalf of its member countries (IIP, 2013). Nowadays, ISO 9001:2008 is implemented by over one million companies and organizations in over 170 countries (ISO, 2013). Quality management principle in ISO 10015 is based on ISO 9000 system, and also the only international criteria for human resource training in the ISO 9000 system (Hockman, Grenville, & Jackson, 1994; Huang, Chang, & Lin, 2006). It is used to enhance the quality of human resource and improve the outcome of education and training in an organization (Cheng, 2005). ISO 10015 is constructed on Plan–Do–Check–Act (PDCA) system, the so-called 'Deming Cycle' or 'Deming Wheel'. Developed by Walter Shewhart in 1920, PDCA was later introduced by Edwards Deming (Fan, 2008). It is now widely used in training quality management (Wu, 2008). There are two core ideas in ISO 10015. One is the strategic connection between investment training and organizational performance. It contributes to precise analysis of performance gaps and the real need for training. The other is setting up the training procedures, which leads organizations to 'learning organizations' by continuous improvements.

2.2. Investors in People UK (IIP UK)

To enhance the competence of enterprises, since 1991 the British Government has set 'IIP' into action. The policy is set by government to provide a standard for enterprises to invest their human resource. It aims to enhance the enterprises' ability to manage and develop human resource and to promote the responsibility of enterprises. It is not only good for workers, but also helps to improve the structure and the performance of the organization (Fan, 2008).

The main purpose of IIP is 'to improve the performance of an organization'. Therefore, IIP follows the three Principles: Plan, Action and Review, which form 10 Indicators (IIP, 2013). To be qualified, each applicant must fit 39 basic requirements from these 10 indicators set by IIP. In addition, to be eligible for bronze honour, based on its own features and development, the applicant needs to qualify at least 65 requirements in which 26 out of 157 can be chosen from advanced requirements to be evaluated. Silver honour requires qualifying at least 115 requirements and 165 for gold honour. Besides, if the applicant who gets the gold honour is willing to promote IIP, give consulting, and accept enterprise visiting, it will be given the certificate of 'leading enterprise' (Yeh & Tsai, 2010) (Table 1).

Table 1. IIP's principles, indicators and requirements.

Divisions		10 Indicators	Requirements	
			Basic	Advanced
Three Principles	Plan	1. Business strategy	6	19
		2. Learning and development strategy	4	13
		3. People management strategy	5	24
		4. Leadership and management strategy	3	10
	Action	5. Management effectiveness	4	21
		6. Recognition and reward	3	16
		7. Involvement and empowerment	3	16
		8. Learning and development	3	17
	Review	9. Performance measurement	5	9
		10. Continuous improvement	3	12
Subtotal			39	157
Total			196	

Note: Adapted from IIP (2013).

2.3. *Taiwan TrainQuali System*

2.3.1. *The origin and structure of TTQS*

In 'Guidelines and Action Plans for Service Industry Development (2004–2008)', the Administrative Council focuses on 'Measurements of Cultivating Talents in Service Industry' to list a certification of the quality of the training. It is mapped by the Labour Committee in order to introduce international standards of training quality. Therefore TTQS, planned by the Labour Committee in 2005, is composed of ISO 10015, vocational training policy in Europe, IIP in UK and active vocational training policy in Australia. The Labour Committee sets up Taiwan's own 'Taiwan TrainQuali Scorecard,' which is a cyclical system to enhance the training quality (Bureau of Employment and Vocational Training Enterprise Network, 2009), shown in Table 2.

2.3.2. *Measurement and evaluation standard of TTQS*

The structure of TTQS mainly follows PDDRO, which means: 'Plan, Design, Do, Review and Outcome', using five aspects with 18 items in total for evaluation. With weighted calculations and circulating training quality modification system, it scores and grades with different counselling. We modify the evaluation items into relative factors in the research questionnaire based on our experience (as shown in Table 3).

Based on 'Bureau of Employment and Vocational Training: TTQS guidelines manual (2010),' items on the score chart are depending on each enterprise. In other words, score of TTQS is judged by whether there is a record or documentation, represented by a value between 1 and 5 and fine scaling by .5. Evaluation of enterprises is based on five aspects with 18 items. Except for No. 19 ('others'), there are 21 scales. The highest score of each square is 5, and the original total score is 105. Under 53 (including) fails; 53.5–63 passes the examination; 63.5–74 is honoured with the bronze medal; 74.5–85 is honoured with the silver medal; over 96.5 is honoured with the gold medal and achieve TTQS standards.

Table 2. Comparison between IIP, ISO 10015 and TTQS.

Items	ISO 10015	IIP	TTQS
Applying country	Switzerland	UK	Taiwan
Starting year	1999	1991	2006
Principles	Plan, Do, Check and Action	Plan, Do and Review	Plan, Design, Do, Review and Outcome
Purpose	1. Make sure the investment in training corresponds with enterprise's target 2. Make sure the application of the training can make a positive return on investment	Through investing in employees' professional development, it can increase employees' devotion towards the organization, and help the organization to reform and improve results	Use strategic training system to improve the training quality and human resource capital, and thus, increase the competitiveness of the organization
Number of indicators	Five main indicators	10	18
Number of detailed catalogues	Nine secondary indicators	39	None
Examining standards	ISO 9001: 2000 training relative standards + ISO 10015 standard requirements + results agree to the business plan	10 requirements	Fundamental standards are based on five aspects of 'PDDRO,' which is the score sheet consisting of 18 standards
Examining mechanism	Qualified for examination (value the process and results)	Qualified for examination and subjective judgment (result-oriented; sampling interview with employees)	Qualified for examination (value the process and results)
Goal	Learning organization with exceptional performance	Team with high commitment and exceptional performance	Learning organization with exceptional performance
Documentary requirement	High (TMIS required)	Low	High
Evaluation type	Qualified for examination (value the process and results)	Qualified for examination and subjective judgment (result-oriented)	Qualified for examination (value the process and results)
Subject	Medium/large-sized training facilities or training centres in groups	Service industry, knowledge industry, high changing industry (technology industry, for example), especially suitable for SMEs	Training departments in organizations and independent training facility
Characteristics	1. Support ISO 9000 system and enforce its training ability 2. Become an independent and complete training system	Condense employees' common consensus and help organization transform. Strong subjectivity, so the quality and the training of verification staffs are very important	1. Verification institute will make sure applicants' quality agrees to the standard of training (1) 2. Assist applicants to conduct counselling and

(*Continued*)

Table 2. (*Continued*).

Items	ISO 10015	IIP	TTQS
	Can be used in enhancing cooperation's competitiveness		view the result of counselling
Sections for Notation	It is the only standard in ISO 9000 international system to regulate human resource training	The first and the only standard of human resource quality in the world	Unique training verification system of Taiwan, based on IIP, ISO 10015 et al.

Note: Adapted from Lin, Kung, and Lin (2009).

2.3.3. *Recent studies of TTQS*

Several recent studies have dealt with the performance, challenges and opportunities of TTQS (Chang & Chen, 2013; Chuang, 2013). Related studies have introduced evaluation models and filled the gaps of Plan and Design structures that TTQS incorporates through data mining in order to help in the future (Lin, Wang, Wu, & Ye, 2011; Lin, Wu, Tung, Huang, & Qin, 2010). Thus, the existing literature notes that introducing TTQS effectively enhances the organization's human capital and this strengthens human capital to be competitive internationally (Hsieh, Lin, & Lee, 2012; Lo, Tsai, Lan, & Lin, 2011). However, most of the studies have focused on the positive implementation of TTQS, reviewing assessment standards and advantages, evaluation models that uplift human capital for the performance of TTQS. Very few have focused on challenges of looking after those assessment items in the limited operational resources to develop competitive advantage (Liu & Liang, 2015), and some even state that is not beneficial at all (Kao et al., 2013; Yeh, 2012). Therefore, very few studies have proposed combinations from a generic algorithm (GA) (Chen, Lin, Chen, Huang, & Chang, 2012a; Huang et al., 2010) and support vector machine (SVM) (Aggarwal, Rani, & Dhir, 2010; Ichihashi, Honda, & Notsu, 2011; Kao et al., 2013; Martens, Baesens, & Van Gestel, 2009) to be GA–SVM algorithm, so they can select the combination of assessment items and identified the most important items to influence the enterprises and be their first priority of execution and to better training target (Kao et al., 2013).

This unique research aims to use 'GRA' to conclude key factors which influence the implementation of TTQS. Empirically, this study introduces nine cases which have passed the TTQS evaluation and honoured with gold medals in Taiwan. It sums up the 12 concrete methods influencing TTQS as references in order to show how to implement the TTQS training quality system for the enterprises in the future, and in turn enhance the quality and performance of human resource training.

2.4. Key success factors

The idea of KSFs is originated in Daniel's article 'Management Information Crisis' (1961) and suggests that 'To succeed, you have to do your job very well.' Later, KSFs were elaborated by Rockart (1979) and Bullen and Rockart (1981) in the perspective of the design management information system. There are three to six KSFs in most industries. For a successful enterprise, these KSFs must be well functioned. Therefore, the concept of KSFs is crucial to the design of information system. Boynton and Zmud (1984) point out the definition of KSFs. It represents that a manager or an enterprise has to offer special and constant notion of success or gain better achievements. It includes

Table 3. Five aspects and evaluation standards in TTQS.

Aspects	Emphatic facets	Evaluation items		Relative factors in this research
Plan	Focus on relativity and the practice of training plan and enterprise's goal	1. Explicitness	1a. Organization's vision	1. Enterprise with visions, missions and strategies.
			1b. Setting of goals and needs	2. Setting clear training goals and needs of the plan
			1c. Clear training policies	3. Enterprises with clear training policies
			1d. Clear core training categories or fields	4. Annual training programs with clear core training categories or fields
		2. Systematic	2a. Management system and documental manual of training quality	5. Enterprises with complete training management system
			2b. Career analysis and applications of training process	6. Clear requirement that the relative staffs should have ability to analyse careers and explain the jobs
		3. Connections	Connections between training plan and business goals	7. Connections between training goals and results of organizations
		4. Ability	4a. Administrative management of training unit	8. Enterprise with professional human resource department or training unit
			4b. Relative ability of training	9. Training staff has relative ability of analysing and also provides career instructions
Design	Focus on the design of training system (including the involvement of beneficial party, standard of course selections, purchase standard)	5. Selection standard of training products and services		10. Products which help training (teachers, teaching materials and training outsourcing)
		6. The involvement of beneficial party		11. Include training party's suggestions in the plan
		7. Combination of training and targets		12. Use training targets as the plans
		8. Systematic design of training		13. Complete training action plan
		10. Purchase process and standardization of training products and services		14. Construct a standardized procedure to cultivate purchase staff or outsourced training
Do	Emphasize the process of training, records	11. Whether the training	10a. Selection of trainees	15. Set up the complete selection

(*Continued*)

Table 3. (*Continued*).

Aspects	Emphatic facets	Evaluation items		Relative factors in this research
	and the standardization of management	follows the plan	corresponds to the plan 10b. Selection of teaching materials corresponds to the plan 10d. Teaching methods correspond to the plan Results transfer to the working environment	system of trainees and teachers 16. Draft the learning and action plan before training 17. Provide job opportunity and sharing after training
		12. Records and information system	11a. Classification and establishing files of training 11b. Procedure of managing information	18. Training unit with the ability to manage files 19. Educational training system
Review	Focus on periodical analysis, monitoring and reaction to accidents	13. Evaluation reports and periodical analysis		20. Periodical evaluation meeting and reports, which will be the adjustment standard next time
		14. Monitoring and reactions		21. Record accidents to be the training review next time
Outcome	Focus on layers, completeness, and sustained improvement of training evaluation	15. Diversity and completeness of training evaluation		22. Full records of diversity and completeness of training evaluation
		16. Results of trainees		23. Training courses can improve the working behaviour effectively 24. Positive rewards after training, and also on their work
		17. Extension Effects of training		25. Combination of training results with salary, evaluation or promotion 26. Training results can enhance competitiveness of enterprises and performance
		18. Results of special training		27. Devote the training results into corporate social responsibility
		19. Others		

Note: Adapted from Bureau of Employment and Vocational Training 'TTQS operating standard of training quality (2011).'

factors about recent and future operating activities of the enterprise. As a result, the enterprises are becoming more promising and also benefit from the 'strategic planning.'

KSFs are thus defined as variables and they are in correspondence with some characteristics such as, for instance the market, which mandates that some skill is a KSF in one but not in another occasion. If an analysis of KSFs is conducted it should not only be a description, but rather contribute to a more general theory about the study area and then it is linked to those particular characteristics.

Thus, KSFs lead into four main research tasks for an empirical analysis in an environment: measuring the perceived KSFs, finding additional hypotheses about the actual causes of superior perceived value and lower relative costs, measuring how businesses score on the potential KSFs, measuring perceived value and relative costs. These four research tasks clearly call for a combination of research methods (Bullen & Rockart, 1981). As this study is different in its nature due to its methodology using GRA to find out the current most vital KSFs of TTQS and future enterprise patterns for implementing TTQS (Chen, 2008), the existing studies have focused on implementation to some case companies and reviewing the general theory of the TTQS. Therefore, it is important to mention that this research shows the most important KSFs that can be utilized when TTQS is implemented.

2.5. Grey relational analysis

GRA is mainly focused on looking for a quantitative method to measure the relevance of factors. It's easier to find out key factors affecting developing trend of the system and then grasp the main features of things. GRA is a measure to analyse the relevance among discrete sequence data without strict demand for numbers of samples. If samples are small, it still accurately analyses the relationship between quantified data and ordering data (Feng & Chiou, 2004; Kung & Yang, 2006). Different from previous methods like regression analysis, related analysis or factor analysis which pose specific demand for numbers and styles of sampling, the method proposed in this paper works beyond the limitation of reliability and normal distribution among huge amount as well as small amount of data and variables. Moreover, it helps to process with uncertainty, multi-variable input, discrete data and non-integrity, compensating for the weaknesses of regression analysis (Chang, Wen, & Chang, 2000). Deng (1987) suggests that the GRA does not strictly require a certain number of samples, and does not require a typical distribution or assumptions for analysis. Therefore, GRA has four main features: (1) the models created are non-function type of series; (2) the calculation is simple and easy to use; (3) it can be calculated based on a small amount of data and (4) the data are not subject to comply with any typical distribution. Grey theory can effectively deal with the 'uncertainty', 'multivariate input', 'discrete data' and 'data incompleteness', and compensate the drawbacks in statistics regression (Wen, Zhao, Zhang, Chen, & Wen, 2009). In conclusion, GRA can overcome the shortage that a quantitative indicator cannot reflect the differences between factors, and make the research more accurate. As a result, GRA fits the purpose to select KSFs after enterprises adopt TTQS evaluation.

Firstly, the procedure of the GRA has to identify possible factors affecting the research target firstly. Then, it collects relevant effecting factors and select major factors as a research target. Next, GRA is processed. In other words, with applications of GRA to determine two sequences of information, the first step is to determine 'sequence analysis' and 'reference analysis.' Next is to calculate grey relational coefficient and figure out grey relation based on the average. After putting grey relation in order, it is possible to get the standard of balancing the importance and select KSFs affecting evaluation. The procedure is explained as follows:

Step 1: Calculate the grey relational coefficient

According to Deng (1989), the calculation of the grey relational coefficient is shown in the following equation:

$$r(x_0(k), x_i(k)) = \frac{\Delta\min + \xi\Delta\max}{\Delta_{0i}(k) + \xi\Delta\max} \tag{1}$$

ξ represents the identified coefficient. Its value lies between 0 and 1. The main function of it is to show a comparison between these two. The value of the identified coefficient can be determined by the reference to different targets. Generally speaking, its average is .5 as its identified value.

Step 2: Calculate grey relation

According to Deng (1989), after grey relational coefficient is derived, grey relation (Equation (2)) is calculated by averaging grey relational coefficient, which also stands for the relationship between two sequences. If two variables are uniform, their simultaneous changes are higher, meaning that their relationship is higher, otherwise is lower. Deng (1989) calculates the values of grey relation based on the following equation:

$$\gamma(x_i(k), x_j(k)) = \frac{1}{n}\sum \gamma(x_i(k), x_j(k)) \tag{2}$$

Step 3: Calculate grey relational sequence

In GRA, grey relation means the relationship between two sequences. The value is not the key, but how to sort each relationship matters the most. It represents values between two sequences. With sorting, the relationship is called 'grey relational sequence'.

Grey relation is a relative concept. Take an *example* of college students fighting with kids in kindergarten and high school students to explain grey relation. When they fight, college students are relatively likely to win. As a result, the college students' value of grey relation is 0.9, while it is .1 for kids in kindergarten. But if college students fight with high school students, the former has to pay much effort. Thus, the college students' value of grey relation is .8 while it is .7 for high school students. The relative concept about grey relation is derived from this example. The closer the value is, the less distance they are. The higher value stands for higher level of importance. The study also calculates the value of grey relation and figure out key factors which values are higher.

3. Questionnaire and sampling

Choices in the questionnaire are based on relevant documents and the authors' adjustments through participating in TTQS evaluation and assisting enterprises. Answers are chosen from one (least important) to five (very important) using five-point Likert scale.

Table 4. Grading scale of ISO system, IIP system and TTQS.

System		Questionnaire
ISO	1.	Executive manager's support and determination
	2.	The staff's participation and recognition
	3.	Educational training and staff quality
	4.	Selection of tutoring units
	5.	Customers' demand orientation and communication
	6.	Usage of information system
	7.	Planning carefully and implement thoroughly
	8.	Current system (ex: ISO) and previous successful experiences
	9.	Implement of document management and document system
	10.	Confirmation of examination and correction
IIP	1.	Service measures to enterprises are market demand
	2.	Connection with enterprises' organizational performance
	3.	Professional reliance on consultants and committees
	4.	Professional personnel in human resource as key members
	5.	Interior strategic communication is clear and deep
	6.	Property and a sense of belongings to the organization. Turnover rate is lowered
	7.	An important management tool for transformation in an organization
	8.	Integration with organizational culture
	9.	Continuous improvements in the enterprises
TTQS	1.	Envision, mission and strategy
	2.	Identified training goals and plans needed
	3.	Clear training policies in an enterprise
	4.	Fields of clear core training in yearly training sessions
	5.	Complete system of educational training management
	6.	Ability to analyse and explain the job is required
	7.	Connections between training plans and organizational performance
	8.	Professional human resource department or training division
	9.	Ability to value the proficiency and offer embodied brochure
	10.	Clear standards of valuation on training products (tutors, materials and outsource training companies)
	11.	Ideas from trainees are taken into training programs
	12.	Planning training demand based on company's training goals
	13.	Detail plans for training programs
	14.	Build up a systematic SOP for training purchasing or expatriates
	15.	Complete selection systems of trainees and tutors are set
	16.	Plans for learning and action-taking on employees are conducted before training
	17.	Practical opportunities and knowledge sharing are provided after training
	18.	Training department has the ability to manage training files
	19.	Information system for educating and training
	20.	Hold meetings of training outcome review and evaluate reports regularly, and adjust the courses according to the results for next time
	21.	Record all the unusual process as the review for the training in the future
	22.	Make sure the evaluations of diverse training results are completely recorded
	23.	Training courses can effectively improve the work behaviour
	24.	After the employees participating the training, they can effectively promote the organizational performance
	25.	Link the employees' training results with salary, evaluation and promotion
	26.	Training can effectively enhance the competitive advantage and business performance
	27.	Companies can put the training results into corporate social responsibility

The questionnaire is divided into two parts. The first part is demographic variables, including gender, age, education, profession and seniority of committee or consultants. Second part is grading scale of TTQS. In order to make the questionnaire and content more clearly, the fourth choice is 'selection of tutoring unit' and the fifth is 'customer demand orientation and communication' in ISO system. In IIP system, the third is 'professional reliance on consultants and committees'. The fifth is 'interior strategic communication is clear and deep' and the seventh is 'an important management tool for transformation in an organization'. The eighth is 'integration with organizational culture' and the ninth is 'continuous improvements in the enterprises' (Table 4).

The survey method has three interrelated advantages. The first advantage is that through this method we aimed to come up with conclusions referring to the main KSFs for implementing TTQS. The most important KSFs are another point that from methodological perspective enterprises answered the survey and that were reliable respondents from different position managers. Therefore, the survey obtained a variety of responses that include different views on KSFs for implementing TTQS. It was important and feasible for such KSFs issues to be accessible to these middle and high-level managers. However, this also related with the theoretical implications and importance of middle and high-level managers in TTQS. The importance of these respondents has been underlined to be crucial. Hence, a survey of a wide range of enterprises offers evidence towards the KSFs to implement TTQS.

The second advantage of the quantitative or surveying approach aimed to provide a wide range of data of KSFs to implement TTQS in different cases and it can obtain two things. One is the comparative analysis of the gathered data and the other is the comparison of KSFs to implement successfully TTQS.

Finally, the third advantage of such approach is that a survey can explore not only firm responses, attitudes and the most important KSFs, but also new conditions introduced by the TTQS approach that enterprises in this country can implement. Through the survey, a variety of information has been collected referring to issues that are directly or indirectly related to TTQS.

The target readers for this research are TTQS committee and consultants. A total of 97 questionnaires were sent. Seventy-five of them were returned. The rate of return was 77.3%. Seventy-three valid questionnaires were analysed while of them were invalid. The rate of validity was 97.3%.

4. Analysis

The analysis is done as follows. First, interpret according to basic information about the interviewees. Next, discuss key factors with analysis of GRA. Third, explain KSFs after enterprises adopt TTQS.

4.1. *Interpretation of collected samples*

The basic profiles of respondents include gender, age, educational level, professional qualification and experience, as shown in Table 5.

The questionnaire has been sent to a total of 97 people and 73 valid answers have been received, which yields 75.3% of the total sample size. A brief analysis is summarized as follows.

Firstly, the respondents were majority male with a total of 61 people, accounting for 83.56% of all the respondents. There were only 12 female respondents, accounting for 16.44%, showing that in the studied region, TTQS assessment members or counsellors are still male dominant.

Table 5. The basic profiles of respondents.

Basic info	Item	Number	Percentage (%)
1. Gender	Male	61	83.56
	Female	12	16.44
2. Age	Below 30	0	0
	31–40	13	17.81
	41–50	31	42.46
	Over 51	29	39.73
3.Educational background	Upper secondary	0	0
	Bachelor	5	6.85
	Master	47	64.38
	Doctor	21	28.77
4.Professional qualification	Assessment committee member	54	73.97
	Counsellor	19	26.03
5. Experiences	1 year	15	20.55
	2–3 years	22	30.14
	4–5 years	21	28.76
	Over 5 years	15	20.55

Secondly, in terms of age, no respondents were under 30 years old. Respondents between 41 and 50 years old were the majority with a total of 31 people, accounting for 42.46%. While there were 29 people over 51 years old, accounting for 39.73%. There were only 13 respondents between 31 and 40 years old, accounting for 17.81%. In overall view, respondents over 41 years old accounted for as much as 82.19%.

Thirdly, the education level of most respondents is master's degree with a total of 47 people, accounting for 64.38%, followed by doctorate with a total of 21 people, accounting for 28.77%. There were only five people with bachelor's degree, representing 6.85%. No respondent has education level of upper secondary or below.

Fourthly, in terms of professional titles, the main categories are assessment committee member and counsellor. In this case study, majority respondents were assessment committee members with a total of 54 people, accounting for 73.97% of all respondents. The other 19 people were counsellors, accounting for 26.03%.

Finally, in terms of experience, the largest group of 22 respondents has two until three years, accounting for 30.14% of all respondents. Second largest group of 21 people has for until five years, accounting for 28.76%. The smallest group has both experience of one year and over five years, each counts 15 people accounting for 20.55%. It can be seen that majority respondents have experience between two and five years with a total of 43 people, accounting for over half.

4.2. *Outcomes of GRA*

Values of grey relation are plotted on the X axis. Similar values are grouped. Fourty-six key factors are divided into six groups. Figure 1 shows the distribution of relation.

Daniel (1961) points out that, 'An enterprise has to achieve 3 to 6 KSFs to survive and succeed in the industry. If it lacks these key factors, it is likely to fail'. As a result, there are three to six factors chosen in all questions based on this principle.

In the first group, there is only one factor, which cannot satisfy the requirement of three to six KSFs. So, we move on to select factors from the next group. In the second group, there are three factors. Including the one factor in the first group, together there are four factors which can satisfy the requirement in total. If we continue to select

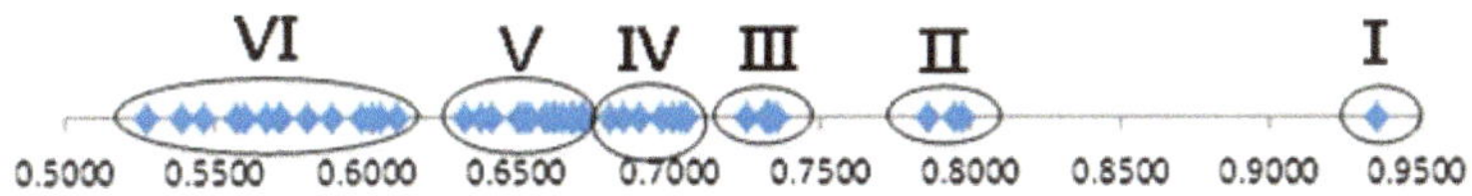

Figure 1. Distribution chart of grey relation among 46 key factors.

factors in the third group, it exceeds the standard of six factors. According to No. 1, 14, 5 and 3 sorted by grey relation in Appendix 1, Table 6 shows ISO System, IIP System and TTQS based on these four key factors.

4.3. Practical analysis of KSFs after TTQS adopted

Nine enterprises that passed the evaluation and won gold medals after adopting TTQS in 2011 are presented as case examples to compare the results analysed with the GRA

Table 6. Lists of KSFs and practical measures.

No.	KSFs	12 Practical ways	Grey relation	Grey relational sequence	System
1	Executive manager's support and determination	1. Show executive managers' support for training 2. Clear training policies and promises 3. Organize professional training personnel and implement training outcomes evaluation. Executive managers take in charge of tracking outcomes	.9361	1	ISO
14	Identified training goals and plans needed	4. Clear KPI and analysis of competency 5. Implement training demand and survey on yearly educational training sessions 6. Quantify training goals and rewards	.7968	2	TTQS
5	Connection with enterprises' organizational performance	7. Build up a complete system of competency qualification 8. Build up diverse training evaluation system 9. Connect training outcomes with salary, checking or promotion	.7945	3	IIP
3	Continuous improvements in the enterprises	10. Implement review sessions which record feedbacks and suggestions from trainees 11. Build up constant checking system, track training outcomes and improve constantly 12. Wind up training programs as the reference for future improvements	.7854	4	IIP

and the four selected key factors. It shows these enterprises which have been awarded gold medals from TTQS have the consistent ideas and relevant actions with the four key factors selected. The KSFs are found through the questionnaire results from the nine firms, and then the results are verified by revisit the nine firms to study the suggested actions, and totally 12 feasible practices have been summarized to achieved the KSFs, which are shown in Table 6.

The highest relative weight is executive manager's support and determination. It is also one of the key factors to reach gold medal after enterprises successfully pass TTQS evaluation. There are four companies whose executive managers greatly support the training and consider that promoting quality of personnel is beneficial to the enterprises. Some of the enterprises even set up personnel in charge of training and inspecting its performance. Secondly, enterprises need to build up identified training goals and plans and have a face-to-face interview with individuals according to the goals. What is more, the yearly individual training plan is needed to form a further yearly educational training plan.

Next, connection with enterprises' organizational performance and continuous improvements in the enterprises are also key factors, for example, to evaluate each index on training programmes based on yearly key performance indicators (KPIs) and call for follow-up meetings to examine the outcomes, or hold routine training conferences to discuss relevant issues as references for future improvements. Especially, the concluding reports after training can be integrated into checking and analysis report. Thus, enterprises can monitor the actual situations of trainees.

The logic connection of these four factors being the KSFs of implementing TTQS is similar to the process of making a strategic plan. Firstly, only if the company is willing to continuously improve the enterprises, they would set the training goals and plans that will lead the way where the company is going. Secondly, management people need to have determination and to support the implementation of all the tactics. Finally, the results should be connected with enterprises' organizational performance evaluation to ensure all the employees will work hard to achieve the goals.

A more elaborated explanation is as follows.

(1) Executive manager's support and determination

For training without the full support of executives, companies will not be able to maintain the continuity of operations and evolution. Therefore, the support of executives is the driving force of training activities. In Section 2 literature review, it can be also found that for any project or action to be successful in a company, the support and determination of top executives is a KSF. This also applies to the successful adaptation and evaluation of TTQS. This finding is in accordance with Chen (2008), Hockman, Grenville, & Jackson, (1994) and Michael (2001).

(2) Identified training goals and plans needed

To ensure the results of training, prior to performing training courses, the training survey needs to be implemented, which includes gap analysis of the organizational performance and current functional status. The curriculum planning is based on discussion with the trainees to clarify the training objectives, assess training effectiveness, set after-training studies or action plans to ensure the effectiveness of the training. From the assessment of nine companies which have received gold medals, the study shows these enterprises or organizations have all set training goals and plans which need improvement to meet their business objectives.

(3) Connection with enterprises' organizational performance

Organizational performance analysis is very important. Only through the analysis it is possible to find out whether the problem is the inadequate capability and understand the gaps in the functions of the staff. The 'plan' indicators of TTQS are connected with 'design' indicators, which expand curriculum design, transfer the training implementation to the training outcome assessments, and connect closely the training and organizational performance.

(4) Continuous improvements in the enterprises

Facing technological change and highly competitive environment, enterprises cannot sustain their operations and will face elimination without continuous improvements (Liu, 2013). Therefore, it is very important to pursue continuous improvement and deeply root such concept and attitude into every member's heart. Most enterprises or organizations which have awarded TTQS gold medals utilize the training to improve the function gaps, and let them understand the development of employees is important. This is because the investment in staff development will result in improving themselves for learning more skills and enhancing expertise, and thus create more reward for the enterprises, which, in turn are willing to spend more time to focus on staff development, forming a virtuous circle. Such action makes the needs of enterprises and employees become more consistent, thus forming common visions and goals.

4.4. Generalization of the results

The research has found that it is possible to generalize through the web platform to guide practitioners to implement the four KSFs by the following three practices:

(1) It provides successful experiences and practices as references to more people. Those who want to use TTQS system can refer themselves and compare their own practices with the references to understand the gaps and make up for deficiencies.
(2) It offers an interactive discussion area for those who are using TTQS system can introduce their current practices and invite other professionals to provide comments or suggestions for improvement.
(3) It offers an automated scoring system, so that people who are using TTQS system can report the practices currently in use which will be scored by the system, allowing the users to understand the performance of the current practices.

When the practitioners find their TTQS performances are not good, they will likely share their own experience on the web platform and discuss with other professionals, who will in turn provide suggestions and guidance, thus lead to a better performance. On the other hand, they can also obtain self-evaluation through the automated scoring system, and once they find their current practices have lower scores they can immediately start improving from the suggested areas, i.e. the four KSFs found in this study. Through these methods, the success rate of implementing TTQS can be improved, and the results of this study can be verified. In summary, the results of this study can be generalized.

5. Conclusions

The contribution of this study is to find out four KSFs with 12 concrete practices verified by nine enterprises which passed the TTQS evaluation and won gold medals as guidelines for other enterprises. In order to strengthen the training quality and enhance labour quality, enterprises are proposed to implement these 12 practices as shown in Table 6. Apart from achieving improved evaluation, decreasing competence gaps of employees, and raising organizational performance and promoting competitiveness can be also achieved. Besides, with the spirit of TTQS, two advices are presented. One is that the four selected KSFs should be focused and emphasized on their relative weights to push the evaluated units to pay more attention to these indexes. If these KSFs can be promoted, it is more efficient to implement TTQS. The other one is that continuous feedbacks and corrections are needed to ensure the successful adoption of TTQS, which will help to strengthen the training quality in enterprises. However, there are still some limitations in this research. The features of SMEs in Taiwan have not been taken into consideration. The sample size is relatively small and only focuses on Taiwanese companies, and this study only covers TTQS, but not IIP or ISO 10015. Therefore, the results may not be directly transferred to other countries. In the future research, analysis of variance may be conducted to analyse the significance of respondent categories as well as the case data, larger samples can be used, and the KSFs in other countries can be compared.

Disclosure statement

No potential conflict of interest was reported by the authors.

ORCID

Tzong-Ru (Jiun-Shen) Lee http://orcid.org/0000-0003-3548-3287
Yang Liu http://orcid.org/0000-0001-8006-3236

References

Aggarwal, A., Rani, R., & Dhir, R. (2010). Recognition of Devanagari handwritten numerals using gradient features and SVM. *International Journal of Computer Applications, 48*, 39–44.

Boynton, A. C., & Zmud, R. W. (1984). An assessment of critical success factors. *Sloan Management Review, 25*, 17–27.

Bullen, C. V., & Rockart, J. F. (1981). *A primer on critical success factors*. Cambridge, MA: Center for Information Systems Research, MIT.

Bureau of Employment and Vocational Training. (2010). *TTQS guidelines manual*. Taipei: Author.

Bureau of Employment and Vocational Training. (2011). *2011 TTQS plan: Operating standards*. Taipei: Author.

Bureau of Employment and Vocational Training Enterprise Network. (2009). TTQS Plan. Retrieved June 7, 2012 from http://training.evta.gov.tw

Chang, W. L., & Chen, S. T. (2013). The performance of Taiwan's training quality excellence system. *Total Quality Management & Business Excellence, 24*, 561–576.

Chang, W. C., Wen, K. L., & Chang, T. C. (2000). *Methods and application of grey relational model*. Taipei: Gau Lih.

Chen, Y. L. (2008). *A study of the key success factors in implementation of strategic training system: The case of Taiwan TrainQuali System benchmarking enterprises*. Taipei: National Taiwan University of Science and Technology, Department of Business Administration.

Chen, T. S., Lin, Y. Y., Chen, J., Huang, C. C., & Chang, H. W. (2012). A study using genetic algorithm and support vector machine to find how the attitude of training personnel of Taiwan TrainQuali System in an enterprise. Information and business intelligence. *Communication in Computer and Information Science Series, 268*, 142–150.

Cheng, Y. C. (2005). *Performance appraisal.* Taipei: Pan Asia.

Chuang, S.-F. (2013). Evaluating training and development practices in Taiwan: Challenges and opportunities. *Human Resource Development International, 16*, 230–237.

Daniel, D. W. (1961). Management information crisis. *Harvard Business Review, 39*, 111–121.

Deng, J. (1987). *Basic methods of grey system* (in Chinese). Wuhan: Huazhong University of Science and Technology Press.

Deng, J. (1989). Introduction to grey system theory. *The Journal of Grey System, 1*(1), 1–24.

Fan, Y. L. (2008). *A case study for enterprises conducting the training quality system of Taiwan TrainQuali Scorecard.* Taipei: Shih Hsin University, Department of Public Policy and Management.

Feng, C. M., & Chiou, Y. C. (2004). *Research and analytical methods.* Taipei: Chien-Tu Cultural Enterprise Company.

Hockman, K. K., Grenville, R., & Jackson, S. (1994). Road map to ISO 9000 registration: A time-tested process for implementing ISO 9000. *Quality Progress, 27*, 39–42.

Hsieh, S. C., Lin, J. S., & Lee, H. C. (2012). Analysis on literature review of competency. *International Review of Business and Economics, 2*, 25–50.

Huang, C. C., Chang, K. C., & Lin, J. S. (2006). *ISO 10015 system visiting report.* Taipei: Bureau of Employment and Vocational Training.

Huang, C. C., Chuang, R. G., Chen, R. C., Chen, T. S., Le, T. N., Hsu, C. J., & Tsai, Y. C. (2010). Finding an optimal combination of key training items using genetic algorithms and support vector machines. *Information Technology Journal, 9*, 652–658.

Ichihashi, H., Honda, K., & Notsu, A. (2011, June 27–30). Comparison of scaling behaviour between fuzzy c-means based classifier with many parameters and LIBSVM. *Proceedings of the IEEE International Conference on Fuzzy Systems* (pp. 386–393), Taipei.

Investors in People. (2013). *Investors in People framework.* London: Author.

ISO. (2013). *ISO 9000 – Quality management.* Geneva: Author.

Kao, Y. H., Chen, T. S., Lee, W. B., Chen, R. C., Huang, Ch. C., Lin, M. C., & Wang, Y. L. (2013). Key training items search of manufacturing assessment based on TTQS and GA-SVM. *Information Technology Journal, 12*, 756–762.

Kung, C. Y., & Yang, P. Y. (2006). Applying grey relational analysis to quality function deployment process – A case study of middle medical centers. *The 11th Grey Theory and Applications Conference* (pp. 7–9), Hsinchu.

Lin, W. T., Kung, C. Y., & Lin, L. L. (2009). Comparing IIP, ISO10015 and TTQS. *Quality Magazine, 45*, 52–56.

Lin, W. T., Wu, Y. C., Tung, C. L., Huang, M. R., & Qin, R. S. (2010). Establishing ISO 10015 accreditation system performance model for domestic enterprises. *Expert Systems with Applications, 37*, 4119–4127.

Lin, W. T., Wang, S. J., Wu, Y. C., & Ye, T. C. (2011). An empirical analysis on auto corporation training program planning by data mining techniques. *Expert Systems with Applications, 38*, 5841–5850.

Liu, Y. (2013). Sustainable competitive advantage in turbulent business environments. *International Journal of Production Research, 51*, 2821–2841.

Liu, Y., & Liang, L. (2015). Evaluating and developing resource-based operations strategy for competitive advantage: An exploratory study of Finnish high-tech manufacturing industries. *International Journal of Production Research, 53*, 1019–1037.

Lo, C. L., Tsai, C. Y., Lan, C. C., & Lin, M. H. (2011). Implementing the Taiwan TrainQuali System (TTQS): A case study of a pharmaceutical company. *World Transactions on Engineering and Technology Education, 19*, 233–238.

Martens, D., Baesens, B., & Van Gestel, T. (2009). Decompositional rule extraction from support vector machines by active learning. *IEEE Transactions on Knowledge and Data Engineering, 21*, 178–191.

Michael, G. S. (2001). *Transition to ISO 9001:2000* (pp. 28–30). Business News Publishing Company.

Rockart, J. F. (1979). Chief executives define their own data needs. *Harvard Business Review, 57*, 81–93.

Small and Medium Enterprise Administration, Ministry of Economic Affairs. (2007). *2007 small and medium enterprise white book.* Taipei: Author.

Wen, K. L., Zhao, Z. X., Zhang, H. Z., Chen, X. Y., & Wen, H. Z. (2009). *Grey theory* (in Chinese). Taipei: Wunan.

Wu, Q. Z. (2008). How to manipulate ISO 10015 to improve the training quality. *Quality Magazine, 44*, 72–76.

Taiwan TrainQuali System. (2012). Retrieved June 17, 2012, from http://ttqs.evta.gov.tw/

Yeh, C. W. (2012, August 13–14). The competency model and training needs for TTQS administrative assistant. *Proceedings of the International Conference on Economics Marketing and Management* (pp. 43–48). Taipei, Taiwan.

Yeh, C. W., & Tsai, S. T. (2010). *IIP system visiting report*. Taipei: Bureau of Employment and Vocational Training.

Appendix 1. List of grey relation analysed by GRA

No.	Key factors	Rank of grey relational sequence	Grey relation
1	Executive manager's support and determination	1	.9361*
14	Identified training goals and plans needed	2	.7968*
5	Connection with enterprises' organizational performance	3	.7945*
3	Continuous improvements in the enterprises	4	.7854*
38	Staff with training to enhance organizational performance	5	.7352
12	The staff's participation and recognition	6	.7333
16	Fields of clear core training in yearly training sessions	7	.7326
37	Training sessions efficiently promote jobs	8	.7318
13	Envision, mission and strategy	9	.7260
19	Connections between training plans and organizational performance	10	.7055
33	Call for checking meetings on training outcomes and evaluation as references for future sessions	11	.7044
40	Training outcomes enhance competitive strengths and managerial performance	12	.7009
8	Interior strategic communication is clear and deep	13	.6973
46	Implement of document management and document system	14	.6903
26	Clear programs of training programs	15	.6848
39	Connection with training outcomes with salaries, checking or promotion	16	.6804
6	Professional reliance on consultants and committees	17	.6723
29	Plans for learning and action-taking on employees are conducted before training	18	.6715
2	Confirmation of examination and correction	19	.6689
36	Complete records of diverse training outcomes	20	.6676
11	Integration with organizational culture	21	.6652
25	Planning training demand based on company's training goals	22	.6621
7	Professional personnel in human resource as key members	23	.6621
15	Clear training policies in an enterprise	23	.6621
17	Complete system of educational training management	25	.6598
10	An important management tool to transformation in an organization	26	.6594
41	Leading to corporate social responsibility	27	.6539
23	Educational training and staff quality	28	.6525
44	Planning carefully and implement thoroughly	29	.6507

(*Continued*)

Appendix 1. (*Continued*)

No.	Key factors	Rank of grey relational sequence	Grey relation
20	Professional human resource department or training division	30	.6417
42	Customers' demand orientation and communication	31	.6378
30	Practical opportunities and knowledge sharing are provided after training	32	.6324
22	Clear standards of valuation on training products (tutors, materials and outsource training companies)	33	.6097
18	Ability to analyse and explain the job is required	34	.6050
31	Training units have the ability to manage files	35	.6018
35	Record unusual process as references for future checking	36	.5982
32	Information system of education training	37	.5888
24	Ideas from trainees are taken into training programs	38	.5799
9	Property and a sense of belongings to the organization. Turnover rate is lowered	39	.5716
21	Ability to value the proficiency and offer embodied brochure	40	.5708
34	Selection of tutoring units	41	.5666
45	Current system (ex: ISO) and previous successful experiences	42	.5594
43	Usage of information system	43	.5572
27	Build up a systematic SOP for training purchasing or expatriates	44	.5457
28	Complete selection systems of trainees and tutors are set	45	.5388
4	Service measures to enterprises are market-demand	46	.5274

10

Influence of surface pretreatment in resistance spot welding of aluminum AA1050

Ihsan K. Al Naimi[a]*, Moneer H. Al Saadi[b], Kasim M. Daws[a] and Niels Bay[c]

[a]Mechanical Department, College of Engineering, Baghdad University, Baghdad, Iraq; [b]Welding Department, University of Kerbala, Baghdad, Iraq; [c]Department of Mechanical Engineering, Technical University of Denmark, Lyngby, Denmark

Resistance spot welding (RSW) of aluminum alloys implies a major problem of inconsistent quality from weld to weld due to problems of varying thickness of the oxide layer. The high resistivity of oxide layer causes strong heat development, which has significant influence on electrode life and weld quality. An experimental study of the influence of pretreatment on weld quality in RSW of AA1050 sheets with three thicknesses, comparing welding of as-received sheet with pretreated sheet by either pickling in NaOH or glass-blasting were investigated. Different weld settings were applied with low-, medium-, and high-energy inputs. The as-received sheet showed higher electrical contact resistance because of thicker oxide layer. Lower values were noticed with pickled surfaces, whereas the lowest electrical contact resistance was obtained when glass blasting, resulting in the roughest surface topography, which facilitated breakdown the oxide layer. Highest strength and smaller scatter in strength were obtained by pickling in NaOH.

Keywords: resistance spot welding; contact resistance; oxide film; surface pretreatment

1. Introduction

The increasing demands for energy savings have implied special focus in automotive industry on lighter vehicles production, among other solutions by introducing skin panels of aluminum instead of steel. Weight savings up to 46% are reported (Wheeler, Sheasby, & Kewley, 1987), and furthermore resistance to corrosion is improved.

Resistance spot welding (RSW) is the most dominant process in sheet metal joining, particularly in automotive industry due to low cost, high productivity, flexibility, easy automation and maintenance, and minimum requirements for skilled labor (Brown, Newton, & Boomer, 1995; Cho, Li, & Hu, 2006; Spinella, Brockenbrough, & Fridy, 2005). The process is also widely applied in other industries of sheet product manufacturing, e.g. other transportation industries and in production of kitchen utensils.

It should, however, be emphasized that RSW of aluminum is more problematic than steel due to higher electrical and thermal conductivity, higher coefficient of expansion, lower melting temperature, and an oxide layer, which has high electrical resistance and high melting temperature (2050 °C). The latter together with the fact that the effective contact resistance grows considerably as the oxide film grows implies large scatter in

*Corresponding author. Email: ihsan_kad@yahoo.com

quality of RSW aluminum sheets, which therefore require close production control (Boomer, Hunter, & Castle, 2003; Kim, Park, Hwang, & Kang, 2009; Mathers, 2002).

The high thermal and electrical conductivity of aluminum require 2–3 times higher current and shorter weld time, typically 25% of that used to spot weld steel. Accurate control and synchronization of current and electrode force is required due to the narrow weld temperature range (*Resistance Welding Manual*, 2003). The problems are especially pronounced when welding unalloyed aluminum AA 1XXX.

Aluminum is highly reactive to oxygen, and after removal of the oxide layer by mechanical or chemical means, a new oxide layer will immediately form on its surface in normal atmosphere. The layer is beneficial as it protects the base metal from corrosion, but creates problems in connection with resistance welding, where it causes severe electrode degradation and scatter in weld quality due to (Han, Thornton, Boomer, & Shergold, 2010; Patrick, Auhl, & Sun, 1984; Sun, 1982). Spot welding in aluminum AA1050 is less stable than in alloyed aluminum, and the electrodes will stick to the sheet after 50 welds because of the oxide film problems (Pedersen, 2010). Contributions to the contact resistance comes from the oxide layer as well as the contaminant film, which may include dirt, lubricant, chemicals, and water vapor (Rashid, 2011; Sun, 2003). Even though maintenance of uniform surface conditions is taken carefully into consideration, the electrical contact resistance will vary from weld to weld implying inconsistent weld quality.

Contact resistance was experimentally investigated by Song, Zhang, and Bay (2005) using a Gleeble system. They showed that interface normal pressure had significant but decreasing influence on the contact resistance with increasing normal pressure, whereas the influence of temperature was less pronounced as pressure increased. Surface roughness together with elastic–plastic properties of the materials also influences the electrical contact resistance (Dzekster & Ismailov, 1990; Zheng, Shan, Hu, & Luo, 2006). Crinon and Evans (1998) showed that the effect of the oxide film is greatest for specimens with the smoothest surface.

Expulsion during RSW may either occur at the faying surfaces or the electrode/workpiece interfaces. The latter may severely affect surface quality and electrode life. The risk of expulsion is especially high in spot welding of aluminum due to the very dynamic and unstable character of the process, related to the requirements of high current and short welding time as compared to welding steels (Mathers, 2002; Senkara, Zhang, & Hu, 2004; Zhang & Senkara, 2012).

A large number of investigations in the literature deal with the influence of surface preparation on RSW of aluminum sheets. Rönnhult, Rilby, and Olefjord (1980) studied the weldability of aluminum alloy AA5252, both as-received and after etching in NaOH and oxalic acid, and proved that removing the oxide layer led to a significant improvement in weld quality. Etched, mill finish, and pretreated/lubricated AA5754 surfaces were examined by Thornton, Newton, Keay, Sheasby, and Evans (1997), they suggested that etched surfaces that had a thin and uniform oxide layer, gave the most consistent surface resistance and weld strength, which was in good agreement with other studies by Pickering and Hart (1994) and Li, Hao, Zhang, and Zhang (2007). Rashid, Fukumoto, Medley, Villafuerte, and Zhou (2007) suggested a special lubricant to lower the tendency to sticking of the workpieces to the electrodes thereby increasing the electrode life. The company Fronius International (2014) developed a technique called DeltaSpot using continuous tape running between the electrodes and the sheets to be joined thereby increasing the electrode life remarkably. Han (2010) reported that frequent, light dressing of the electrodes removing pickup of aluminum increases the electrode life significantly.

Although different techniques and materials were investigated by these researchers, the influence of surface pretreatment before RSW of aluminum AA1050 and the use of mechanical pretreatment by glass blasting have not been clearly identified in any previous work. Blasting with glass beads is easy to automatize and may not cause higher costs than other surface pretreatments in mass production. Furthermore, the treatment is more environmentally friendly than chemical treatment.

The present study focuses on the influence of surface pretreatment prior to RSW of aluminum AA1050, aiming at minimizing variations in weld strength caused by the oxide layer. Mechanical pretreatment by glass blasting and chemical treatment by pickling with NaOH are compared with welding of as-received material.

2. Experimental procedure

2.1. Workpiece material, electrodes, and welding equipment

The experiments were conducted at the Technical University of Denmark (DTU) welding .6, 1.0, and 1.5 mm sheets of unalloyed aluminum AA1050 on a TECNA AC welding machine with specifications, as listed in Table 1. The properties and composition of the workpiece sheets determined by spectrum analysis are shown in Table 2. Samples of dimensions 16 × 115 mm were cut from delivered sheets, with the longitudinal dimension in the rolling direction. They were welded as a lap joint ready for subsequent shear-tensile testing. A hard wood fixture was used to mount the samples in good alignment with the electrodes (see Figure 1).

The electrode tips (Female Cap) were of type A according to ISO 5821-2009 and the American standard RWMA No. FF-25. They were made of zirconium–copper alloy (CuCrZr) with the following chemical composition; Cr: .7–1.2%, Zr: .06–.15%, bal. Cu. The electrodes were of dome configuration, Ø16 mm in diameter with a spherical end

Table 1. Machine specifications, TECNA AC welder.

Specifications	Values	Specifications	Values
Controller	TE-180, 16 functions	Nominal power at 50%	250 kVA
Supply voltage	380 V	Phases	1
Frequency	50 Hz	Supply pressure	6.5 bar
Max. welding current	68 kA	Electrode force per 1 bar	3.14 kN
Max. welding force	18.85 kN	Throat depth	250 mm
Max. welding power	810 kVA	Water cooling	12 l/min

Table 2. Strip material specifications.

Trade name	Thickness (mm)	Tensile (MPa)	Hardness (HV)	Nominal composition (wt%)				
				Fe	Si	Mn	Others	Al
1050	.6	105	30	.255	.173	.021	.051	99.5
1050	1.0	105	30	.378	.100	.018	.004	99.5
1050	1.5	127	45	.350	.070	.010	.070	99.5

(a) (b)

Figure 1. Setup for welding, (a) hardwood-fixture for mounting strips, (b) removing welded strips. Pretreatment of the workpiece surfaces was done by two methods, mechanical and chemical.

surface of radius 40 mm. Close to the tip of the two electrodes, an Ø1.5-mm hole was drilled and copper wires were inserted to measure the secondary voltage over the weld.

The welding current was measured by a Rogowski coil together with a precalibrated TECNA-1430 conditioner, and a Kistler piezoelectric force transducer was used to measure the electrode load. The acquired data were treated on a PC by specially developed software in LabVIEW.

The following parameters of the RSW process were calculated for each experiment, RMS current I (A), RMS voltage U (V), welding time C (s), and the electrode force P (kN). Tensile-shear testing was carried out using a universal testing machine at a deformation rate of 2 mm/min to determine the weld strength S(N). Vickers microhardness measurements were performed with a load of 50-g testing on weld cross-sections in longitudinal direction through the diameter of the nugget at intervals of .5–1.0 mm. Macrographs and micrographs of the welds were made in light optical microscope. Moreover, high-resolution images were made in SEM, using EDS for quantitative chemical analysis.

2.2. *Shot blasting*

The mechanical treatment was done by shot blasting with 100-μm glass beads. Each strip was subjected to 30 s blasting at an air pressure of 200 kPa (see Figure 2). In the main series of experiments, surface treatment was done on both sides of the strips, whereas additional experiments were performed dressing only the outer side of the strips contacting the electrodes.

2.3. *Pickling*

Pickling was performed in an aqueous solution of sodium hydroxide (NaOH) (see Figure 3). During pickling of aluminum alloy strips, the surface oxides or hydroxides

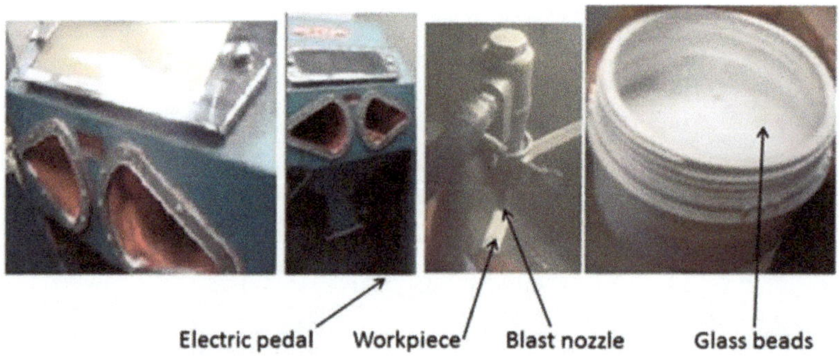

Figure 2. Glass blast surface treatment.

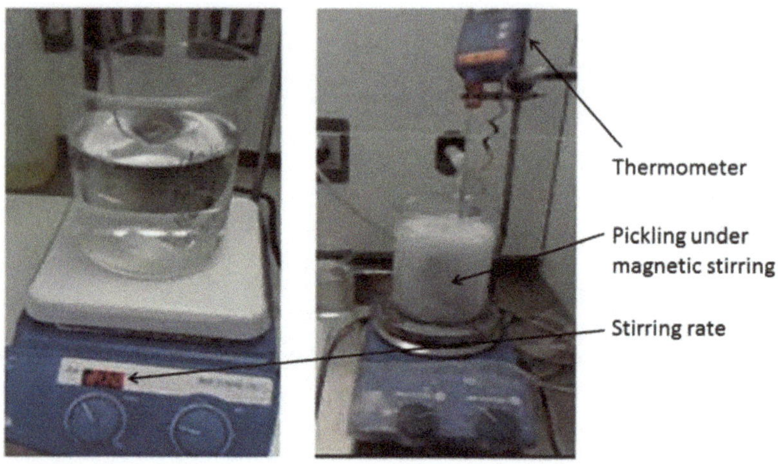

Figure 3. Pickling setup.

are dissolved. The pickling rate is dependent on several variables including; (1) concentration, (2) solution temperature, (3) composition of the surface sheet, and (4) stirring rate of solution.

An aqueous solution of 60 g of NaOH per liter of water was chosen with a bath temperature of 60 °C. Magnetic stirring was done in a fume cupboard at a rate of 200 rpm. The strips were subsequently rinsed in hot water and ethanol. Preliminary experiments with 5-min pickling time resulted in severe sticking of the workpieces to the electrodes as well as expulsions between the electrodes and the strips. Due to this, pickling time was reduced to 2 min thereby minimizing these problems.

As earlier mentioned, an oxide layer is re-established on the aluminum specimens immediately after surface dressing. In order to ensure similar conditions for all experiments with surface dressing, this was carried out within 3 h before spot welding. In this way, a substantially smaller oxide layer than that of the as-received aluminum can be expected.

2.4. *Plan of experiments*

The experiments were planned as a general factorial with three replicates per condition (Cho et al., 2006), and the results were recorded as the average value of each condition. The welding parameters and their associated values are given in Table 3. For each of the three sheet thicknesses, three different weld currents were chosen, representing low, medium, and high welding energy. During welding, possible expulsion and sticking to the electrodes were observed and recorded.

3. Results and discussion

3.1. *Surface roughness*

The two surface treatments were each characterized by five surface roughness measurements on three different strips. Table 4 shows the resulting R_a values. Pretreatment by glass blasting show high values of roughness, in the order of 10 times the value of the as-received surfaces. Chemical surface treatment by pickling gives a bit higher R_a values than the as-received, but the standard deviation (SD) is lower.

Table 3. Experimental setup (welding parameters and values).

	Welding energy					
	Low		Medium		High	
Sheet (mm)	Times (cycles)	Current (kA)	Times (cycles)	Current kA	Times (cycles)	Current (kA)
.6	.04 (2)	20	.1 (5)	23	.18 (9)	26
1.0	.04 (2)	23	.1 (5)	26	.18 (9)	29
1.5	.04 (2)	26	.1 (5)	29	.18 (9)	32

Note: Electrode force 1.85–2.45 kN.

Table 4. Surface roughness values.

Sheet (mm)	As-received	Pickling in NaOH	Glass-blast
.6	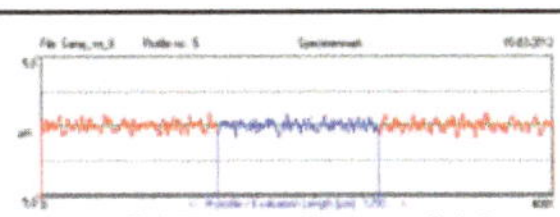 R_a = .287 μm, SD = .046	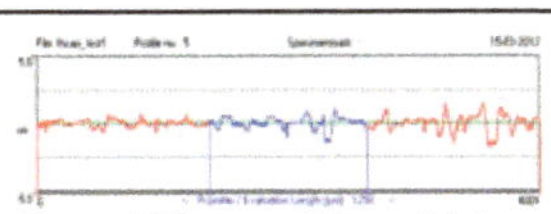 R_a = .339 μm, SD = .011	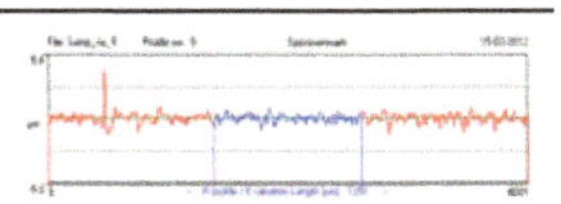 R_a = 2.650 μm, SD = .115
1.0	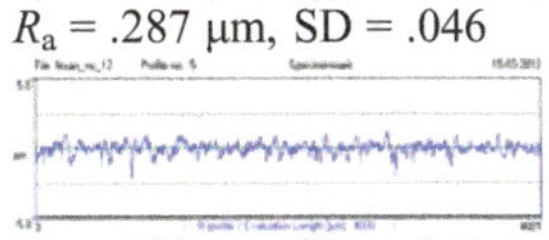 R_a = .300 μm, SD = .051	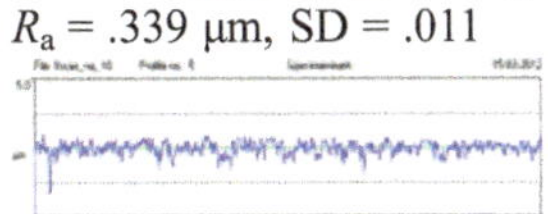 R_a = .385μm, SD = .037	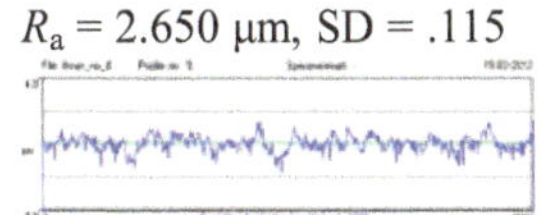 R_a = 4.703 μm, SD = .987
1.5	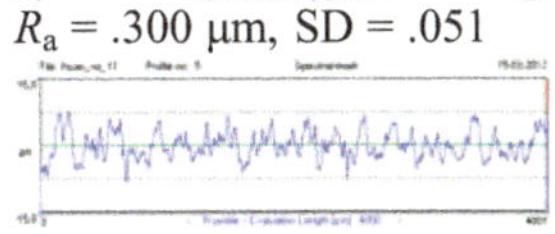 R_a = .238 μm, SD = .039	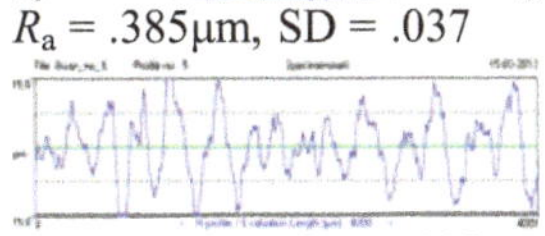 R_a = .411 μm, SD = .028	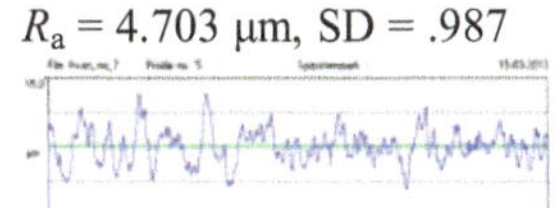 R_a = 3.230 μm, SD = .453

3.2. Electrical resistance as a function of surface condition

In order to study the influence of surface preparation on electrical contact resistance, the total electrical resistance between the two electrodes was determined for each surface dressing condition and each sheet thickness. The resistance was calculated by dividing

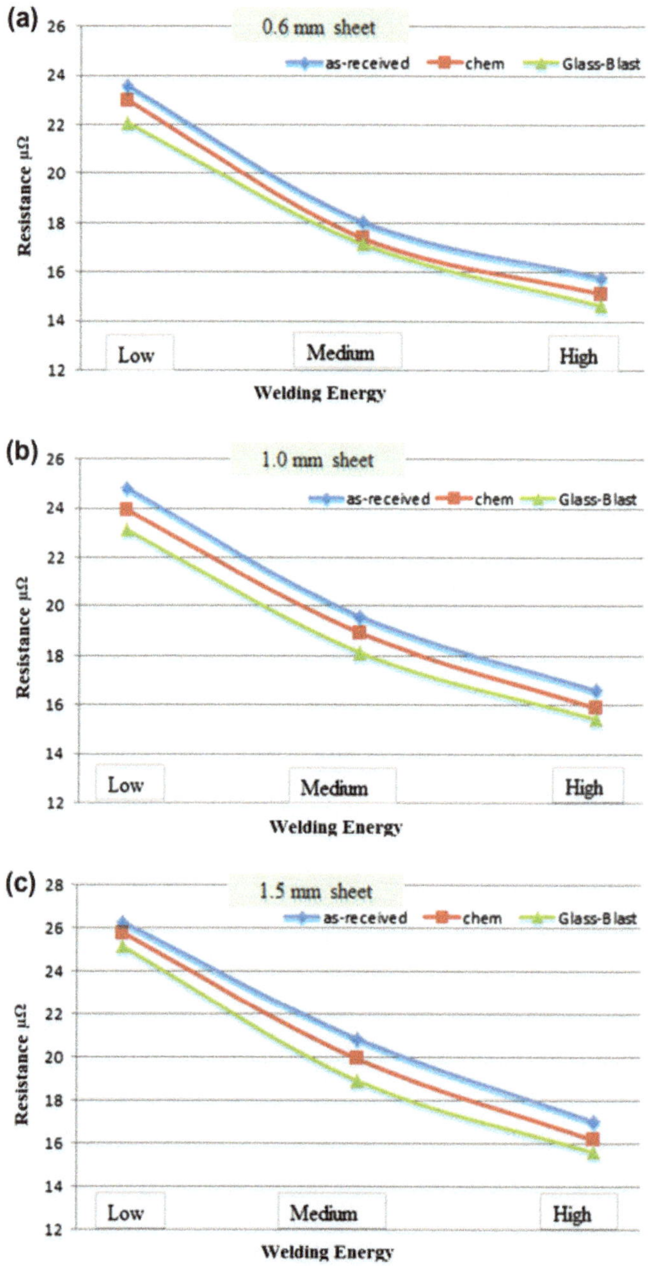

Figure 4. Electrical resistance as a function of energy input and surface dressing for (a) .6 mm; (b) 1.0 mm; (c) 1.5 mm sheets.

the RMS voltage by the RMS current determined by sampled current and voltage data using the LabVIEW software program. The error due to induction of emf was minimized by twisting the wires, when measuring the voltage between the electrodes. Although it is the entire electric resistance between the electrodes, which is determined in this way, a comparison between the individual surface treatments gives a good idea of their influence on the accumulated contact resistance between workpieces and between workpieces and electrodes.

Figure 4 shows the total electrical resistance between the two electrodes as a function of the energy input for each surface condition and each sheet thickness. It is noticed that the as-received sheet has the highest electrical resistance for all sheet thicknesses due to the thicker oxide layer increasing the contact resistance. Medium contact resistance is obtained with chemical pickling, which lowers the contact resistance by decreasing the oxide layer thickness. The lowest electrical resistance is obtained with glass-blasted sheets, which besides removing the thick oxide film creates a rough surface causing easy breakup of the newly formed oxide layer, when the applied electrode load deforms the contacting surface asperities. It is, furthermore, noticed that the electrical resistance decreases with increasing welding energy (increased current and welding time). This might be explained by easier breakdown of the layer of oxides and contaminant film at high temperature as also observed by others (Crinon & Evans, 1998; Han et al., 2010; Song et al., 2005). Finally, a small increase in the electrical resistance is observed with increasing sheet thickness, as to be expected.

3.3. Microhardness tests and microstructure examinations

Figure 5 shows the microhardness profiles for RSW of 1.0-mm sheets with the three different surface treatments. The hardness of the fusion zone FZ and the heat-affected zone (HAZ) is noted to be lower than that of the base metal for all cases due to annealing of the work-hardened base metal during welding. A few points indicate relatively low microhardness due to presence of void defects nearby, e.g. points 5 and 7 in as-received sheets and point 4 in pickled sheets.

The comparison of hardness profiles in Figure 5(d) shows the hardness in the FZ and the HAZ to be highest for welds with pretreatment by glass blasting and lowest for the case of as-received sheets. This is explained by the influence of electrical contact resistance on heat generation discussed earlier. Hardness profiles for welded 1.5-mm sheets showed the same trend.

Macrographs also showed the nugget size when welding as-received sheets to be larger than pretreated sheet surfaces. This is due to the smaller electrical contact resistance in the faying surfaces generating less heat in the latter cases. For example, the nugget sizes were 5.1, 5.0, and 4.7 mm in 1.0-mm sheets, and 7.0, 6.54, and 5.35 mm in 1.5-mm sheets for as-received, pickled, and glass-blasted sheets, respectively. The minimum nugget size was obtained by glass-blasted sheets, due to minimum electrical contact resistance.

Figure 6 shows micrographs of the 1.0-mm glass-blasted strips welded with 29-kA welding current and .1-s (five cycles) welding time. In the oval nugget region, recrystallized, small, equiaxial grains, and insoluble particles of $FeAl_3$ (black) are noticed as well as a narrow zone of columnar grains along the nugget edge nugget. Next to this zone is the HAZ, which consist of dendritic grains. Moreover, some porosities (large, black areas) are evident. No significant difference in the microstructures was found in welds between the as-received and the pretreated sheets.

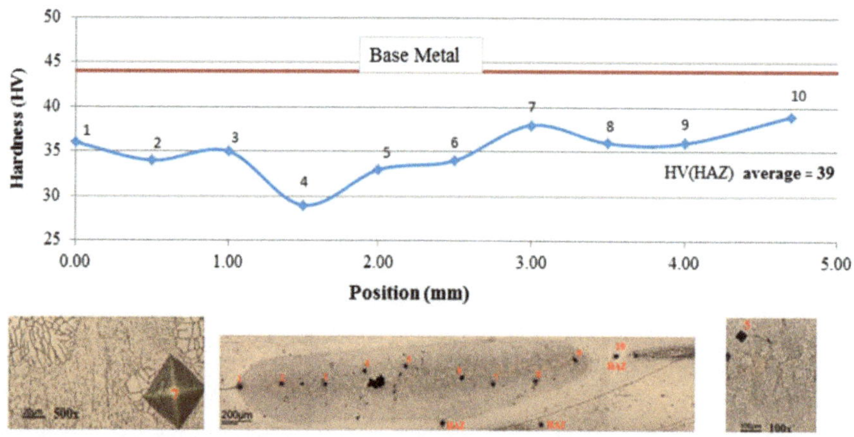

(a) Glass-blasted pretreatment surfaces

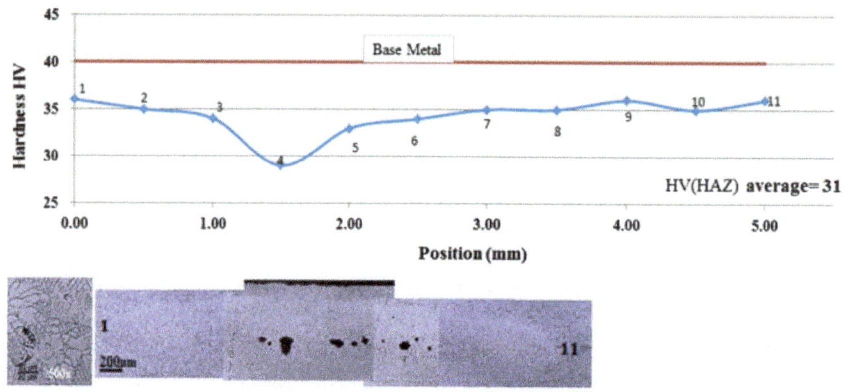

(b) Pickling in NaOH pretreatment surfaces

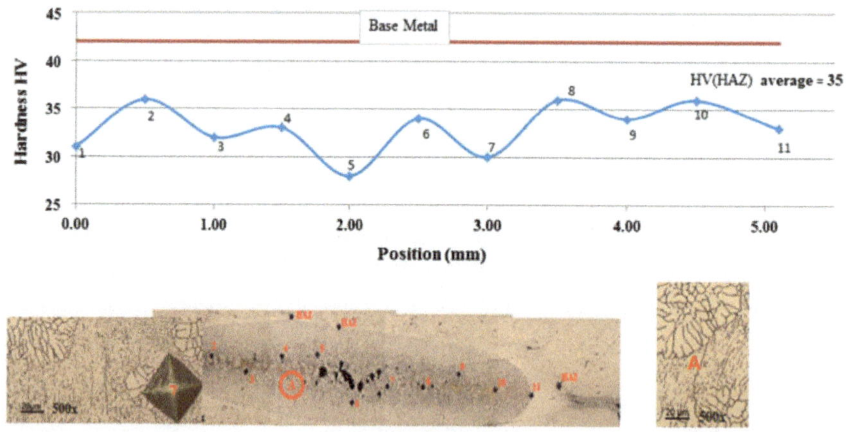

(c) As-Received sheet

Figure 5. Microhardness profiles of 1.0-mm sheet, welding parameters 29 kA, .1 s (five cycles).

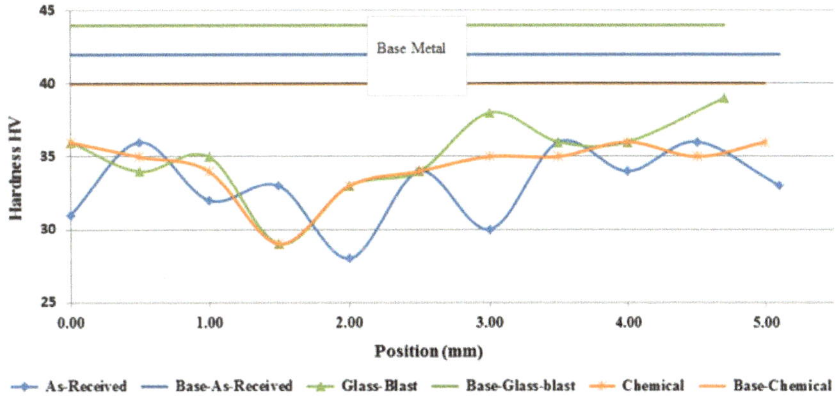

(d) All surface dressing conditions

Figure 5. (*Continued*).

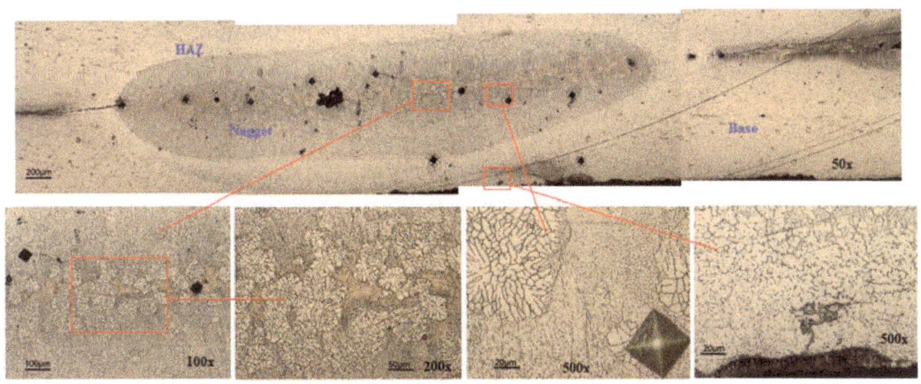

Figure 6. Micrographs of RSW of 1.0-mm sheet, welding parameters 29 kA, .1 s (five cycles).

3.4. Tensile-shear tests

Tensile-shear tests of the welded joints indicate their strength and the failure mode. Figure 7 shows the strength as function of the welding energy input specified in Table 3 for the three different surface treatments (as-received, pickled, and glass blasted). The surface treatment is generally done on both sides of the strips, but for the case of glass blasting and medium energy input, one-sided treatment of the outer sides contacting the electrodes is also investigated. The data represented by an average value and a scatter bar are based on an average of three identical tests. It is seen that an increase in energy input increases the strength. The mode of failure was recorded in each test and classified as one of the following three types: (1) interfacial failure (nugget fracture in shear), (2) plug failure (nugget pull-out), and (3) failure in the HAZ. Failure mode 1 typically occurred at low input energy indicating insufficient heating, mode 2 at medium input energy indicating satisfactory heating, and mode 3 at high input energy indicating over-heating and softening of the region near the nugget. A few abnormal test results were observed to fall outside these types of failure. In line with other studies (Mathers, 2002;

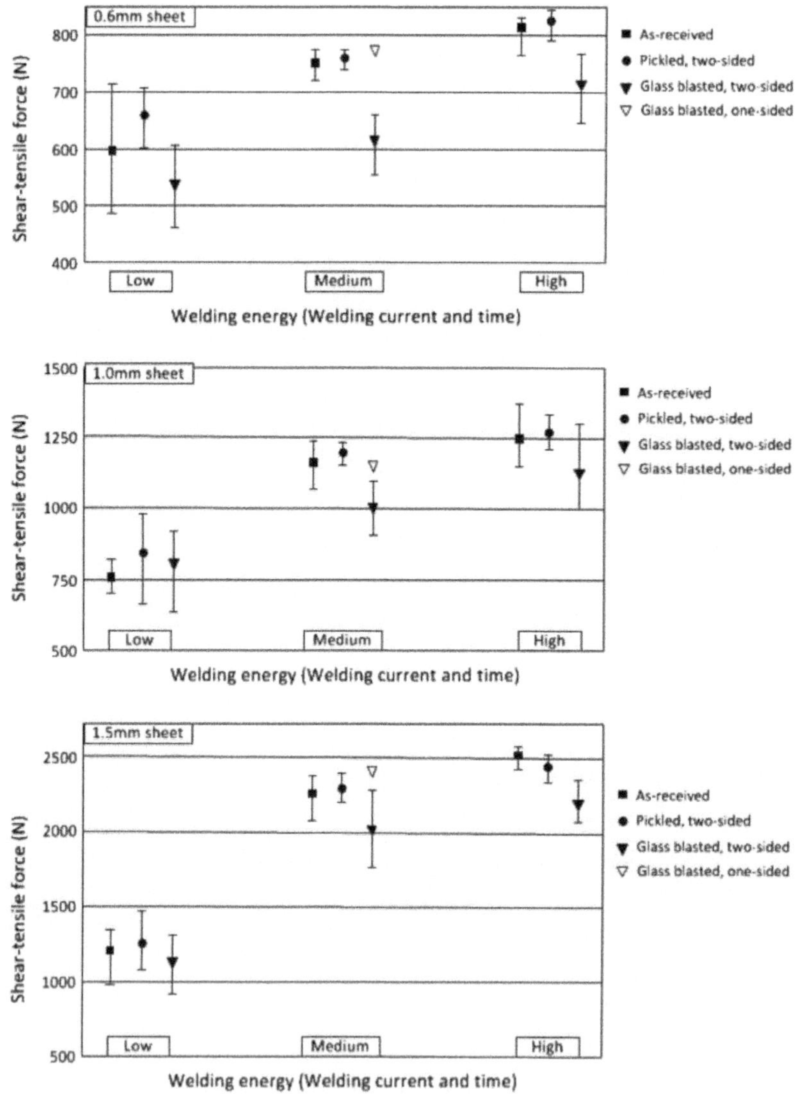

Figure 7. Shear-tensile force vs. welding energy input for different surface treatments and sheet thicknesses.

Newton, Browne, Thornton, Boober, & Keay, 1994; Rashid, 2011; Senkara et al., 2004; Shi & Guo, 2013), expulsion and severe sticking of the workpiece to the electrodes occurred at high-energy input. Furthermore, fracture in mode 3 was usually in the form of ductile tearing around the nugget.

The scatter in weld strength is less when surface treating by pickling than for the as-received strips. Largest scatter is generally noticed for the two-sided, glass-blasted sheets, which may be attributed to the large surface roughness, low interface resistance, and difficulties in controlling this manually operated surface dressing. The treatment by pickling is also best regarding average weld strength independent of the sheet thickness.

The values for medium energy input are 760, 1193, and 2283 N for .6, 1.0, and 1.5-mm sheets, respectively. The lowest strength is obtained for the glass-blasted sheets due to the lowest values of electric contact resistance; see Figure 4, which leads to small nugget size. The values for medium heat input are 616, 1008, and 2020 N for .6, 1.0, and 1.5-mm, respectively. The low electrical contact resistance of the glass-blasted, rough surfaces may be attributed to easier breaking up of the oxide layer as also observed by other authors (Dzekster & Ismailov, 1990; Han et al., 2010; Song et al., 2005).

If only the surfaces contacting the electrodes are glass blasted leaving the faying surfaces untreated, a significant increase in strength occurs, i.e. 775, 1147, and 2408 N for .6, 1.0, and 1.5-mm sheet thickness, respectively, for medium-energy input. The reason for this is higher heat generation due to higher electrical contact resistance at the faying surfaces. This treatment is recommendable, since it also ensures less sticking to the electrodes, and thus increased electrode life. The fact that pickling improves weld quality as regards level and consistency compared to the as-received surface condition even if the surface roughness is similar indicates that the latter is not of overriding importance controlling the electric contact resistance although increasing roughness may facilitate breakdown of the oxide layer. The main importance is the thickness of the oxide layer as also stated by other researchers (Patrick et al., 1984; Song et al., 2005; Sun, 1982).

3.5. SEM and EDS examinations

Micrographs were also made in SEM to study microstructure and surface profile and carry out high-resolution mapping determining chemical composition of areas of special interest by X-ray spectroscopy (EDS). When welding as-received sheet (with thick oxide layers), considerable indentation into the aluminum sheets was observed, as well as rapid degradation of the electrodes. This was due to pitting of the electrode surfaces and dissolving or diffusion of copper into the aluminum workpiece material and vice versa (see Figure 8(a)).

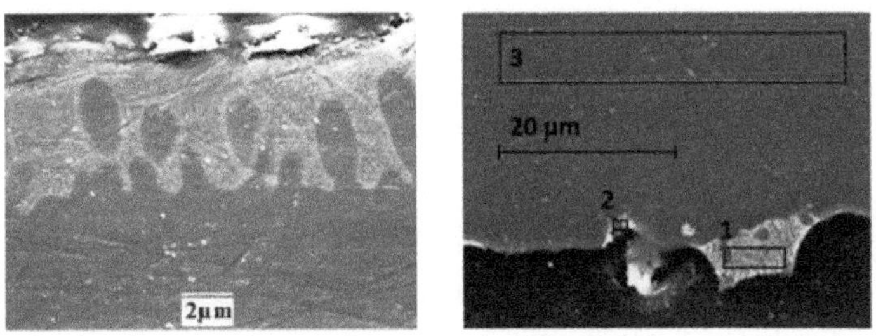

(a) 1.0 mm as-received sheet **(b)** 1.5 mm chemical pretreatment

Al O Cu Total				
1	69.4	1.0		100
2	66.5	2.2		100
3	100			100
Max.	100	2.2	31.3	
Min.	66.5	1.0	29.6	

(**c**) EDS analysis

Figure 8. SEM images RSW and measured composition referring to Figure 5(b).

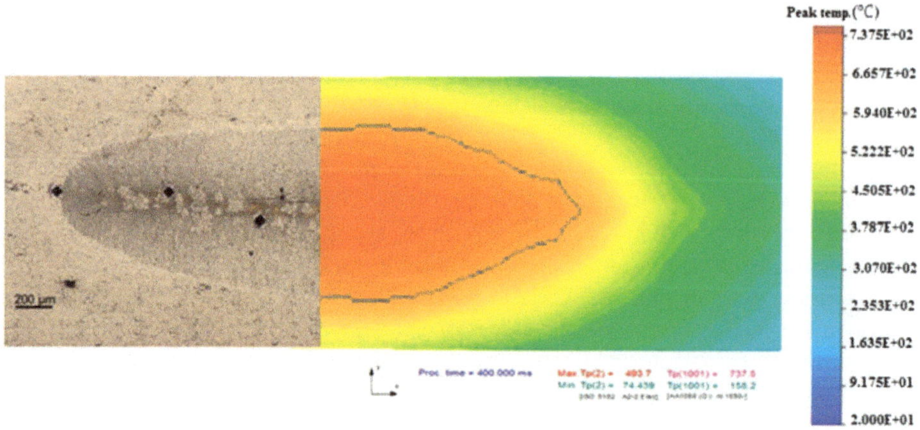

Figure 9. Comparison of experimental and predicted nugget shape.

Welding of pretreated sheets showed less indentation and dissolving of electrode material in the workpiece due to a thinner oxide layer causing less heating (Figure 8(b)). The reason for a few defects in this case may be pitting of the electrode surface. The white colored areas in the SEM images indicate the presence of oxide as well as dissolving of copper into the workpiece, as shown in the EDS analysis table in Figure 8(c).

3.6. Numerical simulation

The commercial finite element program SORPAS® developed especially for resistance welding (SWANTEC Software & Engineering ApS. SORPAS®) was used in the present work. It combines mechanical, electrical, thermal, and metallurgical models analyzing the development of heat and distributions of current, voltage, temperature, stresses, and strains (Bay & Zhang, 2007; Zhang, 2003). Using a new feature called Weld Schedule Specifications optimized weld current, electrode force, welding time, and holding time were calculated, see Table 5, for welding of as-received as well as surface-dressed sheet adjusting the input value of the electrical resistance at 20 °C in the materials database of the program to lower value in case of surface dressing according to Figure 4. It is clearly noted that especially the proposed welding current for the pretreated surfaces is lower than that for as-received sheets.

Figure 9 shows a comparison between predicted and actual weld nugget with as-received strip surface, in 1.0-mm sheet for 27-kA welding current and .1-s five cycles welding time. It is noticed that the nugget shape is rather well predicted.

Table 5. WSS of weld planning optimization SORPAS.

Sheet	As-Received (mm)			Pretreatment (mm)		
	0.6	1.0	1.5	0.6	1.0	1.5
Welding Current (kA)	23.9	26.27	31.64	22.87	24.58	27.26
Welding Time (cycles)	2	4	5	2	4	4
Electrode Force (kN)	1.16	1.44	1.78	1.04	1.44	1.49

4. Conclusions

The experiments showed that consistency in weld quality is improved when the oxide layer of the as-received aluminum surfaces is properly removed. However, the resistance of an oxide layer is beneficial at the interface between the faying surfaces in order to increase local heating to form the weld nugget, whereas the resistance at the contact between electrodes and workpiece should be as low as possible to avoid sticking and poor electrode life. The most significant conclusions drawn from this experimental work are as follows:

(1) The surface condition of the aluminum sheets has significant influence on the weldability and the electrode lifetime.
(2) The highest value of electrical contact resistance is obtained with as-received sheet due to a thicker oxide layer.
(3) Pretreatment by pickling in 60 °C NaOH gives highest weld strength and lowest scatter. Optimum pickling time is 2 min.
(4) Lowest contact resistance is obtained by glass blasting, which besides removing the original oxide film provides a rough surface decreasing the contact resistance due to breakdown of the oxide layer by asperity deformation.
(5) One-sided surface treatment by glass blasting of the surfaces contacting the electrodes is favorable leading to less electrode sticking, and thus increased electrode life and furthermore better weld strength than two-sided glass blasting.
(6) The microhardness in the FZ and HAZ is lower than in the base metal for all cases due to annealing of the as-received, workhardened material.
(7) SEM and EDS examinations confirm the presence of copper (electrode material) in the welded aluminum strips especially when welding as-received sheet indicating degradation of the electrodes due to dissolving and diffusion, which may furthermore cause pitting and accelerated electrode wear due to the pitting on the surface of the electrodes.
(8) Numerical simulation with SORPAS® gives good prediction of appropriate welding parameters and nugget size.

Acknowledgments

This work is part of cooperation on independent research between Baghdad University and Technical University of Denmark (DTU). The authors would like to thank co-workers at the Department of Mechanical Engineering, DTU, especially Peter Sanderhoff, Steffen Munch, Rolf Jensen, and Chris V. Nielsen for helping with the experimental work. First author, Ihsan Al Naimi, is grateful for the financial support by Iraqi Ministry of Higher Education and Scientific Research.

Disclosure statement

No potential conflict of interest was reported by the authors.

Funding

The work of Ihsan Al Naimi was supported by Iraqi Ministry of Higher Education and Scientific Research.

References

Bay, N., & Zhang, W. (2007, May). Thermomechanical modelling of resistance welding. *Proceedings of International Conference on Computational Methods for Coupled Problems in Science and Engineering* (162–165). Ibiza, Spain.

Boomer, D. R., Hunter, J. A., & Castle, D. R. (2003). *A new approach for robust high-productivity resistance spot welding of aluminum* (SAE Technical Paper -01-0575).

Brown, D. J., Newton, C. J., & Boomer, D. (1995). Optimization and validation of a model to predict the spot weldability parameter lobes for aluminum automotive body sheet. Advanced Technologies & Processes. In *Proceedings of International Body Engineering Conference, IBEC* (pp. 100–106).

Cho, Y., Li, W., & Hu, S. J. (2006). Design of experiment analysis and weld lobe estimation for aluminum resistance spot welding. *AWS Welding Journal, 85*, 45s–51s.

Crinon, E., & Evans, J. T. (1998). The effect of surface roughness, oxide film thickness and interfacial sliding on the electrical contact resistance of aluminum. *Material Science and Engineering Journal, A242*, 121–128.

Dzekster, N. N., & Ismailov, V. V. (1990). Some methods for Improving Aluminum Contacts. In *Proceeings of 36th Annual Holm Conference on Electrical Contacts* (pp. 518–520). Chicago, IL, Illinois Institute of Technology.

Fronius International. (2014, September). Retrieved from http://www.fronius.com/cps/rde/xchg/SID-3AC27A9B-F495FC44/fronius_international/hs.xsl/79_33390_ENG_HTML.htm#.VBmeZaPU-70

Han, L. (2010, September 22–24). Development of RSW Al for LWV manufacture. In *Proceedings of the 6th International Seminar on Advances in Resistance Welding* (pp. 55–70). Hamburg.

Han, L., Thornton, M., Boomer, D., & Shergold, M. (2010). Effect of aluminium sheet surface conditions on feasibility and quality of resistance spot welding. *Journal of Materials Processing Technology, 210*, 1076–1082.

Kim, D. C., Park, H. J., Hwang, I. S., & Kang, M. J. (2009). RSW of aluminum alloy sheet 5J32 Using SCR type and inverter type power supplies. *International Scientific Journal, 38*, 55–60.

Li, Z., Hao, C., Zhang, J., & Zhang, H. (2007, April). Effects of sheet surface conditions on electrode life in resistance welding aluminum. *Welding Journal*, 81–89.

Mathers, G. (2002). *The welding of aluminum and its alloys*. Cambridge: Woodhead.

Newton, C. J., Browne, D. J., Thornton, M. C., Boober, D. R., & Keay, B. F. (1994). *The fundamental of resistance spot welding aluminum*. Sheet Metal Conference VI; Paper E2.

Patrick, E. P., Auhl, J. R., & Sun, T. S. (1984). *Understanding the process mechanism is key to reliable resistance spot welding aluminum auto body components* (SAE Technical Paper 840291).

Pedersen, K. R. (2010). *Resistance welding of aluminum alloys* (Master thesis). Department of Mechanical Engineering, Technical University of Denmark.

Pickering, E. R., & Hart, C. J. (1994). *Optimizing resistance spot welding on aluminium alloy 6111 auto body sheet* (SAE Technical Paper, 940662).

Rashid, M. (2011). Some tribological influences on the electrode-worksheet interface during resistance spot welding of aluminum alloys. *Journal of Materials Engineering and Performance, 20*, 456–462.

Rashid, B. M., Fukumoto, S., Medley, J. B., Villafuerte, J., & Zhou, Y. (2007, March). Influence of lubricants on electrode life in resistance spot welding of aluminum alloys. *Welding Research,* 62–70.

Resistance Welding Manual. (2003). Bridgeport, NJ: RWMA Resistance Welding Manufacturers Association.

Rönnhult, T., Rilby, U., & Olefjord, I. (1980). The surface state and weldability of aluminium alloys. *Materials Science and Engineering, 42*, 329–336.

Senkara, J., Zhang, H., & Hu, S. J. (2004). Expulsion prediction in resistance spot welding. *AWS, Welding Journal, 83*, 123s–132s.

Shi, Y., & Guo, H. (2013). Fatigue performance and fatigue damage parameter estimation of spot welded joints of aluminium alloys 6111-T4 and 5754. *Fatigue & Fracture of Engineering Materials & Structures, 36*, 1081–1090.

Song, Q., Zhang, W., & Bay, N. (2005). An experimental study determines the electrical contact resistance in resistance welding. *AWS Welding Journal, 84*, 73s–76s.

Spinella, D. J., Brockenbrough, J. R., & Fridy, J. M. (2005). Trends in aluminum resistance spot welding for the auto industry. *AWS Welding Journal, 84*, 34–41.

Sun, M. A. (2003). *Fundamental study of contact resistance behavior in RSW aluminum* (PhD dissertation). School of The Ohio State University, Ohio State University.

Sun, T. S. (1982). *Electrode deterioration mechanisms in resistance spot welding of aluminum* (AlCOA Internal Report No. 53-82-3).

SWANTEC Software and Engineering ApS. SORPAS® (Version 10.6). Retrieved from http://www.swantec.com

Thornton, M. C., Newton, C. J., Keay, B. F. P., Sheasby, P. G., & Evans, J. T. (1997). Some surface factors that affect the spot welding of aluminium. *Transactions of Institute of Metal Finishing, 75*, 165–170.

Wheeler, M. J., Sheasby, P. G., & Kewley, D. (1987). *Aluminum structured vehicle technology. A comprehensive approach to vehicle design and manufacturing aluminum* (SAE Technical Paper 870146).

Zhang, H., & Senkara, J. (2012). *Resistance welding – Fundamentals and applications* (2nd ed.). Boca Raton, FL: Taylor & Francis Group, LLC.

Zhang, W. (2003). Design and implementation of software for resistance welding process simulations. Welding & Joining, SP-1766, SAE. In *World Congress* (pp. 105–113). Detroit, MI, USA.

Zheng, Z. T., Shan, P., Hu, S. S., & Luo, Z. (2006). Resistance spot welding of aluminum alloy with great resistance medium. *Transactions of the China Welding Institution, 27*, 5–8.

11

Criteria definition and approaches in green supplier selection – a case study for raw material and packaging of food industry

Narges Banaeian[a]*, Hossein Mobli[a], Izabela Ewa Nielsen[b] and Mahmoud Omid[a]

[a]Department of Agricultural Machinery Engineering, Faculty of Engineering and Technology, University of Tehran, Karaj, Iran; [b]Department of Mechanical and Manufacturing Engineering, Aalborg University, Aalborg, Denmark

This paper formulates an integrated framework for deciding about the green supplier selection criteria in food supply chain and also proposes different methods that account for single and multiple sourcing of supplier selection. Green supplier selection relies on green criteria, so determination of suitable set of criteria will affect decision-maker results directly. In this research, an operational model including combination of general and environmental criteria is introduced for green supplier selection criteria in raw material and packaging of food industry. This model reviewed a literature on general and environmental criteria that help us to know and make a set of common green criteria. Afterward, weighting criteria and collecting of sub-criteria are done by an expert team using the Analytic Hierarchy Process and Delphi method. The expert team tried to propose important and practical sub-criteria which are well fitted to food industry. Finally, in the section of supplier selection, two kinds of qualitative and quantitative data are discussed when single and multiple sourcing are required, respectively. Fuzzy Grey Relational Analysis is proposed to ranking suppliers in presence of qualitative and imprecise data. Also linear programming is used to present a model which can select the best suppliers and allocate the orders among them optimally.

Keywords: green supplier selection; food industry; raw material; packaging supplier

1. Introduction

Enterprises realize that a large amount of direct and indirect profits can be obtained from effective and efficient supply chain management practices. Supplier selection has great impact on integration of the supply chain relationship. Effective and accurate supplier selection decisions are significant components for productions and logistics management in many firms to enhance their organizational performance.

The issue of supply chain (management of networks, decision-makings, scheduling, etc.) is the area of science that has been strongly developing since the 80s of the last century (Bocewicz, Wójcik, & Banaszak, 2012; Sitek & Wikarek, 2014). A definition of a supply chain can be proposed as: the global network of players involved in executing the production and distribution of a predefined good from raw materials through to delivery to satisfy a consumer demand (Chen, Simon, Reich-Weiser, & Woo, 2013). The aim of supply chain management is to increase sale, reduce costs, and take full advantage of business (Sitek & Wikarek, 2014). Green supply chain is similar while

*Corresponding author. Email: nrgsbanaei@ut.ac.ir

managers simultaneously consider environmental impacts and work to reduce those impacts over time. Namely according to Sirvastava (2007), Green Supply Chain Management (GSCM) can be defined as "integrating environmental thinking into supply chain management, including product design, material sourcing and selection, manufacturing process, delivery of the final product to the consumers as well as end-of-life management of the product after its useful life." GSCM is to eliminate or minimize waste in the form of energy, emission, hazardous, chemical, and solid waste (Olugu, Wong, & Shaharoun, 2011).

Deciding whether to make or buy a new product or service is the first decision that must be made in managing supply chain. It determines where and how the supply chain will be managed (Schoenfeldt, 2008). If you decide to buy, so next step of SCM implementation includes the purchase, movement, and storage of raw materials (Harland, 1996).

Having seen the significance of collaborative supply chain and centralized purchasing in environmental management performance, it is now obvious that deciding which suppliers to collaborate with and how to select suppliers is a very crucial decision for the organization performance.

Fundamentally, supplier selection is divided into two kinds as single sourcing and multiple sourcing. In single sourcing, one supplier can satisfy all the buyer's requirements, but in multiple sourcing, company needs to select a collection of selected suppliers and their allocations (Wikipedia contributors, 2014).

Decision-making on supplier selection is based on criteria, so definition and selection of criteria have an important role in your selection. Supplier selection is a critical managerial decision-making problem, which requires considering various qualitative and quantitative criteria (Toloo, 2014). In conventional supply chain, general criteria are used; general selection criteria ordinarily intend to cover issues such as quality, capacity in terms of finance, services, and equipment, quantity, responsiveness, and others. In GSCM, environmental criteria are joined to general; environmental supplier selection criteria arise from an organization inclination to respond to any existing trends in environmental issues related to business management and processes. A good combination of conventional and environmental criteria will show the exact image of green supplier selection.

Building accepted definitions and characteristics of these criteria before their implementation in decision models is necessary for the acceptance of the methodologies, and, most importantly, for the decisions that derive from this methodologies. Despite the large amount of literature to supplier selection, there is narrow literature on green supplier selection and especially criteria definition, most of which are qualitative. Also, the methods for identifying, defining, grouping, and filtering criteria would greatly enhance the research, development, and application of models and their criteria. These areas are important and necessary directions (Govindan, Rajendran, Sarkis, & Murugesan, in press). The selected criteria are systematically integrated from supplier selection papers (general criteria) and green supplier selection (environmental criteria).

Given this complexity, this study presents a compound approach, first capable to integrate Delphi and AHP for criteria selection and then, according to supplier's data and company's requirements, defines two different kinds of GSS: (1) single sourcing with qualitative supplier's data using FUZZY GRA to rank suppliers and select just one green supplier and (2) multiple sourcing with quantitative supplier's data using linear programming to select and allocate green suppliers of raw ingredients and packaging material food industries.

Generally, this study is organized as follows. Sections 2 and 3 explain greening food industry and reviews pertinent literature on green supplier selection of food industry. Next, Section 4 introduces the applied model and related methods. Section 5 describes the applicability of the proposed approach by food industry experts. Conclusions are finally drawn in Section 6, along with recommendations for future research.

2. Greening food industry

When compared to other, industries of food and agricultural products have some unique characteristics related to public health. Food supply chains can be distinguished into "fresh agricultural products" (e.g. vegetables or fruits) and "processed food products" (e.g. convenience food or soft drinks) (Apaiaha, Linnemanna, & van der Kooi, 2006) which here we focused on it. Resulting product-related issues include but are not limited to "shelf life constraints, variable quantity and quality due to biological variations, seasonality, random factors connected with weather and pests and other biological hazards" (van der Vorst, van Dijk, & Beulens, 2001).

Efforts have consequently been intensified to enable tracking and tracing of materials and ingredients in food supply chains, for example, food safety and traceability standards by the European Union (EU) requiring every ingredient to be traceable (EU, 2002). In supply chain management, to build strategic and strong relationships, food industries should select best suppliers by applying appropriate method and selection criteria.

Food companies are closely monitored by their customers who want to be confident that the food being purchased is safe. In this regard, they are usually asked to pass an independent audit to check whether good manufacturing practice, good hygiene practice, and HACCP (Hazard Analysis and Critical Control Points) principles are fully implemented and strictly followed in each step of the process (Losito, Visciano, Genualdo, & Cardone, 2011). Quality and safety of food ingredients and packaging are paramount to company success, so choosing a good supplier is a critical business decision. Generally, companies in food industry try to develop their supplier assessment form according to HACCP, ISO 9000, and other related regulation and special requirements.

Something that has become apparent is that in the near future, the design and operation of agricultural supply chains will be subject to more stringent regulations and closer monitoring, in particular, those for products destined for human consumption (Bosona & Gebresenbet, 2013). This implies that the traditional supply chain practices may be subject to revision and change. One of the aspects that may be the subject of considerable scrutiny is the supplier selection along the supply chains of food products (Gendron & Audet, 2012).

Beside the conventional factors like quality, safety, and hygienic factors, it should be noted that the food industry is one of the world's largest industrial sectors, and hence has large environmental impacts (Grunert, Hieke, & Wills, 2014). Greenhouse gas (GHG) emission, which has increased remarkably due to tremendous input and energy use, has resulted in global warming, perhaps the most serious problem that humankind faces today (Ilbery & Maye, 2005). With growing worldwide awareness of environmental protection, green food production has become an important issue for almost every producer and will determine the sustainability of producers in the long term. It is the responsibility of the food industry to develop and implement green supplier selection in the first step of green supply chain establishment.

3. Literature review

A food producer needs to buy raw ingredients and packaging material, so suppliers in food industries are divided in two groups. A number of literature reviews have been completed to summarize supplier selection criteria adopted and decision models published (Aissaoui, Haouari, & Hassini, 2007; Chai, Liu, & Ngai, 2013; Govindan et al., in press). Unfortunately, there are few studies which are focused on food supplier selection and none of them suggested environmental criteria. Some information about their criteria and techniques are summarized in Table 1.

In the field of package printing industry, Vachon and Klassen (2006) studied a comparative green supplier selection problem and determined green supplies in Canadian and United States Package Printing Industry. They purposed Green projects partnership as environmental criteria to select green supplier. Also Kumar, Hong, and Haggerty (2011) and Magdalena (2012) studied a global supplier selection process for packaging. Kumar et al. (2011) noted risk of disruption, volume and product flexibility, and innovation as specific criteria for packaging suppliers. This paper attempts to construct a practical model for purchase manager and decision-maker in green supplier selection of food industry. In the latest research of Banaeian, Nielsen, Mobli, and Omid (2014) on GSS in food industry, they used method of Data Envelopment Analysis (DEA) to rank and compare green suppliers, different criteria of price, and carbon footprint as inputs, and delivery and quality as outputs of models.

4. Material and method

4.1. General model (guideline)

Figure 1 shows the structured model. In this paper, we discussed widely about green supplier selection criteria and formed a selection model step by step:

Phase I

- Reviewing general & green supplier selection criteria
- Selection of relevant criteria for green supplier selection

Phase II

- Determining the weights of relevant criteria using AHP
- Determining branches for each relevant criterion and selecting the most effective branch based on Delphi group opinion in raw material and package suppliers

Phase III

- Identifying company requirements and supplier characteristics
- Decision-making for selection of best green supplier in case of single sourcing
- Modeling for selection and allocation of best green suppliers in case of multiple sourcing

Table 1. Supplier selection criteria in food industry.

Authors/Year	Technique	Criteria/Sub-criteria	
Cebi and Bayraktar (2003)	AHP+LGP	Logistic	Lead time, Supply lots, Flexibility, Delivery condition
		Technologic	Capacity, Involvement. Improvement efforts, problem solving
		Business	Reputation, Financial strength, Management,
		Relationship	Communication, Past experience, Sales representation
Shen, Liu, and Tzeng (2012)	DEMATEL+ANP	Cost	Unit cost, operating cost, warranty cost
		Delivery	Delivery performance, delivery speed, fill rate
		Quality	Product quality, steadiness of quality
		Service	Follow-up, process flexibility, professionalism, supply chain responsiveness
		Technology	Ability to solve technical problems, machinery, outlook for the future
		Company image	Attitude, relationship closeness, reputation and financial position
Prusak, Stefanow, Niewczas, and Sikora (2013)	AHP	Finance	Price, Price difference, Time limit for payment
		Assortment	Possible changes, Diversity
		Logistic	Buffer stock, Flexibility, Promptness of supply
		Service	Innovativeness, contact, Acceptable procedure, Customer service, On-line platform
		Quality	The use of SPC (Statistical process control), ISO standards, Technical support, Reaction on problem
Liao, Fu, Chen, and Chih (2012)	Fuzzy MSGP	Financial Stability Product quality Reputation in industry R&D capability Service quality	
Shen, Olfat, Govindan, Khodaverdi, and Diabat (2013)	DEA	Inputs Synthetic cost, Geographical position Outputs	
		Quantitative	Qualitative
		Product qualification ratio, Market occupancy ratio, Order changing received ratio, Deliver punctual Ratio, R&D success rate, R&D return rate, Total assets turnover ratio, Profit increasing ratio	Quality system, enterprise reputation, compatibility, upstream firm's management, traceability system, information level

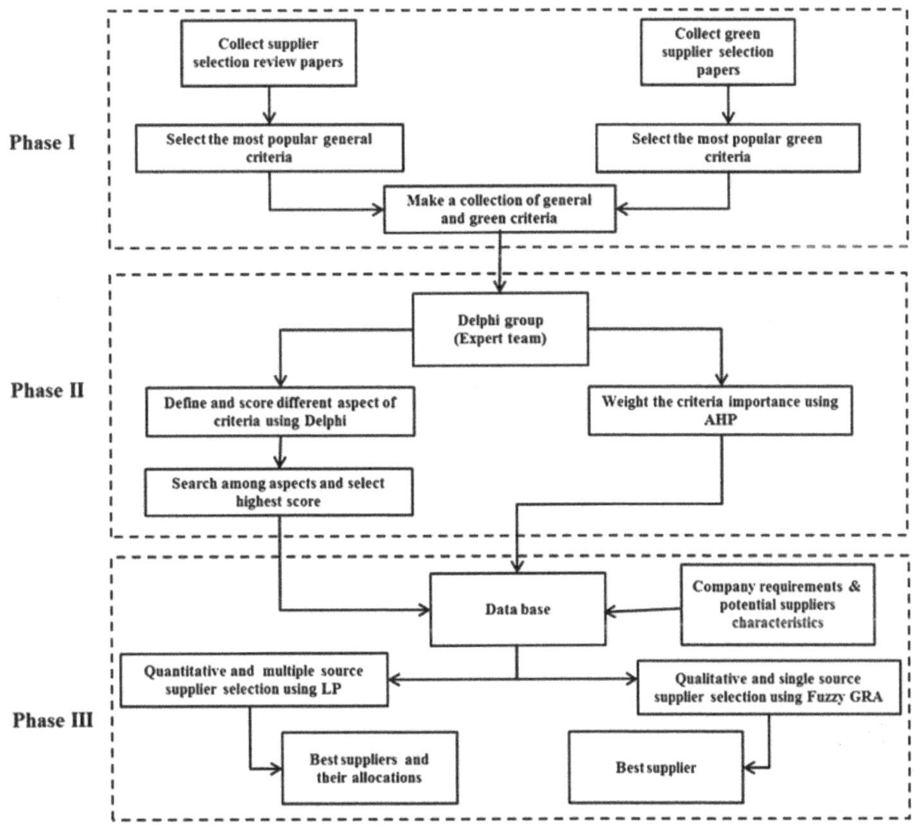

Figure 1. Structure of green criteria definition and supplier selection model.

4.2. Analytical hierarchy process

AHP is a structured technique for organizing and analyzing complex decisions based on paired comparisons of both projects and criteria (Saaty, 1990). This method compares weights in pairs and is more straightforward and easier to use for the decision-makers. The strength of the AHP method lies in its ability to construct complex, multi-person, multi-attribute, and multi-period problem hierarchically. Thus, it has been discussed in a wide variety of decision situations in fields such as business planning, resource allocation, priority setting, and selection among alternatives.

First, under the designed hierarchy model, pairwise comparisons are classically carried out by asking the decision-maker how valuable a criterion (C1) is when compared to another criterion (C2) with respect to overall goal. The verbal judgments of the decision-maker are then transformed into numerical values using the scale presented in Table 2, then a pairwise comparison matrix is constructed.

The pairwise comparison of criteria can be summarized in an $(n*n)$ evaluation matrix $A=(a_{ij})$ for n criteria, where every element $(i, j=1,2,\ldots,n)$ is the quotient of weights of the criteria. This pairwise comparison can be shown by a square and reciprocal matrix, (see Equation (1)).

Table 2. Verbal judgments and numerical rate.

Verbal judgment of preference	Numerical rate
Equal importance	1
Weal importance of one over another	3
Essential or strong importance	5
Demonstrated importance	7
Absolute importance	9
Intermediate values between the two adjacent judgments	2,4,6,8

$$A = (a_{ij}) = \begin{pmatrix} 1 & a_{12} & \cdots & a_{1n} \\ a_{21} & 1 & \cdots & a_{2n} \\ \vdots & \vdots & \ddots & \vdots \\ a_{n1} & a_{n2} & \cdots & 1 \end{pmatrix}; \tag{1}$$

$$a_{ij} = \left\{ \begin{matrix} 1 & i = j \\ a_{ij},\ a_{ij}^{-1} & i \neq j \end{matrix} \right\}; \tag{2}$$

In order to measure the subjective evaluation of the decision-maker, (Saaty, 1990) proposes a consistency index (CI). CIs are calculated for each pairwise comparison matrix and to check consistency using a random index (RI). If the calculated ratio is significantly small (10% or less), the priorities will accepted otherwise, the DM is asked to revise the pairwise comparisons.

4.3. Delphi method

Delphi is a method of popular survey technique that brings consensus of opinions among a set of experts or panelists (informed individuals) by maintaining the unanimity among them (Prusty et al., 2010). In this paper, Delphi method is used for describing sub-criteria based on food industry necessity and selects the most important sub-criteria. Each Delphi panelist was asked to answer two main questions:

First round – According to criteria collection, describe sub-criteria which are important for your supplier selection.

Second round – Score each sub-criteria between 1 (not important) and 9 (very important).

Finally, Delphi score is proposed by Listone and Turoff (2002) as illustrated in Equation (2):

$$\text{Delphi Score} = \frac{\text{Lowest Score} + \text{Highest Score} + (4 \times \text{Average Score})}{6} \tag{3}$$

4.4. Qualitative and single source green supplier selection

In decision-making process of supplier selection, the decision-makers try to gather as much information as possible through surveys, investigations, sampling, etc. To reach the aspired decision, sometimes obtaining all the quantitative information is impossible; therefore, decisions are often made in uncertain process, i.e. without complete or quantitative information. In these situations, decision-makers try to apply qualitative models.

Fuzzy theory and sets, first introduced by Zadeh (1965), have been recognized as a powerful tool for modeling systems with imprecise input data, such as the decision-maker opinions. Linguistics terms are used to express the variables by means of fuzzy sets and membership function. Triangular fuzzy number (TFN) has been the most popular form to present fuzzy numbers which represented with three points; $\tilde{A} = (l, m, u)$. A model based on subjective preferences of a decision-maker is not always accurate as it demands a great deal of knowledge, expertise, and experience (Li, Yamaguchi, & Nagai, 2008). This is where "fuzzy" approaches are used in MCDM problems. Incorporating fuzzy sets in decision-making models can produce more realistic outcomes as they aim to transform uncertain human knowledge into a mathematical formula (Zadeh, 1965).

Numerous MCDM techniques have been proposed (Chai et al., 2013) to tackle the problem of supplier selection in order to rank them in the presence of qualitative data. Grey Relational Analysis (GRA), originally proposed by Dong (1982), is applied to imprecise information in the form of interval values and also aims to show the degree of similarity or difference of development trends between an alternative and the reference (ideal) alternative (Kuo & Liang, 2011). Grey system theory has shown to be able to handle both incomplete information and ambiguities (Hu, Zhang, Hu, & Jiang, 2010).

Therefore, combination of Fuzzy logic and GRA is a good choice for single sourcing when quantitative and precise data are not available; so Fuzzy GRA would allow for effective qualitative assessment.

4.5. Quantitative and multiple source Green Supplier Selection

The incorporation of environmental criteria into supplier selection practices is gaining increasing importance. By increasing GSS studies and development of its concept, the applications of GSS in companies are growing quickly. Therefore in future, measuring of many quantitative criteria is possible. Future GSS researches can consider other important environmental criteria (like Carbon Footprint, GHG emissions, Life Cycle Assessment of products, etc.) which currently have been focused on. These kinds of environmental subjects will provide more practical results in GSS, provided that suppliers trace its environmental footprint accurately.

The purchaser only needs to make one decision, which supplier is the best. However, in multiple sourcing, purchasers need to decide which are the best suppliers and how much order should be placed on each selected supplier. It is based on firm's condition and depends on manager's decision directly. Linear programming is the easiest way which helps managers to select multiple source suppliers and their allocations, otherwise by changing constraints, managers are able to select single source. Linear programming (LP; also called linear optimization) is a method to achieve the best outcome (such as maximum profit or lowest cost) in a mathematical model whose requirements are represented by linear relationships. This method was proposed by Ghodsypour and O'Brien (1998) for the first time to indicate best suppliers and their allocations in multiple sourcing models.

5. Results

5.1. Criteria collection (Phase I)

One of the most significant business decisions faced by purchasing managers in a supply chain is the selection of appropriate suppliers while trying to satisfy multiple criteria based on price, quality, customer service, and delivery (Arikan, in press). Hence, supplier selection by its nature involves the need to trade-off multiple criteria, as well as the presence of both quantitative and qualitative data. A popular area of research in the field of supply chain was establishing the criteria for supplier selection and evaluation since the 1960s. If we look into the history of popular criterion identified in the literature, we can conclude three primary categories of emphasis: in the late 1970s and early 1980s, cost was the main focus; then in the early 1990s, cycle time and customer responsiveness were considered, and, finally, in the late 1990s, the focus shifted to flexibility. Now, environmental factors are a key issue which gives rise to the new paradigm of focusing on green supply chains (Haung & Keskar, 2007).

5.1.1. General criteria

Traditionally, companies only focus on their internal environmental measures of waste reduction, pollution control, and hazardous material control to comply with existing regulations and address their social responsibilities. These days, companies have started to realize that they have to do something more than these internal controls, so they are responsible for environmental concerns about supplier selection, product distribution, etc. (Chen et al., 2013).

Many previous studies on supplier selection and evaluation defined numerous evaluation criteria and selection models for supplier selection. In consideration of the criteria for supplier selection, the pathfinder work by Dickson (1966) has been one of the most cited studies.

Weber, Current, and Benton (1991) re-examined Dickson's work by reviewing published articles between 1966 and 1990. Twenty-three distinct criteria are identified in various supplier selection problems in Dickson (1966) and Weber et al. (1991) Among the 23 criteria identified for vendor selection, the product quality was ranked as most important, it was followed by on-time delivery, performance history of supplier and warranties and claimed policies, and so on.

In a later work, Cheraghi, Dadashzadeh, and Subramanian (2004) continued to extend these key players initial work to obtain the current perspective of supplier selection by analyzing articles published between 1990 and 2001. Cheraghi et al. (2004) provided an update of Dickson's seminal work with 13 more criteria. Also, Thiruchelvam and Tookey (2011) reviewed supplier selection studies and comprise the selection criteria between two periods of 1966–2001 and 2001–2010.

Previous studies indicate that quality, price, and delivery are general (traditional) key criteria that consistently appear for vendor selection (Wu, Zhang, Wu, & Olson, 2010).

Chang, Chang, and Wu (2011) used DEMATEL to find influential factors in selecting SCM suppliers and found that "technology ability", "stable delivery of goods", "lead-time", and "production capability" criteria are more influential than other evaluation criteria. They also suggested extending the scope of this study and exploring the addition of a green supply chain in future studies.

Table 3 shows summary results of review paper on most important general criteria on supplier selection. Often each of these important criteria contains several sub-criteria which are described completely in past literature.

The past few years have led researchers to investigate the environmental concepts in management and supply chains. We comprehensively collected the literature associated with the descriptors "Green/Environmental supplier selection," "vendor selection," and

Table 3. Most important general criteria on conventional supplier selection.

Authors	The most important criteria as a result of review papers
Dickson (1966)	Quality, delivery, performance history
Weber et al. (1991)	Price, delivery, facilities and capacity, geographic location, technology capability
Cheraghi et al. (2004)	Quality, delivery, price, repair service, technical capability
Ho et al. (2010)	Quality, delivery, price/cost, manufacturing capability, service, management, technology
Thiruchelvam and Tookey (2011)	Quality, delivery, price

from academic databases including Science Direct, Emerald, Springer-Link Journals, IEEE Xplore, Academic Search Premier, and World Scientific Net. After a methodological decision analysis of all collected articles, we reviewed 59 international journal articles published to 2014. This review of literature provides the basis for identification and definition of green supplier selection criteria.

As you can see in Figure 2, most of empirical studies on green supplier selection are concentrated on electronic (42%) and automotive industries (19%).

So for exact studies on other industries, specialized consultant is necessary to choose suitable criteria. Unfortunately, agriculture and food industry is neglected from the vision of researchers. Regarding to high environmental impacts on agriculture and food industry (Pluimers, Kroeze, Bakker, Calla, & Hordijik, 2000), it is necessary to pay attention more on this field.

Based on reviewing papers, widespread set of criteria were purposed. Unfortunately, lack of consistent studies about basic green criteria caused widespread number of green criteria in each study. Jabbour and Jabbour (2009) concluded there is no standard for criteria or for use on the part of analyzed companies. Nonetheless, evolving criteria have

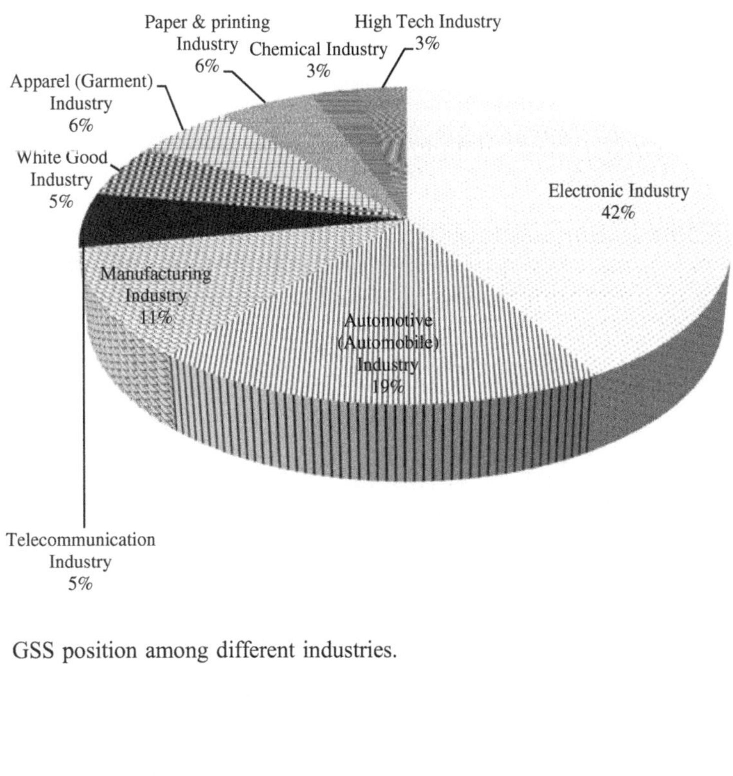

Figure 2. GSS position among different industries.

emerged due to competitive market as number of suppliers has increased over the years. Therefore, inception of new criteria is essential to narrow and qualify not only deserving but exceptional suppliers. Also under the current rapid GSCM transformation edge, reliability and flexibility of each supplier are considered as key contributing factors (Chang et al., 2011).

Govindan et al. (in press) reviewed 33 papers on green supplier and introduced relevant selection methods and criteria. The most popular approach in green supplier selection was AHP followed by ANP, DEA, also EMS was the most popular environmental criteria (Table 4).

In two comprehensive review paper (Govindan et al., in press; Nielsen, Banaeian, Golinska, Mobli, & Omid, 2014), researcher collected environmental criteria on supplier selection literature (Table 4). The most important green criteria derived from 57 review articles are: environmental management system (EMS) (20 frequencies), green image (8 frequencies), environmental competences (7 frequencies), design for environment (5 frequencies), environmental improvement costs (5 frequencies), and environmental performance (5 frequencies).

5.2. Criteria selection (Phase II)

According to Figure 1, Phase II investigate calculate the weights of criteria collected in previous section using expert team, also sub-criteria which are more important and influential in edible oil production are introduced.

5.2.1. Expert team

Due to many factors to be considered, purchasing managers might choose an easier path of evaluating based on certain important criteria and ignoring the rest. A larger number of attributes also contribute to the complications of being assigned with consistent and meaningful weights. The selection decision also becomes more difficult as the number of business rules that must be considered increases.

The Expert team is applied first to weighting criteria, second to define related sub-criteria and third for determination of most effective sub-criteria. Handfield, Walton, Sroufe, and Melnyk (2002) developed an expert team of environmental experts and noted that for better results in future studies, group should include purchasing managers who may not be directly familiar with environmental metrics. A good combination of experts will help to ensure that the tool is practical from an actual user's standpoint (Handfield et al., 2002). In order to obtain a rational framework, the model was developed through an expert team (10 people) described in Table 5. However, experts had a variety of titles, but were responsible for materials management and waste reduction activities within food processing organizations.

5.2.2. Weighting criteria

Weight of each criterion can be concluded from the paired comparison between the criteria using the AHP method, which represents the importance of criteria, is determined by means of a nine-point Likert scale.

The AHP method is often combined with Fuzzy Linear Programming in the decision-making process (Cebi & Bayraktar, 2003; Tektas & Aytekin, in press, etc.); the difference is that in relevant studies, AHP is used for estimating criteria situation in each supplier, but here, we used it for weighting criteria and tried to use quantified data for each supplier.

A questionnaire, which is based on the format proposed by Humphreys, Wong, and Chan (2003), ought to be carried out to determine the number of criteria. According to verbal judgement (Table 1), pairwise comparison of criteria are done by expert team. Finally using AHP method, criteria weights in raw material and package are calculated and shown in Table 6.

5.2.3. Sub-criteria description

The most relevant criteria are selected from reviewed papers (Cheraghi et al., 2004; Dickson, 1966; Ho et al., 2010; Thiruchelvam & Tookey, 2011; Weber et al., 1991) on supplier selection criteria (quality, price, and finally service and delivery). Main green criteria is selected based on reviewing green supplier selection papers (EMS), as (Govindan et al., in press) determined EMS as the most popular criteria on green supplier selection.

Since the lack of sufficient information about green supplier selection in food industries, expert team (Delphi group) determines sub-criteria based on the nature of industry. They also emphasized that in a practical model, measures should be relatively easy to assess, as commodity strategy development teams often do not have access to all the requisite information. Finally, each criterion is described by several sub-criteria as mentioned follow (Table 7). For example in criteria A, Capital and financial power of supplier company (A1) proposed raw material price (A2) and Transportation cost to the geographical location (A3) were the most important sub-criteria from the financial point of view.

Quality for food package material depends on durability, shelf life, interaction with content, density, etc., but based on expert opinion, the highest difference between package suppliers is providing precision and intact items to reduce leakage and waste rate. Also preserving nutritive and qualitative characters (Shelf life) is determined as quality factor for raw ingredients.

5.2.4. Conclusive criteria selection

Determining sub-criteria will help the Delphi group to decide more precision. The group was asked to score each sub-criteria for raw material and packaging supplier separately (It can be seen in Table 8). Delphi results of financial criteria shows that, capital and financial power of supplier in selection of raw ingredients and proposed price in

Table 4. Frequency of most important environmental criteria on conventional supplier selection.

	Govindan et al. (in press)	Nielsen et al. (2014)
Duration	1996–2011	1996–2014
Number of reviewed paper	33	57
Criterion		
Environmental Management System	11	20
Green Image	4	8
Environmental Competences	3	6
Design for environment	3	5
Environmental Improvement Costs	2	5
Environmental Performance	3	5

selection of packaging material were the most important sub-criteria in financial aspect. On-time delivery (lead time) and quality are most important in service and qualitative criteria in both raw ingredients and packaging materials. Mean Delphi score shows that there are some differences between raw ingredients and packaging materials from the aspect of EMS criteria. Environmental prerequisite and environmental friendly material are the most important for the raw ingredients and packaging materials, respectively. The greenest concerning subject in choosing packaging material is easily recycling capability; but in the food ingredient, training special courses for staff training are the most important criteria. For suppliers claiming to be green, the most tangible activity is awarding at least one environmental certificate or labels (like ISO14000, eco-friendly label and carbon footprint label, etc.), so we consider this term as first provision for green supplier selection.

5.3. Supplier selection (Phase III)

After identification of criteria and their weight (the main goal of this paper), decision-makers have to decide about supplier selection methodology. According to Sections 4.3 and 4.4, methods are different and should select based on company requirements and data characteristics.

5.3.1. Ranking and supplier selection by Fuzzy GRA

Fuzzy GRA works as follows. Assume that m potential suppliers are determined, n selected criteria (here based on the previous section, four selected sub-criteria in Table 8), and k decision-makers (here based on Table 5, 10 decision-makers). According to Table 9, the appropriate linguistic ratings can determine for suppliers with respect to criteria ($\tilde{x}_{ij}$) as TFN. It means that each decision-maker has to rank importance degree of n criteria in m suppliers (Overall $n \times m \times k$ linguistic variables) for raw material and packaging supplier separately.

Table 5. Summary of Delphi group.

Expert occupation	Experience	Total expert
Active member of Food Industry Association (president)	Professional and academic	4
Purchase manager in food industry	Professional and academic	4
Environmental technician	Professional and academic	2

Table 6. Criteria weight in raw material and package.

	Raw material	Package
Financial (w_1)	0.04	0.09
Service (w_2)	0.21	0.12
Quantitative (w_3)	0.65	0.51
EMS (w_4)	0.10	0.28

Table 7. Sub-criteria description based on expert team opinion.

Criteria	Sub-criteria
A- Financial	A1- Capital and financial power of supplier company A2- Proposed raw material price A3- Transportation cost to the geographical location (Availability)
B- Delivery & service	B1- Communication System (willing to trade, attitudes, acceptance of procedures and flexibility) B2- On time delivery (lead time) B3- After sales service (policy, quality assurance, and damage ratings) B4- Production capacity
C- Qualitative	C1- Quality (the supplier ability to access the quality characteristics) C2- Operational control (reporting, quality control, inventory control, research and development) C3- Expert labor, technical capabilities and facilities C4- Business experience and Position among competitors
D- Environmental management system	D1- Environmental prerequisite (environmental staff training) D2- Environmental planning (program to reduce environmental impacts, green research and development) D3- Environmental friendly material (low waste: easily recycling and reuse capability) D4- Environmental friendly technology (emission of pollutant: CO_2 equivalent, VOC, BOD and COD content and etc.)

Table 8. Mean Delphi score of experts.

		Delphi score	
Criteria	Sub-criteria	Raw ingredients	Packaging material
A	A1	24.5	21.5
	A2	22.5	23.5
	A3	21	20.5
B	B1	19.8	18.8
	B2	27.2	26.6
	B3	19.5	21.5
	B4	20.5	22.7
C	C1	30.9	18.9
	C2	11.3	11.2
	C3	14.6	16.4
	C4	15.4	17.3
D	D1	6.11	9
	D2	5.5	7.3
	D3	4.4	12.3
	D4	5.8	8.3

In the initial step of the selected multi-criteria supplier selection methods, aggregated ratings of suppliers (decision matrix) are constructed. The aggregated fuzzy rating $\tilde{r}_{ij}$ of alternative A_i under criterion C_j is:

$$\tilde{x}_{ij} = \frac{l}{k}\left[\tilde{x}_{ij}^{1} + \tilde{x}_{ij}^{2} + \cdots + \tilde{x}_{ij}^{k}\right] \tag{4}$$

The aggregated decision matrix can then be constructed as follows:

$$\tilde{X} = \begin{matrix} A_1 \\ A_2 \\ \vdots \\ A_m \end{matrix} \begin{bmatrix} \tilde{x}_{11} & \tilde{x}_{12} & \cdots & \tilde{x}_{1n} \\ \tilde{x}_{21} & \tilde{x}_{22} & \cdots & \tilde{x}_{2n} \\ \vdots & \vdots & \vdots & \vdots \\ \tilde{x}_{m1} & \tilde{x}_{m2} & \cdots & \tilde{x}_{mn} \end{bmatrix}; \quad (5)$$

Decision matrix ($\tilde{X}$) is converted into the normalized decision matrix $\tilde{R}$. Given $\tilde{r}_{ij} = (l_{ij}, m_{ij}, u_{ij})$, the normalized performance rating can be calculated as (Gumus, Yayla, Celik, & Yildiz, 2013; Zhang & Liu, 2011):

$$\tilde{r}_{ij} = \left(l_{ij}/u_j^+, m_{ij}/u_j^+, u_{ij}/u_j^+\right), i = 1, \ldots, m, \quad j = 1, \ldots, n \quad (6)$$

$$\text{where} \quad u_j^+ = \max_i\{u_{ij}\} \quad \forall ii = 1, \ldots, m \quad (7)$$

The reference series can be determined from (Gumus et al., 2013):

$$\tilde{R}_0 = [\tilde{r}_{01}, \tilde{r}_{02}, \tilde{r}_{03}, \tilde{r}_{04}] \text{ where } \tilde{r}_{0j} = \max(\tilde{r}_{ij})\, j = 1, \ldots, 4 \quad (8)$$

The distance matrix is then established. The distance δ_{ij} between the reference value and each comparison value is computed from Equation (7) of TFNs. The distance between $\tilde{A}$ and $\tilde{B}$ is defined as follows (Krohling & Campanharo, 2011; Zhang & Liu, 2011):

$$d(\tilde{A}, \tilde{B}) = \sqrt{\frac{1}{3}\left[(l_1 - l_2)^2 + (m_1 + m_2)^2 + (u_1 - u_2)^2\right]} \quad (9)$$

The grey relational coefficient (ξ_{ij}) is defined as (Gumus et al., 2013):

$$\xi_{ij} = \frac{\delta_{\min} + \rho\delta_{\max}}{\delta_{ij} + \rho\delta_{\max}}, \ \delta_{\max} = \max(\delta_{ij}), \ \delta_{\min} = \min(\delta_{ij}) \text{ and } \rho \text{ resolving coefficient } \rho \in [0, 1] \quad (10)$$

Defuzzification formula (Definition 4) is applied to $\tilde{w}_j$; then the grey relational grade γ_i is estimated by the relation (Gumus et al., 2013):

$$\gamma_i = \sum_{j=1}^{n} w_j \xi_{ij}, \quad i = 1, \ldots, m \quad (11)$$

Table 9. Linguistic variables for supplier ratings.

Number	Linguistic variables (degree of importance)	TFNs
1	Very Poor (VP)	(0, 1, 2)
2	Poor (P)	(1, 2, 3)
3	Medium Poor (MP)	(2, 3.5, 5)
4	Fair (F)	(4, 5, 6)
5	Medium Good (MG)	(5, 6.5, 8)
6	Good (G)	(7, 8, 9)
7	Very Good (VG)	(8, 9, 10)

where w_j is the weight of the jth criterion, $\sum_{j=1}^{n} w_j = 1$ which are presented in Table 5. The alternatives can now be ranked in accordance with the value of grey relational grade. The bigger the value, the better would be the alternatives.

5.3.2. *Supplier selection and allocation by programming*

An environmental objective function was added to the proposed model by Ghodsypour and O'brien (2001). Thus, our model is a multi-objective model considering four objectives of financial, quantity, service, and EMS issues for green supplier selection in food industry. The presented model can select the best suppliers and allocate the orders among them optimally too. Consequently, one objective function and five constrains are considered to calculate Total Purchase Value (TPV). The elements of model and constraints are presented below:

All suppliers have limited capacities and minimum order quantity (13, 9).

Total order quantity must be equal the amount of demand (14).

Summation of supplier's capacity must be above the amount of demand (15).

Since managerial matter, also there is a limitation on number of selected suppliers (16).

The parameters of the proposed model are defined as follows:

$$\text{Max TPV} = \sum_{i=1}^{n} \left[Y_i Z_i \left(w_1 F_i^* + w_2 S_i + w_3 Q_i^* + w_4 E_i^*\right)\right] * X_i \tag{12}$$

$$ST : X_i \le C_i Y_i i = 1, 2, \ldots, n \tag{13}$$

$$\sum_{i=1}^{n} X_i = D \tag{14}$$

$$\sum_{i=1}^{n} C_i \ge D \tag{15}$$

$$n_{\min} \le \sum_{i=1}^{n} Y_i \le n_{\max} \tag{16}$$

$$X_i \ge Q_{i\min} Y_i i = 1, 2, \ldots, n \tag{17}$$

Base on criteria selection process, the model includes four objective functions to optimize the four criteria for green supplier selection: financial, service, qualitative, and EMS of suppliers are expressed as follows:

$$F_i^* : \begin{cases} \text{for raw ingredients} = \frac{F_i}{\text{MaxF}}; F_i : \text{Capital of supplier} \\ \text{for packaging materials} = \frac{\text{MinF}}{F_i}; F_i : \text{Proposed price by supplier } i \end{cases} \tag{18}$$

$$E_i^* : \begin{cases} \text{for raw ingredients} = \dfrac{\text{Number of trained staff}}{\text{Number of total staff}} \\ \text{for packaging material} = \dfrac{\text{Amount of collected package from consumer}}{\text{Total amount of production}} \end{cases} \tag{19}$$

$$Q_i^* : \begin{cases} \text{for raw ingredients} = \frac{Q_i}{\text{Max } Q}; Q_i = \text{Shelf life of items provided by supplier } i \\ \text{for packaging material} = \text{percent of intact items delivered by supplier } i \end{cases}$$

$S_i = \frac{S_i}{\text{Max}} S$; S_i: Lead time (day) items provided by the supplier i

n: Number of suppliers

D: The amount of demand for the period

X_i: Order quantity for the supplier i

$Q_{i\ \min}$: Minimum order quantity of the ith supplier

C_i: The capacity of the supplier i

Y_i: 1 if the supplier is selected; 0 otherwise

Z_i: 1 if the supplier award environmental certificate; 0 otherwise

$W_{1,2,3,4}$: Represent the weight of each criterion using the AHP method as described completely in previous section which is different for raw ingredients and packaging materials.

6. Conclusion

GSS literature reviewed in this paper showed that most of papers are focused on automotive, electronics, and manufacturing industries. The principal benefit of this paper is to provide an extended GSS guideline (from the definition of criteria to GSS) to help decision-makers in various industries often neglected in previous GSS studies.

Food and agricultural products have some unique characteristics related to public health, which has not been properly investigated in GSS literature. Green purchase is an inclusive concept and people may attach different meanings to it, especially among multifarious number of environmental management actions in food industry. So in this study, an expert team tried to define easy understanding criteria and sub-criteria and also select easy access ones in a food company.

As the main goal, this study tried to show the single platform in order to choice criteria and sub-criteria in GSS and calculate the weights of criteria, applied in an edible oil company. In addition, an example of different ways to GSS (single and multiple sourcing) is also briefly illustrated.

In this paper, general and environmental criteria are systematically selected: general and green supplier selection criteria are reviewed comprehensively, and relevant criteria for green supplier selection are selected. In order to obtain a rational framework, the model was developed through an expert team (including Active members of Food Industry Association, purchase managers and environmental technicians) to find most suitable criteria and their weights. The weights of relevant criteria using AHP are determined, also branches for each relevant criterion and selecting the most effective branch based on Delphi group opinion in raw material and package suppliers are calculated.

In the section of supplier selection, two kinds of qualitative and quantitative data are discussed when single and multiple sourcing are required, respectively. Fuzzy GRA is proposed to ranking suppliers in presence of qualitative and imprecise data. Also programming is used to present a model which can select the best suppliers and allocate the orders among them optimally.

There are certain limitations for this study that can provide opportunities for further research in this area. While it is believed that the presented model provides operational value in criteria selection, there are also further points in supplier selection that can be included. Application of single and multiple sourcing of GSS (phase III) will be study in future papers. Additional research is required to identify and define more clearly specific criteria in different products and industries.

Acknowledgments

The financial support provided by University of Tehran is duly gratefully acknowledged. Thanks and appreciations of authors also go to colleagues in developing the research and people who have willingly helped us with their abilities.

References

Aissaoui, N., Haouari, M., & Hassini, E. (2007). Supplier selection and order lot sizing modeling: A review. *Computers & Operations Research, 34*, 3516–3540.

Apaiaha, R. K., Linnemanna, A. R., & van der Kooi, H. J. (2006). Exergy analysis: A tool to study the sustainability of food supply chains. *Food Research International, 39*(1), 1–11.

Arikan, F. (in press). An interactive solution approach for multiple objective supplier selection problem with fuzzy parameters. *Journal of Intelligent Manufacturing*. doi:10.1007/s10845-013-0782-6

Banaeian, N., Nielsen, I. E., Mobli, H., & Omid, M. (2014). Green supplier selection in edible oil production by a hybrid model using Delphi method and Green Data Envelopment Analysis (GDEA). *Management and Production Engineering Review, 5*, 3–8.

Bocewicz, G., Wójcik, R., & Banaszak, Z. (2012). Cyclic scheduling for supply chain network. *Trends in practical applications of agents and multiagent systems* (pp. 39–47). Springer Berlin Heidelberg.

Bosona, T., & Gebresenbet, G. (2013). Food traceability as an integral part of logistics management in food and agricultural supply chain. *Food control, 33*, 32–48.

Cebi, F., & Bayraktar, D. (2003). An integrated approach for supplier selection. *Logistics Information Management, 16*, 395–400.

Chai, J., Liu, J. N. K., & Ngai, E. W. T. (2013). Application of decision-making techniques in supplier selection: A systematic review of literature. *Expert Systems with Applications, 40*, 3872–3885.

Chang, B., Chang, C.-W., & Wu, C.-H. (2011). Fuzzy DEMATEL method for developing supplier selection criteria. *Expert Systems with Applications, 38*, 1850–1858.

Chen, Y., Simon, R., Reich-Weiser, C., & Woo, J. (2013). Green supply chain. In D. A. Donrfeld (Ed.), *Green manufacturing: Fundamental and applications* (pp. 83–105). Springer.

Cheraghi, S. H., Dadashzadeh, M., & Subramanian, M. (2004). Critical success factors for supplier selection: An update. *Journal of Applied Business Research, 20*, 91–108.

Dickson, G. W. (1966). An analysis of vendor selection system and decisions. *Journal of Purchasing, 2*, 5–17.

Dong, J. L. (1982). Control problems of grey systems. *Systems and Controls Letters, 5* 288–294.

EU. (2002). Regulation (EC) no 178/2002 of the European parliment and of the council of 28 January 2002. *Official Journal of the European Commiunities, L31*, 1–24.

Gendron, C., & Audet, R. (2012). Key drivers of the food chain. In J. L. Boye & Y. Arcand (Eds.), *Green technologies in food production and processing* (pp. 23–39). Food Engineering Series. New York, NY: Springer.

Ghodsypour, S. H., & O'brien, C. (2001). The total cost of logistics in supplier selection, under conditions of multiple sourcing, multiple criteria and capacity constraints. *International Journal of Production Economics, 73*, 15–27.

Ghodsypour, S. H., & O'Brien, C. (1998). A decision support system for supplier selection using an integrated analytic hierarchy process and linear programming. *International Journal of Production Economics, 56–57*, 199–212.

Govindan, K., Rajendran, S., Sarkis, J., & Murugesan, P. (in press). Multi criteria decision making approaches for green supplier evaluation and selection: A literature review. *Journal of Cleaner Production*. doi:10.1016/j.jclepro.2013.06.046

Grunert, K. G., Hieke, S., & Wills, J. (2014). Sustainability labels on food products: Consumer motivation, understanding and use. *Food Policy, 44*, 177–189.

Gumus, A. T., Yayla, A. Y., Celik, E., & Yildiz, A. (2013). A combined fuzzy-AHP and Fuzzy-GRA methodology hydrogen energy storage method selection in Turkey. *Energies, 6*, 3017–3032.

Handfield, R., Walton, S. V., Sroufe, R., & Melnyk, S. A. (2002). Applying environmental criteria to supplier assessment: A study in the application of the analytical hierarchy process. *European Journal of Operational Research, 141*, 70–87.

Harland, C. M. (1996). Supply chain management, purchasing and supply management, logistics, vertical integration, materials management and supply chain dynamics. *Blackwell encyclopedic dictionary of operations management*. Oxford, UK: Blackwell.

Haung, S. H., & Keskar, H. (2007). Comprehensive and configurable metrics for supplier selection. *International Journal of Production Economics, 105*, 510–523.

Ho, W., Xu, X., & Dey, P. K. (2010). Multi-criteria decision making approaches for supplier evaluation and selection: A literature review. *European Journal of Operational Research, 202*, 16–24.

Hu, L., Zhang, C., Hu, C., & Jiang, G. (2010). Use of grey system for assessment of drinking water quality: A case study of Jiaozuo City, China. In S. Liu & J. Y. L. Forrest (Eds.), *Advance in Grey Systems* (pp. 469–480). Berlin Heidelberg, Germany: Springer.

Humphreys, P. K., Wong, Y. K., & Chan, F. T. S. (2003). Integrating environmental criteria into the supplier selection process. *Journal of Materials Processing Technology, 138*, 349–356.

Ilbery, B., & Maye, D. (2005). Food supply chains and sustainability: Evidence from specialist food producers in the Scottish/English borders. *Land Use Policy, 22*, 331–344.

Jabbour, A. B. L., & Jabbour, C. J. (2009). Are supplie selection criteria going green? Case studies of companies in Brazil. *Industrial Managemant & Data Systems, 109*, 477–495.

Krohling, R. A., & Campanharo, V. C. (2011). Fuzzy TOPSIS for group decision making: A case study for accidents with oil spill in the sea. *Expert Systems with Applications, 38*, 4190–4197.

Kumar, S., Hong, Q. S., & Haggerty, L. N. (2011). A global supplier selection process for food packaging. *Journal of Manufacturing Technology Management, 22*, 241–260.

Kuo, M. S., & Liang, G. S. (2011). Combining VIKOR with GRA techniques to evaluate service quality of airports under fuzzy environment. *Expert Systems with Applications, 38*, 1304–1312.

Li, G. D., Yamaguchi, D., & Nagai, M. A. (2008). A grey-based rough decision-making approach to supplier selection. *The International Journal of Advanced Manufacturing Technology, 36*, 1032–1040.

Liao, C. N., Fu, Y. K., Chen, Y. C., & Chih, L. (2012). Applying Fuzzy-MSGP approach for supplier evaluation and selection in food industry. *African Journal of Agricultural Research, 5*, 726–740.

Listone, H. A., & Turoff, M. (2002). *The Delphi Method: Techniques and Applications. College of Computing Sciences*. Newark, NJ: New Jersey Institute of Technology.

Losito, P., Visciano, P., Genualdo, M., & Cardone, G. (2011). Food supplier qualification by an italian large-scale-Distributor: Auditing system and non-conformances. *Food Control, 22*, 2047–2051.

Magdalena, R. (2012). Supplier selection for food industry: A combination of Taguchi loss function and fuzzy analytical hierarchy process. *The Asian Journal of Technology Management V, 5*, 13–22.

Nielsen, I. E., Banaeian, N., Golinska, P., Mobli, H., & Omid, M. (2014). Green supplier selection criteria: From a literature review to a flexible framework for determination of suitable criteria. In P. Golinska (Ed.), *Logistic operations, supply chain management and sustainability* (pp. 79–100). Springer.

Olugu, E. U., Wong, K. Y., & Shaharoun, A. M. (2011). Development of key performance measures for the automobile green supply chain. *Resource, Conservation and Recycling, 55*, 567–579.

Pluimers, J. C., Kroeze, C., Bakker, E. J., Calla, H., & Hordijik, L. (2000). Quantifying the environmental impact of production in agriculture and horticulture in The Netherlands: Which emissions do we need to consider? *Agricultural Systems, 66*, 167–189.

Prusak, A., Stefanów, P., Niewczas, M., & Sikora, T. (2013). Application of the AHP in evaluation and selection of suppliers, 57th EOQ congress quality renaissance – co-creating a viable future, June 17–20, 2013, Tallinn, Estonia.

Prusty, S. K., Pratap, K. J., Mohapatra, C. K., & Mukherjee (2010). GOS tree (Goal–Objective–Strategy tree) approach to strategic planning using a fuzzy-Delphi process: An application to the Indian Shrimp Industry. *Technological Forecasting and Social Change, 77*, 442–456.

Saaty, T. L. (1990). How to make a decision: The analytic hierarchy process. *European Journal of Operational Research, 48*, 9–26.

Schoenfeldt, T. I. (2008). Make or buy: The first step in supply chain management. In P. O'Mara (Ed.), *A practical application of supply chain management principles* (pp. 53–59). Milwaukee, WI: American Society for Quality Press.

Shen, J. L., Liu, Y. M., & Tzeng, Y. L. (2012). The cluster-weighted DEMATEL with ANP method for supplier selection in food industry. *Journal of Advanced Computational Intelligence and Intelligent Informatics, 16*, 567–575.

Shen, L., Olfat, L., Govindan, K., Khodaverdi, R., & Diabat, A. (2013). A fuzzy multi criteria approach for evaluating green supplier's performance in green supply chain with linguistic preferences. *Resources, Conservation and Recycling, 74*, 170–179.

Sirvastava, S. K. (2007). Green supply-chain management: A state-of-the-art literature review. *International Journal of Management Reviews, 9*, 53–80.

Sitek, P., & Wikarek, J. (2014). Hybrid solution framework for supply chain problems. Distributed computing and artificial intelligence (DCAI 2014). *Book Series: Advances in Intelligent Systems and Computing, 290*, 11–18.

Tektas, A., & Aytekin, A. (in press). Supplier selection in the international environment: A comparative case of a Turkish and an Australian company. *IBIMA Publishing*. doi:10.5171/2011.598845

Thiruchelvam, S., & Tookey, J. E. (2011). Evolving trends of supplier selection criteria and methods. *International Journal of Automotive and Mechanical Engineering, 4*, 437–454.

Toloo, M. (2014). Selecting and full ranking suppliers with imprecise data: A new DEA method. *The International Journal of Advanced Manufacturing Technology, 74*, 1141–1148.

Vachon, S., & Klassen, R. D. (2006). Green project partnership in the supply chain: The case of the package printing industry. *Journal of Cleaner Production, 14*, 661–671.

van der Vorst, J. G. A. J., van Dijk, S. J., & Beulens, A. J. M. (2001). Supply chain design in the food industry. *The International Journal of Logistics Management, 12*, 73–86.

Weber, C., Current, J., & Benton, W. (1991). Vendor selection criteria and methods. *European Journal of Operation Research, 50*, 2–18.

Wikipedia contributors, S. C. M. (2014). Wikipedia, The Free Encyclopedia. Retrieved 30 June, 2014, from http://en.wikipedia.org/wiki/Supply_chain_management#cite_note-11

Wu, D. D., Zhang, Y., Wu, D., & Olson, D. L. (2010). Fuzzy multi-objective programming for supplier selection and risk modeling: A possibility approach. *European Journal of Operational Research, 200*, 774–787.

Zadeh, L. A. (1965). Fuzzysets. *Inform Control, 8*, 338–353.

Zhang, S. F., & Liu, S. Y. (2011). A GRA-based intuitionistic fuzzy multi-criteria group decision making method for personnel selection. *Expert Systems with Applications, 38*, 11401–11405.

12

Yogurt production line: reliability analysis

Panagiotis H. Tsarouhas[a]* and Ioannis S. Arvanitoyannis[b]

[a]Department of Standardization & Transportation of Products – Logistics, Technological Educational Institute of Central Macedonia, Kanellopoulou 2, Katerini 60100, Greece; [b]School of Agricultural Sciences, University of Thessaly, Fytokou Street, 38446 Nea Ionia Magnesias, Volos, Greece

A reliability analysis of an automated yogurt production line was carried out for all machines and the entire line level. Trend and serial correlation tests validated the assumption of independence and identical distribution of the failure data was conducted. The determination of the theoretical distribution parameters that had the best fit to the failure data was made. Furthermore, the reliability and failure rate models for the entire production line were developed. The models could prove to be a useful tool, both to assess the current conditions and to predict the reliability for upgrading the operations management policies of the yogurt line. It was found that: (a) the Weibull distribution provided the best fit for the yogurt production line to describe the time between failures, (b) the failure rate of the production line increased implying that the currently applied maintenance strategy is not adequate and must be upgraded in the near future, and (c) to avoid losses related to quality and productivity, the reliability must be improved initially on the pasteurizing boiler, and secondly on the filling machine.

Keywords: yogurt production line; reliability models; performance evaluation; quality; operations management

Introduction

The dairy sector is one of the most important in terms of economic growth within the food market. Manufacturers are focusing on applied reliability analysis of their production lines to increase the productivity and the efficiency for improving competitiveness. Although ensuring an adequate reliability level is of great importance for industrial applications on food production lines, reliability itself is quite complex and there are numerous interlocking variables to be taken into account when one has to evaluate and achieve various reliability levels. Inadequate reliability leads to several problems such as high maintenance costs, unsafe conditions, and eventual downgrading of the quality of the products. It takes a long time for a company to establish a reputation for reliability but only a short time for its reputation to be ruined (Croarkin & Tobias, 2005).

A number of methods, all with reference to reliability, availability, maintainability, and safety analysis, have been developed and can make a significant contribution to the performance improvement of both industrial and non-industrial complex systems (Regattieri, Manzini, & Battini, 2010). The literature includes a huge number of interesting methods, linked for example to preventive maintenance models. These

*Corresponding author. Email: ptsarouh@logistics.teithe.gr

models can determine the best frequency of maintenance actions, or the optimization of spare parts consumption or the best management of their operating costs (Manzini, Regattieri, Pham, & Ferrari, 2009).

In this study, a reliability analysis was conducted for the yogurt production line. The best fit of the failure data among the common theoretical distributions was found and the corresponding parameters were identified. Moreover, the reliability and hazard rate models for the entire production line were calculated. The latter can potentially be useful tools to assess the current situation, and, more importantly, to predict reliability for upgrading the operation management of the yogurt production line and other relevant production lines.

The rest of this paper is organized as follows: (a) in Section 2, the literature review was presented, (b) in Section 3, the yogurt production system was described, (c) the research methodology was presented in Section 4, (d) in Section 5, the results with the reliability analysis were reported, (e) in Section 6, the results of the case study are discussed, and finally, (f) the conclusions were drawn in Section 7.

Literature review

There is a vast literature on reliability, maintainability, and maintenance, including many articles and books (Blischke & Murthy, 2003; Gupta & Bhattacharya, 2007; Jalali & Forouhandeh, 2011; Usher, Roy, & Parsaei, 1998). Wang, Loman, and Vassiliou (2004) reported that one of the main purposes of reliability and availability analysis is to identify the weaknesses of the system and to quantify the potential impact of failure. Reliability and availability are appropriate approaches for measuring product quality, product performance, and cost effectiveness of the product itself (Goetz & Villa, 1998). Tsarouhas (2012) reviewed reliability, availability, and maintainability analysis in the food industries, and identified the critical points of the production systems that should improve the operational performance and the maintenance effectiveness. Hajeeh and Chaudhuri (2000) worked on reliability and availability assessment of reverse osmosis by analyzing its failure behavior and downtime patterns. Markeset and Kumar (2001) discussed the application of reliability, maintainability, and risk analysis methods to minimize the life cycle cost of the system. Kiureghian, Ditlevson, and Song (2007) analyzed the availability, reliability, and downtime of systems with repairable components. Singh and Swaminathan (2002) reported a methodology for determining sample sizes for system availability.

The serial configuration is the most commonly encountered reliability block diagram in engineering practice i.e. industrial systems, production lines, etc. In a serial configuration, all the consisting components of the system should be operating to maintain the required operation of the system (Blischke & Murthy, 2000; Dhillon, 2006; Meeker & Escobar, 1998). Thus, failure of any one component of the system will cause failure of the whole system (Tsarouhas, 2010). The system reliability in a series configuration is less than the reliabilities of each component (Stapelberg, 2009). Kardon and Fredendall (2002) developed a model that allows consideration of not only the total maintenance cost but also the overall probability of a serial system breakdown when determining the time intervals between preventive maintenance activities. Ruan and Sun (2006) presented an exact method for cost minimization problems in series reliability systems with multiple component choices.

The available literature referring to the food industry is quite limited. Tsarouhas, Varzakas, and Arvanitoyannis (2009) and Tsarouhas, Arvanitoyannis, and Varzakas

(2009) developed reliability and maintainability analysis for strudel and feta cheese production line at machine, workstation, and entire line level. Moreover, Tsarouhas, Arvanitoyannis, and Ampatzis (2009) investigated the reliability and maintainability in a Greek juice bottling medium size enterprise. Wang, Hu, and Li (2010) presented a study to determine the buffer capacity in dairy filling and packing lines through transient analysis. Tsarouhas and Arvanitoyannis (2010a) assessed the operation management of a beer-packaging line based on field failure data. Furthermore, Tsarouhas and Arvanitoyannis (2010b) analyzed the reliability and maintainability of a bread-production line. Tsarouhas and Arvanitoyannis (2011) conducted quantitative analysis on the management of a peach-canning production line. They pointed out that the failure times follow the logistic distribution whereas the repair times comply with the Weibull distribution, and the repair rate of failure increases, thereby implying that the maintenance staff expertise increases with time. In a recent study, Tsarouhas and Arvanitoyannis (2012) carried out reliability and maintainability analysis on a limoncello production line by applying statistical techniques on field failure data.

Description of the system

Yogurt production process

The yogurt production process begins with receiving milk from local or other cooperating regional producers and its transportation in a tank (see Figure 1). The milk is weighed and then transferred to an auto-cooling storage tank (M1). The entire process occurs as a continuous flow system. A pre-heating stage follows where the milk is pasteurized within few seconds at 63–65 °C in the heat exchanger (M2) and then moves to a separator (M3) to standardize the fat content using centrifugation. Then, the milk moves to a pasteurizing boiler (M4) – production tank – where it is boiled to kill any contaminating micro-organisms. At the same time, yogurt rennet is added to speed up the coagulation. The filling of plastic containers (cups) in the filler machine (M6) with milk-yogurt takes place using a filling pump (M5). The last step for yogurt production is the ripening-cooling where the filled cups move to the ripening/incubation chamber for three hrs and then to the rapid cooling chamber for two more hours. Once the production run is completed, the cleaning of all machines involved starts automatically using *cleaning in place*-CIP (M7). The latter lasts for 5–7 min.

Operation management of the yogurt production line

The yogurt production line contains seven machines in series. In a serial configuration, all machines of the production line should be functioning to maintain the proper operation of the system. Thus, the failure of any machine may potentially cause failure in the entire system. Failure and repair data for the yogurt production line were collected from the files of the maintenance staff (mechanical and electrical) at the end of each shift. The records included the failures occurring per shift, the action taken to repair the failure, the downtime, and the exact time of failure. Apart from the failures occurring, the line stopped whenever alarm signals (related to quality and/or potential safety problem) were recorded. Therefore, the times related to these activities were also included in the failure data.

The failure data for the yogurt production line were also collected from the files of the technical department at the end of each shift, and a total of 759 failures were recorded. The line operates in one eight-hour shift each workday to include 6 h for

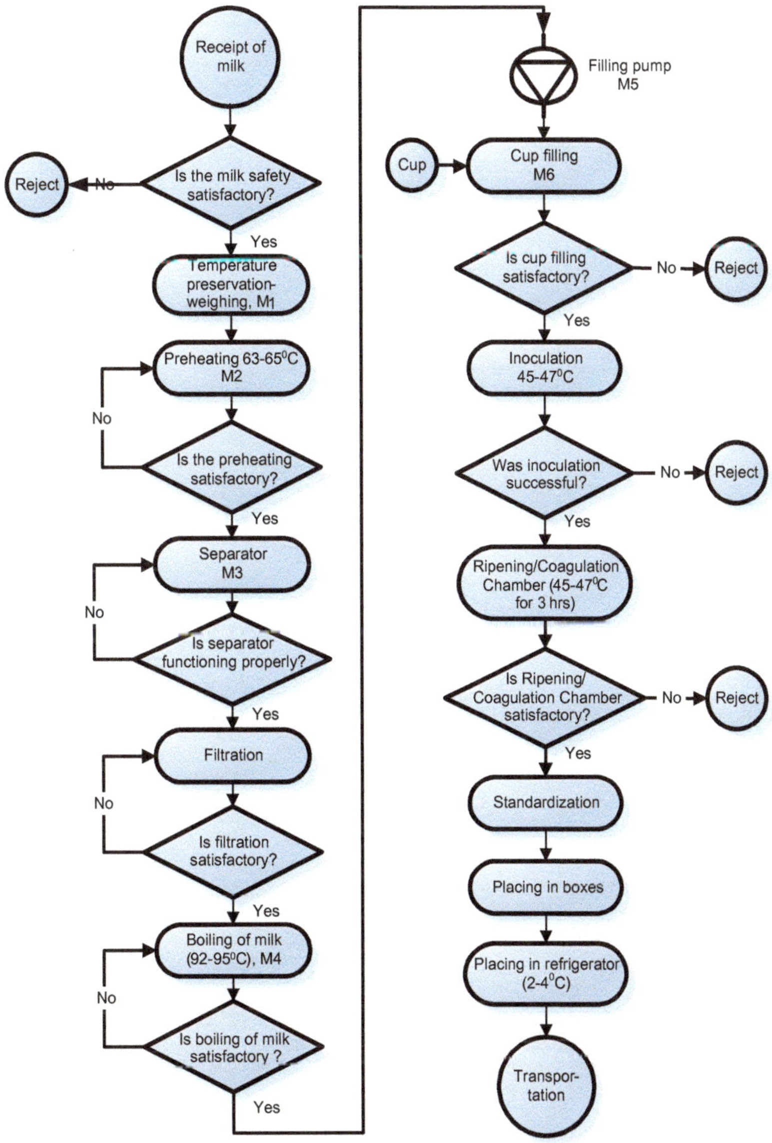

Figure 1. Flow diagram of yogurt production line where CIP (M7) is applied for cleaning purposes to all machines involved in this process.

yogurt production. For the rest of the time (2 h), the line is cleaned and prepared for the following day (i.e. raw materials, packing materials, etc.).

The currently applied maintenance policy of the yogurt line is both corrective (unscheduled and carried out in case of failure) and preventive maintenance (scheduled and performed periodically); this procedure requires immediate action by the maintenance staff to restore the machine into an operational state. This maintenance policy may include any or all the following steps: identification, localization and diagnosis of faults, correction actions (disassemble, remove, replace, reassemble, and adjust), and operation checkout. The corrective maintenance is applied while the line is running.

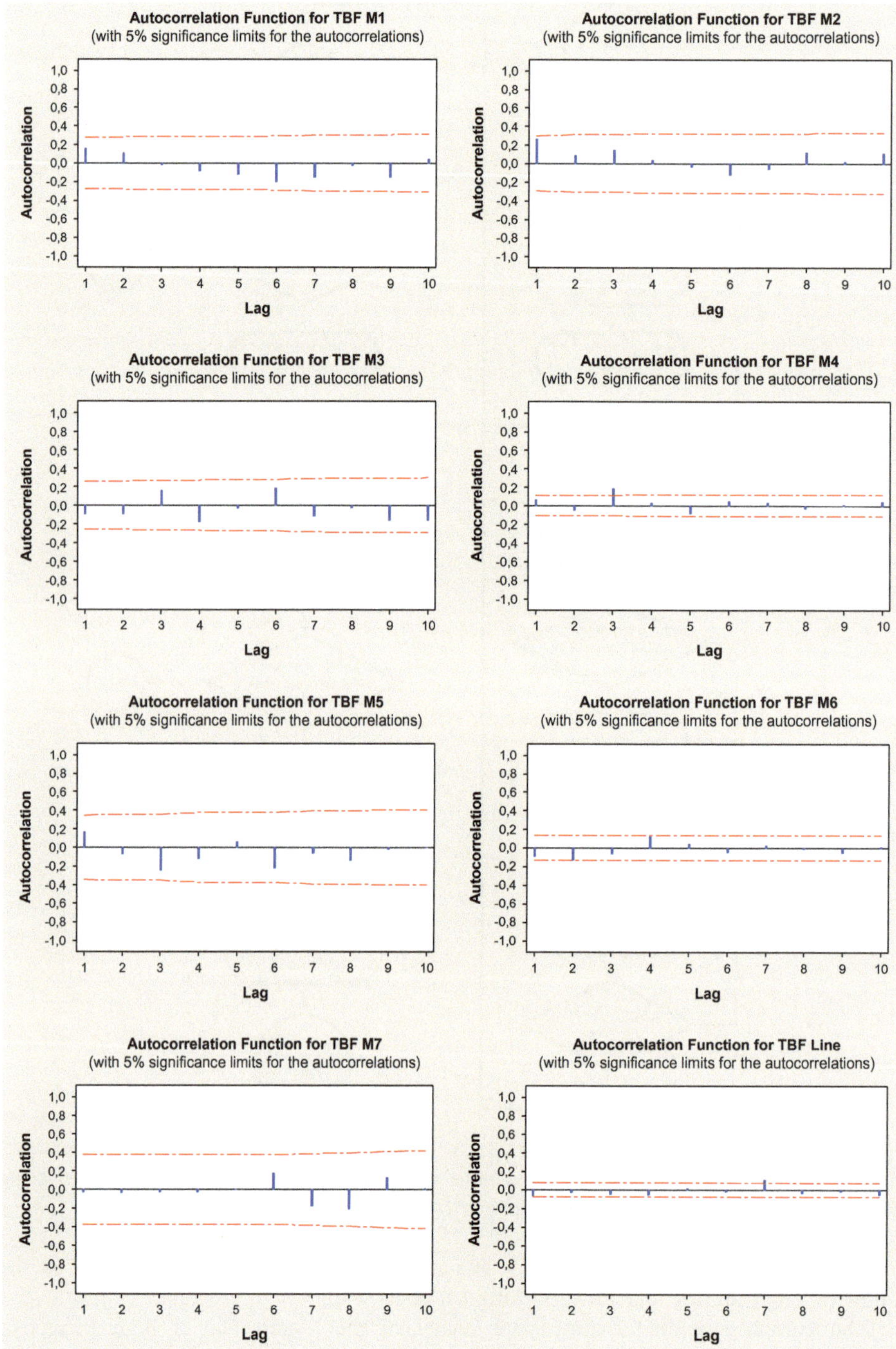

Figure 2. Correlation diagrams for the TBFs at machine and line level for yogurt production line.

Methodology

Failures data i.e. *time between failures* (TBFs) are the backbone of reliability studies because they provide invaluable information to concerned professionals such as reliability engineers, design engineers, and managers (Dhillon, 1999). Thus, the statistical analysis to describe the basic features of the failure data for TBFs for all the machines and the entire production line were analyzed with respect to theoretical distributions. The maximum likelihood and goodness-of-fit test was used to determine the best theoretical distribution to represent each machine and line failure data. The TBF of equipment is defined as the time that elapses from the moment the equipment is turned on and starts operating after a failure, until the moment it goes down again and stops operation due to a new failure. The TBFs are recorded in minutes. A quantitative analysis (i.e. descriptive statistics) of the failure data for the production line was obtained. The descriptive statistics have been done using the software package MINITAB. Thus, it is possible to extract the minimum and the maximum value of the sample, mean, standard deviation (StDev), coefficient of variation (CoefVar), skewness, and kurtosis of the failure data at machine and line level. The StDev of the random variable is defined as the square root of the variance, and is often used in place of the variance to describe the distribution spread. Since the CoefVar of a random variable is defined as the ratio of the StDev over the mean of the random variable, it is a dimensionless measure of the variability of the random variable. Skewness and kurtosis are statistics that characterize the shape and symmetry of the distribution. Skewness is a measure of the degree of asymmetry of a distribution while kurtosis is a measure of whether the data appear as peaks or are flat. A normal distribution will have kurtosis and skewness values equal to zero.

After collection, sorting, and classification of the 'data', the validation of the assumption for independent and identically distributed (IID) nature of the TBF data of each machine and the entire line must be checked. The validations of the trend for the TBF at machine and line level are displayed in Table 2. Moreover, the correlation of the failure data is shown in Figure 2.

Results

Reliability analysis

Reliability is the probability that the machine or the entire line will perform a required function, under specific operating conditions, for a given period of time t. The TBF of the line is defined with T (or machine); $T \geq 0$, then the reliability can be expressed as, $R(t) = P(T \geq t)$. Moreover, the unreliability function, $Q(t)$, is defined as the probability of failure in t, $Q(t) = 1 - R(t) = P(T \leq t)$. In reliability theory, the hazard or failure rate function is indicated as $\lambda(t) = f(t)/R(t)$, where $f(t)$ is the probability density function (PDF) of the failure distribution.

The yogurt production line for TBF follows the Weibull distribution (see Table 3), and it is fair to indicate T as the continuous random variable representing the failure time. The PDF of the Weibull distribution is:

$$f(t;\beta,\theta) = \frac{\beta}{\theta}\left(\frac{t}{\theta}\right)^{\beta-1} \cdot \exp\left(-\frac{t}{\theta}\right)^{\beta} \tag{1}$$

where the parameter β is the shape parameter. A value of $\beta > 1$ signifies an increasing *failure rate* (or *hazard rate*) function, whereas a value of $\beta < 1$ signifies a decreasing *failure rate* function. When $\beta = 1$, the failure rate function is constant and the Weibull distribution is identical to the exponential distribution (Ebeling, 1997). The scale parameter of the Weibull distribution, denoted by θ, influences both the mean and the spread of the distribution. As θ increases, the reliability at a given point in time increases, whereas the slope of the hazard rate decreases.

The reliability and the unreliability of the yogurt production line are depicted by Equations (2) and (3), respectively

$$R(t) = \Pr(T \geq t) = \int_T^{\infty} f(t;\beta,\theta)dt = \exp\left(-\frac{t}{\theta}\right)^{\beta} \tag{2}$$

$$Q(t) = 1 - R(t) = 1 - \exp\left(-\frac{t}{\theta}\right)^{\beta} \tag{3}$$

The hazard or failure rate function of the line is given by

$$\lambda(t) = \frac{f(t)}{R(t)} = \frac{\beta}{\theta}\left(\frac{t}{\theta}\right)^{\beta-1} \tag{4}$$

Consequently, the Equations (2)–(4) are used to calculate the reliability, unreliability, and the failure rate of the yogurt production line based on the Weibull distribution per time t. The parameters β and θ are 1.41453 and 207.909, respectively (see Table 4). Therefore, the reliability model and the failure rate model of the yogurt production line from Equations (2) and (4) are as follows:

$$R_{Line}(t) = \exp\left(-\frac{t}{207.909}\right)^{1.41453} \tag{5}$$

$$\lambda_{Line}(t) = \frac{1.41453}{207.909}\left(\frac{t}{207.909}\right)^{1.41453-1} \tag{6}$$

Thus, these models were used to determine the operational behavior as a line performance evaluation indicator.

Discussion

To obtain quantitative analysis of the failure data for the yogurt production line, the descriptive statistics of the basic features of the TBF data are given in Table 1, and the following observations were made: (a) in the yogurt production line, the mean TBF is about 190 min (or approximately 3 h), meaning that about every 3 h of operation there is a failure which occurs on the production line. (b) The coefficient of variance at machine level is around one, whereas at line level it is less than one (.69). In other words, although the machines have the greatest StDevs, the production line has less variability relative to its mean. (c) The minimum mean TBFs are observed at the pasteurizing boiler (M4) with 467.6 min, and at the filling machine (M6) with 650 min. Therefore, the maintenance staff must initially concentrate on these machines, and (d) the distribution of TBF at line level display positive skew values, this indicates that the tail on the right side is longer than on the left side. Thus, many failures in the production line have low values for TBFs, while few failures have high values for TBFs. The distribution of TBF presented borderline mode < median < mean.

In Table 2, the trend test results for the yogurt production line are shown, with the correlation diagrams for the failure data exhibited in Figure 2. These show that all

Table 1. Descriptive statistics: the minimum and the maximum value of the sample (Count), mean, standard deviation (StDev), coefficient of variation (CoefVar), skewness, and kurtosis of the failure data for TBF (in min) at machine and entire line levels.

Variable	Count	Mean	StDev	CoefVar	Minimum	Maximum	Skewness	Kurtosis
TBF M1	54	2738	2591	.9463	225	13,677	2.23	6.13
TBF M2	48	3080	3390	1.1006	237	16,019	2.05	4.71
TBF M3	59	2505	2338	.9335	90.0	8876	1.24	.70
TBF M4	313	467.6	383.2	.8196	15.0	2395.0	1.71	3.83
TBF M5	35	4225	4540	1.0744	76.0	20,078	1.76	3.36
TBF M6	226	650.1	590.4	.9082	27.0	4311.0	2.23	8.41
TBF M7	30	4932	5582	1.1319	158	24,378	2.17	4.95
TBF Line	759	189.88	131.06	.6902	1.00	822.00	.92	1.51

Table 2. Calculation of the test statistic χ^2 for TBFs at machine and line level.

Unit	Variable	*df*	χ^2 statistic	$x^2_{a.df}$	Decision for H_0
M1	TBF	106	111.63	83.2403	Not reject
M2	TBF	94	111.45	72.6398	Not reject
M3	TBF	116	131.49	92.1338	Not reject
M4	TBF	624	697.24	567.051	Not reject
M5	TBF	68	69.37	50.0202	Not reject
M6	TBF	450	496.66	401.817	Not reject
M7	TBF	58	68.06	41.492	Not reject
Line	TBF	1516	1662.84	1426.58	Not reject

machines and the entire line are free from trend and serial correlation. Therefore, the failure data for the yogurt production line are IID.

To identify the distributions of the trend-free failure data between several theoretical distributions (i.e. Weibull, lognormal, exponential, normal distribution etc.), the maximum likelihood estimation method was applied per each distribution and assessed using the Anderson–Darling goodness-of-fit test. The Anderson–Darling statistics of several theoretical distributions for TBF based on failure data from all machines and the entire line are summarized in Table 3. The smallest statistic value indicates the distribution which best fits the data, thereby making clear that the TBF at line level follows the Weibull distribution.

Table 3. The Anderson–Darling statistics for TTB at machine and line level. The smaller the statistic value, the better the model fitting (i.e. in the row for M1, the lowest value is .462 belonging to the loglogistic distribution).

	Weibull	T Lognormal	B Exponential	F Loglogistic	Normal	Logistic
M1	.958	.471	1.541	.462[a]	3.77	2.669
M2	.888	.538[a]	.886	.585	3.842	3.21
M3	.671	.733	.651	.640[a]	3.318	2.908
M4	6.826[a]	9.202	11.946	7.958	16.281	13.053
M5	.509[a]	.812	.626	.602	2.414	2.101
M6	3.025[a]	4.633	4.689	4.04	10.919	8.069
M7	.864[a]	1.25	.903	.992	2.948	2.238
Line	5.53[a]	17.819	25.931	13.368	10.238	10.935

[a]Indicates the lowest value.

The hazard plots for the yogurt production line at machine and the entire line level are displayed in Figure 3. An increasing failure rate at the pasteurizing boiler (M4), the

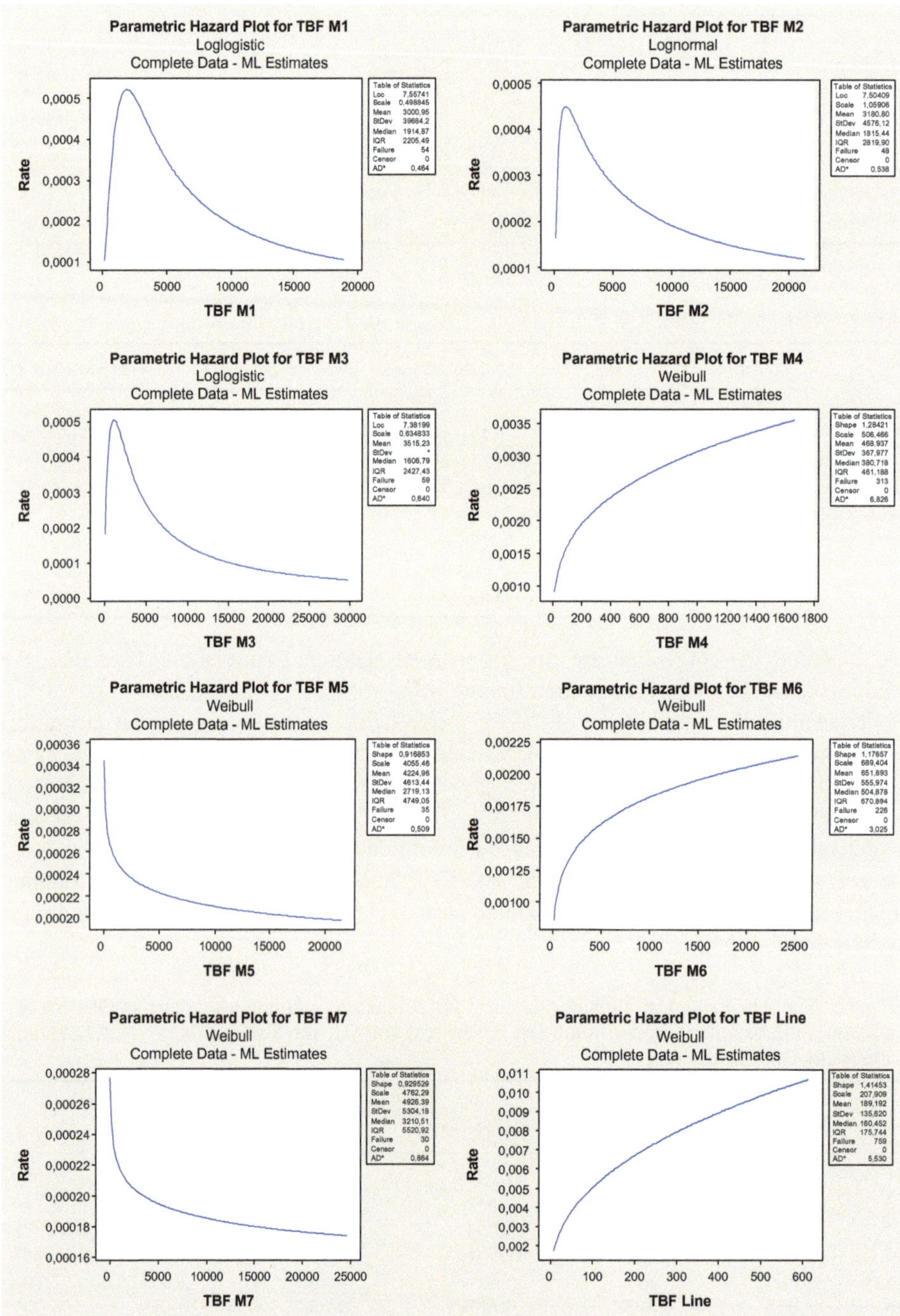

Figure 3. Hazard function for time between failures (TBFs) at machine and line level are shown graphically in conjunction with the appropriate distributions and their statistics.

filling machine (M6), and at the entire production line was observed. Therefore, this suggests that the maintenance policy should be initially focused on these machines to improve the reliability of the entire production line.

Table 4 displays the results derived for TBF at line level based on Weibull distribution with shape parameter $\beta = 1.41453$ and scale parameter $\theta = 207.909$. Quartiles are values that divide the failure data into four equal parts. With them, one can quickly evaluate a data-set's spread and central tendency. From this table, the following observations were made; (a) 25% of the failures (first quartile, Q_1) are less than or equal to 86.16 operating minutes, 75% (third quartile, Q_3) of the failures are less than or equal

Table 4. Distribution analysis of TBF applying the Weibull distribution for the yogurt production line, with 95% confidence interval (CI).

Variable: TBF Line
Uncensored value 759
Estimation Method: Maximum Likelihood
Distribution: Weibull

Parameter Estimates

Parameter	Estimate	Standard Error	95.0% Normal CI Lower	Upper
Shape (β)	1.41453	0.0411963	1.33605	1.49762
Scale (θ)	207.909	5.59975	197.218	219.179

Goodness-of-Fit
Anderson-Darling (adjusted) = 5.530

Characteristics of Distribution

	Estimate	Standard Error	95.0% Normal CI Lower	Upper
Mean (MTBF)	189.192	4.90704	179.814	199.058
Standard Deviation	135.620	4.69849	126.717	145.148
Median	160.452	4.82921	151.261	170.202
First Quartile (Q1)	86.1694	3.65914	79.2879	93.6481
Third Quartile (Q3)	261.913	6.73168	249.046	275.445
Interquartile Range (IQR)	175.744	5.17271	165.893	186.180

Table of Percentiles

Percent	Percentile	Standard Error	95.0% Normal CI Lower	Upper
1	8.04488	0.853129	6.53511	9.90344
5	25.4656	1.88272	22.0304	29.4364
10	42.3602	2.55231	37.6418	47.6700
20	72.0040	3.36539	65.7010	78.9116
30	100.312	3.91516	92.9247	108.287
40	129.312	4.37186	121.021	138.171
50	160.452	4.82921	151.261	170.202
60	195.449	5.38148	185.181	206.286
70	237.064	6.17041	225.273	249.471
80	291.059	7.48352	276.755	306.102
90	374.922	10.1939	355.465	395.444
99	612.002	21.1960	571.838	654.988

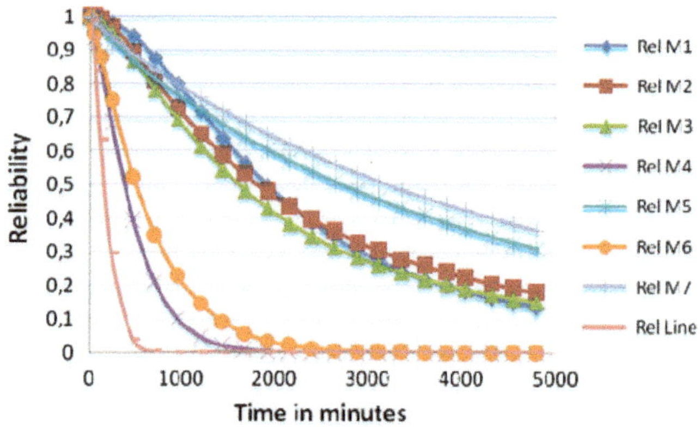

Figure 4. Reliability diagram for all machines and the entire system for yogurt production line.

to 261.91 operating minutes whereas the time when it spans the middle 50% of the failures (interquartile range: $IQR = Q_3 - Q_1$) is 175.74 operating minutes. (b) From the percentiles with 95% confidence interval, it is evident that 30% of all the failures which are expected to occur amounts to 100 min.

The reliability diagrams for all the machines and the entire line are shown in Figure 4. The optimal maintenance interval with the expected reliability level can be estimated, for example to achieve 80% reliability for the heat exchanger (M2), the

Table 5. Reliability at machine and entire line level for different time intervals.

Time	Rel M1	Rel M2	Rel M3	Rel M4	Rel M5	Rel M6	Rel M7	Rel Line
5	.99999	1	.999888	.997346	.99785	.996966	.998299	.994884
10	.99997	1	.999665	.993549	.995946	.993154	.996763	.986421
30	.99976	.99995	.998113	.97382	.988938	.975291	.991037	.937373
60	.99903	.99936	.994397	.937432	.979217	.945016	.982998	.84164
120	.99614	.99484	.983486	.854396	.961124	.880001	.967867	.631548
240	.98468	.97197	.95235	.681644	.927872	.749049	.939688	.293722
480	.94124	.89546	.870252	.393216	.868195	.52041	.888267	.038165
720	.87662	.80874	.779798	.207816	.814664	.349092	.841375	.003042
960	.79966	.72629	.692392	.102975	.76579	.228471	.797982	.000166
1200	.71846	.65207	.612973	.04843	.720776	.146667	.757536	.000007
1440	.63907	.58658	.543051	.021788	.679091	.092655	.719666	0
1680	.56521	.52918	.48246	.009428	.640343	.057733	.684101	0
1920	.49866	.47892	.430324	.00394	.604227	.03554	.650626	0
2160	.43992	.43483	.385548	.001595	.570492	.021642	.619061	0
2400	.38872	.39605	.34705	.000627	.538925	.013049	.589257	0
2640	.3444	.36183	.313854	.00024	.509345	.007798	.561082	0
2880	.30615	.33152	.285116	.00009	.481594	.004621	.534422	0
3120	.27316	.30457	.260127	.000033	.455531	.002717	.509174	0
3360	.24468	.28053	.238296	.000012	.431032	.001586	.485246	0
3600	.22003	.259	.219133	.000004	.407984	.00092	.462554	0
3840	.19863	.23967	.202234	.000001	.386285	.00053	.441022	0
4040	.18293	.22503	.189638	.000001	.369168	.000333	.423915	0
4320	.1637	.20651	.173945	0	.346578	.000173	.401166	0
4560	.14939	.19225	.162045	0	.328407	.000098	.382718	0
4800	.13679	.17929	.151371	0	.311263	.000055	.365182	0

maintenance must be carried out in less than 720 min, thereby meaning that the maintenance must be carried out every two working days. Moreover, Rel_{Line} (240) = .2937 means the probability that the system will not fail in 240 min of operation is .2937.

The reliability is calculated for all machines and the entire line at different time intervals in Table 5. Some conclusions from Table 5 include the following; (a) the reliability of the line after 10 min of operation is 98.64%, after 60 min is 84.16%; whereas after 120 min of operation, it drops sharply to 63.15%; (b) the highest reliability values are recorded for CIP (M7), the filling pump (M5), and the auto-cooling tank (M1); and (c) the lowest reliability values are shown at the pasteurizing boiler (M4), the filling machine (M6), and the separator (M3).

Conclusions

The main research findings can be summarized as follows:

(1) The Weibull distribution provided the best fit for the yogurt production line to describe the TBFs.
(2) The reliability must be improved both on the pasteurizing boiler (M4), and the filling machine (M6), and the entire production line as well. The importance of these machines is crucial and their maintenance must be meticulous to avoid any losses related to quality, safety, and productivity.
(3) The failure rate of the production line is increasing ($\beta = 1.41453 > 1$), implying that the currently applied maintenance strategy is not adequate and must be upgraded immediately.
(4) the highest reliability values were determined for the CIP (M7), the filling pump (M5), and the auto-cooling tank (M1), and
(5) Both reliability and failure rate models at line level were developed, thereby ensuring that line operation forecasting, at least in the short term, is feasible.

In conclusion, it can be claimed that the current maintenance policy is not adequate for the entire yogurt production line and more focused effort is required to improve the line operation management.

References

Blischke, W. R., & Murthy, D. N. P. (2000). *Reliability: Modeling, prediction and optimization.* New York, NY: Wiley.

Blischke, W. R., & Murthy, D. N. P. (2003). *Case studies in reliability and maintenance* (pp. 351–445). Hoboken, NJ: Wiley.

Croarkin, C., & Tobias, P. (2005). *NIST/SEMATECH e-Handbook of Statistical Methods.* Retrieved from http://www.itl.nist.gov/div898/handbook

Dhillon, B. S. (1999). *Design reliability: Fundamentals and Applications.* Boca Raton, FL: CRC Press LLC.

Dhillon, B. S. (2006). *Maintainability, maintenance and reliability for engineers.* Boca Raton, FL: Taylor & Francis Group, LLC.

Ebeling, C. E. (1997). *An introduction to reliability and maintainability engineering.* New York, NY: McGraw Hill.

Goetz, W., & Villa, S. D. (1998). Reliability improvement through the use of plant operating data and the ORAP database. *O&M Conference Modern Power Systems*, London.

Gupta, S., & Bhattacharya, J. (2007). Reliability analysis of a conveyor system using hybrid data. *Quality and Reliability Engineering International, 23*, 867–882.

Hajeeh, M., & Chaudhuri, D. (2000). Reliability and availability assessment of reverse osmosis. *Desalination, 130*, 185–192.

Jalali, S. E., & Forouhandeh, S. F. (2011). Reliability estimation of auxiliary ventilation systems in long tunnels during construction. *Safety Science, 49*, 664–669.

Kardon, B., & Fredendall, D. L. (2002). Incorporating overall probability of system failure into a preventive maintenance model for a serial system. *Journal of Quality in Maintenance Engineering, 8*, 331–345.

Kiureghian, A. D., Ditlevson, O. D., & Song, J. (2007). Availability, reliability & downtime of system with repairable components. *Reliability Engineering and System Safety, 92*, 231–242.

Manzini, R., Regattieri, A., Pham, H., & Ferrari, E. (2009). *Maintenance for industrial systems*. London: Springer London Ltd.

Markeset, T., & Kumar, U. (2001). R&M and risk-analysis tools in product design, to reduce life cycle cost and improve product attractiveness. *Proceedings of annual reliability and maintainability symposium* (pp. 116–121). New York, NY: IEEE.

Meeker, W. Q., & Escobar, L. A. (1998). *Statistical methods for reliability data*. New York, NY: Wiley.

Regattieri, A., Manzini, R., & Battini, D. (2010). Estimating reliability characteristics in the presence of censored data: A case study in a light commercial vehicle manufacturing system. *Reliability Engineering and System Safety, 95*, 1093–1102.

Ruan, N., & Sun, X. (2006). An exact algorithm for cost minimization in series reliability systems with multiple component choices. *Applied Mathematics and Computation, 181*, 732–741.

Singh, V. P., & Swaminathan, S. (2002). Sample sizes for system availability. *Proceedings of annual reliability and maintainability symposium* (pp. 51–55). New York, NY: IEEE.

Stapelberg, F. R. (2009). *The handbook of reliability, availability, maintainability and safety in engineering design*. London: Springer-Verlag London Limited.

Tsarouhas, P. (2010). Measuring the efficiency of medical equipment. *International Journal of Applied Science, Engineering and Technology, 6*, 15–20.

Tsarouhas, P. (2012). Reliability, availability, and maintainability (RAM) analysis in food production lines: A review. *International Journal of Food Science & Technology, 47*, 2243–2251.

Tsarouhas, P., & Arvanitoyannis, I. (2010a). Assessment of operation management for beer packaging line based on field failure data: A case study. *Journal of Food Engineering, 98*, 51–59.

Tsarouhas, P., & Arvanitoyannis, I. (2010b). Reliability and Maintainability analysis of bread production line. *Critical Reviews in Food Science and Nutrition, 50*, 327–343.

Tsarouhas, P., & Arvanitoyannis, I. (2011). Quantitative analysis for peach production line management. *Journal of Food Engineering, 105*, 28–35.

Tsarouhas, P., & Arvanitoyannis, I. (2012). Reliability and maintainability analysis to improve the operation of the limoncello production line. *International Journal of Food Science & Technology, 47*, 1669–1675.

Tsarouhas, P., Arvanitoyannis, I., & Ampatzis, Z. (2009). A case study of investigating reliability and maintainability in a Greek juice bottling medium size enterprise (MSE). *Journal of Food Engineering, 95*, 479–488.

Tsarouhas, P., Arvanitoyannis, I., & Varzakas, T. (2009). Reliability and maintainability analysis of cheese (Feta) production line in a Greek medium-size company: A case study. *Journal of Food Engineering, 94*, 233–240.

Tsarouhas, P., Varzakas, T., & Arvanitoyannis, I. (2009). Reliability and maintainability analysis of strudel production line with experimental data: A case study. *Journal of Food Engineering, 91*, 250–259.

Usher, J. M., Roy, U., & Parsaei, H. R. (1998). *Integrated product and process development*. Hoboken, NJ: Wiley.

Wang, J., Hu, Y., & Li, J. (2010). Transient analysis to design buffer capacity in dairy filling and packing production lines. *Journal of Food Engineering, 98*, 1–12.

Wang, W., Loman, J., & Vassiliou, P. (2004). Reliability importance of components in a complex system. In *Proceedings of the annual reliability and maintainability symposium (RAMS)* (pp. 6–11), Los Angeles, CA.

13

Sequencing jobs in an engineer-to-order engineering environment

Douglas H. Grabenstetter and John M. Usher*

Department of Industrial and Systems Engineering, Mississippi State University, MS 39762, USA

Engineer–to-order (ETO) firms produce complex – one of a kind – products and desire shorter lead times as a key component to cost competitiveness. In ETO firms, the engineering process is the largest controllable consumer of lead time. Given that lead time is a function of completion rate and scheduling policy, one critical process is to accurately sequence jobs in front of the engineering function. However, unlike other manufacturing models, such as make–to-stock or make-to-order models, the design for an ETO product is not realized until after the engineering process has been completed. Hence, the only information available does not include data normally required by most sequencing algorithms. Therefore, the problem becomes the determination of an accurate schedule within a complex transactional process for jobs which have not even been designed yet. This paper investigates this topic in the context of the engineering process within the ETO model. Based on research conducted in conjunction with multiple firms, common factors are identified which drive complexity, and a new framework and algorithm are presented for using these factors to sequence jobs. Using discrete event simulation, the performance of this new algorithm is found to be a significant improvement over current industry and published methods.

Keywords: engineer-to-order; sequencing; design

1. Introduction

The engineer-to-order (ETO) model is used by a significant number of manufacturers across multiple sectors. Indeed, ETO firms comprise approximately one-fourth of all North American manufacturing and are growing at a rate of 20% (Cutler, 2005). One reason for this level of growth is the increased demand for customized products which, in turn, has caused markets to become more and more fragmented. Use of the ETO model provides firms with increased agility and flexibility thereby, allowing them to be ideally positioned to respond to rapid market shifts through production of one of kind products. Despite these advantages, the very nature of the ETO environment drives complexity. For instance, ETO products are manufactured and assembled in low quantities to individual customer specifications. In addition, the goods manufactured using ETO models tend to be complex in nature, resulting in deep and complicated multilevel product structures and routings which often contain job-specific and engineered-to-order components (Hicks & Braiden, 2000; Hicks, Song, & Earl, 2007). In the ETO environment, the engineering process is the largest controllable consumer of lead time taking up to one half of the total lead time (Little, Rollins, Peck, & Porter, 2000; Pandit & Zhu, 2009). Since ETO firms produce products which are often used in large projects, it

*Corresponding author. Email: usher@ise.msstate.edu

is not unusual for their customers to impose large cost penalties for lateness. Further, given that lead time is a function of completion rate and scheduling policy, one critical process is to accurately sequence jobs in front of the engineering function and therefore is an important topic for these companies.

Yet, despite the aforementioned need, there has not been much sequencing research focused on the ETO engineering environment. Therefore, this paper investigates sequencing approaches used in the literature in order to generate good (not necessarily optimal) job sequences. A new complexity sequencing approach is introduced that has been tested using real-world data from a typical ETO industrial firm and found to meet the needs of ETO firms in that it is quick, easy to use, accessible, and accurate. Indeed, the approach has been found to significantly outperform traditional sequencing methods used in the ETO environment.

2. Background

When discussing ETO firms, it is important to consider a couple of important constraints. The first of these is that the engineering process is a bottleneck resource per Gosling and Naim (2009), Cutler (2005, 2009), and Pandit and Zhu (2009). Indeed, it is a core process and is never outsourced (Hicks, McGovern, & Earl, 2000). Further, competition from other firms has caused some firms to reduce their costs by 50% (Hicks, McGovern, & Earl, 2001). Since resources for the engineering processes are costly (Qi & Tu, 1998), these staffs in turn have been scaled down (Hicks et al., 2001). As a result, engineering staffs are now often resource constrained. Secondly, most ETO firms use enterprise resource planning (ERP) systems, per Stevenson, Hendry, and Kingsman (2005), which are substantial investments per Worthen (2002). Hence, trying to convince these firms to throw out their ERP systems likely will be a hard sell! Unfortunately, despite the need for research on the engineering process, the vast majority of research performed for ETO firms by such authors as Hicks and Pongcharoen (2006, 2009), Hicks and Braiden (2000), Hicks et al. (2000, 2001, 2007), Pongcharoen, Hicks, and Braiden (2004), Pongcharoen, Hicks, Braiden, and Stewardson (2002), and Song, Hicks, and Earl (2001, 2002, 2006) recommends complex global shop floor scheduling approaches which seek to replace ERP systems. Coates, Ritchey, Duffy, Hills, and Whitfield (2000) provide the only discussion found which focuses on the ETO engineering process. They present an agent-based design decision evaluation tool for large make-to-order products. This tool chooses the design concept which takes the least amount of projected time with minimum resource use. The purpose of the agents is to perform this task while dealing with a diverse set of design tools using data from heterogeneous distributed computing systems. Seven agents are identified with each of the agents fulfilling a specific role performing several activities. A resource manager agent optimizes the utilization of the resources. If the utilization of a particular resource falls below a given target, or if an alternative resource's efficiency level is better than the current resource, the resource manager triggers a scheduling agent to create a task matrix which contains task dependencies and task durations. The scheduling agent determines in which order certain design candidates are evaluated. The scheduling problem is defined by the authors as a job shop problem (as opposed to a single-machine setting) which has been proven to be NP-Hard (Brucker, Drexl, Moehring, Neumann, & Pesch, 1999). Therefore, the authors make use of a multicriteria genetic algorithm (GA) which "optimizes" multiple objective functions simultaneously. The GA produces a pareto-optimal set of schedules using the task matrix and chooses the schedule which maximizes the utilization of the resources, minimizes the number of resources, and minimizes the time to complete design concept evaluations. The main problem with the GA approach lies with its complexity as driven by the scheduling problem definition. Since the scheduling problem is defined as a job shop problem, computation time will likely be an issue due to problem logic complexity.

In contrast to what has been proposed by others, this paper offers a more robust approach which uses the output of ERP and focuses on scheduling just the engineering process not each process of the entire system. The new methodology addresses the inherent scheduling complexity of the ETO environment through the use of dispatching rules (DR). Compared to the aforementioned methods, DR are low in computational complexity, having only to compute which job to process next as opposed to having to determine the entire sequence of all the jobs in the queue. While use of DR will not guarantee optimal solutions, good solutions are often achieved for either static or dynamic, single or multiple machine environments. As well, DR are very easy to implement. Therefore, it is no surprise to find widespread use of DR by ETO firms.

There are three basic types of sequencing approaches found in the literature which use due dates as an input: allowance-based, slack-based, and ratio-based. In his study, Baker (1984) investigates approaches from each of these three categories and finds that the performance of slack-based approaches is equivalent to the allowance-based approaches. Referring to Table 1, Baker and Trietsch (2009) find that shortest processing time (SPT) performs quite robustly for the proportion tardy (PT) and mean tardy (MT) objective functions. Additionally, Baker reports that modified due date (MDD) performs well for the single-machine dynamic maximum lateness (T) problem and that early due

Table 1. Brief summary of dispatching rule metrics (adapted from Baker & Trietsch, 2009).

Metric	Best rule(s)	Other rules compared	Authors
PT	S/OPN	EDD/MST	Conway (1965a), Conway, Maxwell, and Miller (1967)
	SPT	MST,EDD,S/OPN	Elvers (1973)
	SPT	EDD, ODD, MST, SCR, SCR,OST	Muhlemann, Lockett, and Farn (1982)
	SCR, A/OPN	S/OPN	Miyazaki (1981)
	SPT	MST, OST, EDD, ODD, SCR, OCR	Kanet and Hayya (1982)
	SPT, S/OPN, MST	EDD	Elvers and Taube (1983)
MT	MST	S/OPN, SPT	Gere (1966)
	SPT	S/OPN, EDD	Conway (1965a,1965b), Conway and Maxwell (1962), Carroll (1965)
	S/OPN	SPT	Weeks (1979)
	SPT, ODD	MST, OST, SCR, OCR	Kanet and Hayya (1982)
	A/OPN	SCR, S/OPN	Miyazaki (1981)
	SPT, EDD	MST, SCR	Baker and Bertrand (1981)
	MST, SPT, EDD	OST, S/OPN, ODD, SCR	Muhlemann et al. (1982)
CMT	S/OPN	SPT, EDD	Conway (1965a,1965b), Conway and Maxwell (1962)
	SCR	SPT, EDD, ODD, MST, S/OPN, OST	Muhlemann, Lockett, and Farn (1982)
	OCR	SPT, EDD, ODD, MST, OST, SCR	Kanet and Hayya (1982)

date (EDD) maximizes the minimum service level in the stochastic single machine environment. Further, EDD has been found by this author to be the method of choice among ETO firms.

3. Sequencing research in the ETO engineering process

As previously mentioned, the available research in the literature is not directly applicable to the ETO environment. In addition, very little information exists about the ETO engineering process and how engineering due dates are calculated. Therefore, subject-matter experts (SMEs) from five ETO firms of various industrials (listed in Table 2) were consulted remotely via telephone, or, where feasible, face-to-face. With each of the ETO firms, research was conducted regarding the front-end processes and flows they employed including due-date setting methods.

The engineering process is distinct from traditional shop floor processes as its inputs and outputs are purely intellectual and not physical in nature. Hence, in order to gain an idea of how sequencing is performed in ETO firms, SMEs at each site were interviewed about their individual process flows including the inputs and outputs, as well as the details of the tasks at each step. Not surprisingly, the challenges of the ETO environment have forced each of the firms to adopt similar process steps. The data gathered are summarized in the form of a generic IDEF0 diagram (see Figure 1).

In step A1, an initial job due date is set. The lead times are set intentionally wide enough to cover the amount of uncertainty present. Once the order is obtained, the firm must provide a committed job due date in step A2. The job information is forwarded on to the engineering step (A3). Research found that several constraints exist in regard to sequencing for the ETO design environment. First, the ETO environment is complex, and each job is unique and often complicated. In addition, there is a limited amount of information available. Referring to Figure 2, the first step in the engineering process (A3.1) is a review and sequencing process where the job information is scanned for long lead time non-inventoried components. In addition, the job data are assessed to ensure that vital data required by the engineering discipline in play (e.g. wiring diagrams, ladder diagrams, and other details) do not contain clean-order defects. These clean-order defects encompass missing information not provided by the sales force, incorrect information in that it is wrong, infeasible or non-sensical, and conflicting information between two or more pieces of data. For instance, a ladder diagram may call for logic which does not translate properly into the expected components.

Clean-order defects are a major source of variability in the ETO environment and cause rework loops in the engineering process (steps A3.2 and A3.4). These issues in

Table 2. Description of the ETO industrial firms studied.

Firm	Market	Product	Annual sales	Employees
A	Industrial	Motor control center	$60MM	256
B	Construction	Switchboards	$25MM	20
C	Industrial	Process instrumentation	$38MM	80
D	Construction	Switchgear	$81MM	317
E	Industrial	Busway	$63MM	232

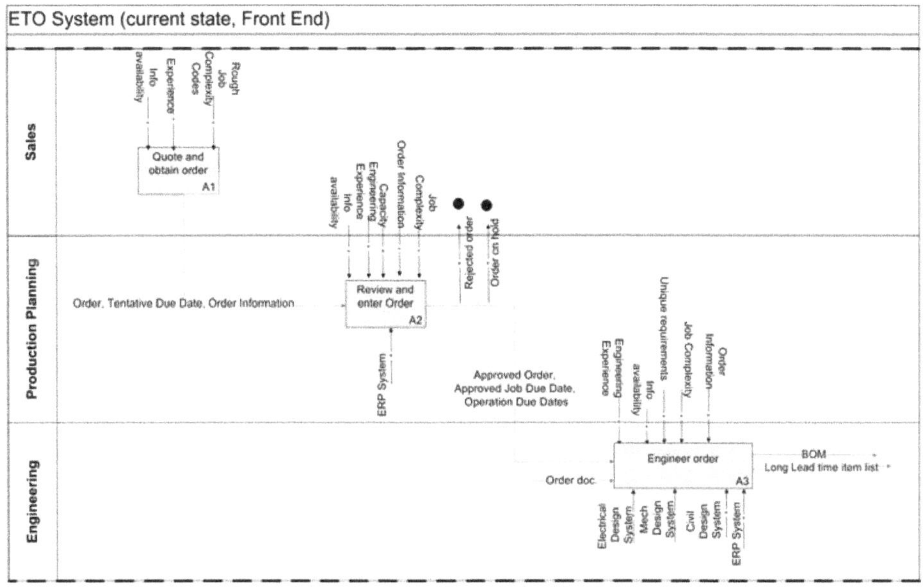

Figure 1. ETO engineering flow of front-end process.

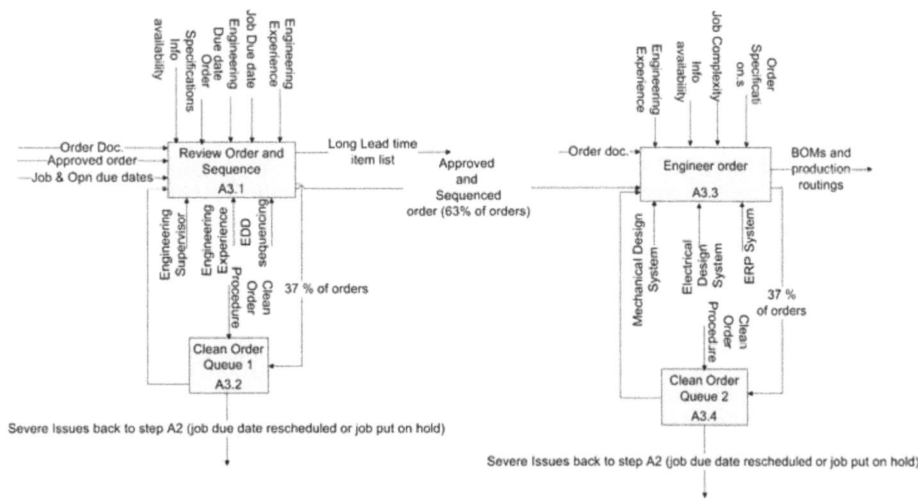

Figure 2. ETO engineering process.

ETO firms are problematic. Indeed, a four-year study conducted by Firm A found that 37% of orders contain some sort of clean-order error. Normally, the cycle time to solve clean-order defects may take up to 2 weeks. However, if the customer is found to be the source of the defect, the ETO sales force is contacted and negotiations then occur, which can drive up the resolution cycle time. According to Firm A, such a defect occurrence can take 60 days or more to resolve. Additionally, if the customer is at fault or if a change is initiated by the customer, the due date could be moved out. If the clean-order negotiations become intractable, it is also possible that the order will be put on hold or even canceled.

Once the clean-order review has been completed, the job is then sequenced in step A3.1. As previously mentioned, research with the five partner firms has revealed that the EDD method of sequencing is used as a general practice. The main reason for using EDD is its simplicity. Each of the firms stated that they prefer simple, easy to use methods that can be run using a spreadsheet. This follows the findings of Stoop and Wiers (1996) that simple and robust approaches are the most broadly used in practice. Otherwise, the firms will likely not use the methodology (O'Neil, 1991). The scheduling process is typically conducted on a weekly basis for several reasons. First, Pandit and Zhu (2009) and Little et al. (2000) point out that the ETO engineering process can consume up to one-half of the production lead time. Hence, the engineering task is considered to be a long duration process. Second, the engineering step is a complex, intellectual process and interruptions to the process can lead to an inflated error rate. Thus, when an engineer begins work on a job, if at all possible, it is desired that they be allowed to complete the engineering task without interruption (non-preemption). Otherwise, constant rescheduling is known to cause schedule nervousness.

Upon the job being sequenced and assigned to an engineer, the next step in the process (A3.3) is the actual engineering task where the job inputs are transformed into a tangible product design. Given the high complexity of ETO products, the engineers themselves typically possess a high degree of experience and are difficult to replace. In addition, due to the high product complexity present, each of the partner ETO firms makes use of a proprietary computer-aided job engineering configuration software system (CAJECSS). When engineering the job, the engineer chooses predefined job characteristics found in the system. The CAJECSS then uses logic tables to generate the appropriate engineering output (e.g. wiring diagrams and assembly drawings). In addition, it also generates a top-level bill of material which is used in the subsequent production planning step to order materials. Only when the engineering process is finished, can the remainder of lower level discrete components be procured. Once the components are on-hand, the manufacturing process itself can begin.

These findings lead to an overriding question: How can scheduling capabilities which co-exist with existing ERP systems be developed for the highly complex ETO engineering environment while, at the same time, remaining realistic, robust, acceptable, and easily implemented? One might conjecture that the answer simply lies in applying what has been researched already in the job shop literature. However, much of this research presupposes that the process time of each job is known or can be easily estimated. However, because each job is unique, in the ETO environment, this simply is not the case. Restated, the literature survey and ETO firm research have shown that what is needed is a sequencing policy which:

- robust and relatively simple to implement and use (spreadsheet driven),
- accurate (an improvement over EDD), and
- complementary to an ERP system (Recall that many firms have invested heavily in their ERP systems and are in no hurry to replace them).

4. Regression-driven complexity-based flow time prediction

Based upon the previous discussions, the engineering process can be modeled as a dynamic job shop with the following assumptions:

- arrivals are dynamic,
- machines process one job at a time,
- setup times are independent of job sequence,
- machines are continuously available, and
- neither idle time nor preemption is allowed.

SPT is well known to be a robust sequencing algorithm. However, given the nature of the ETO engineering environment, nothing is known ahead of time in regards to job processing time. To overcome this problem, Grabenstetter and Usher (2013) identified seven engineering factors that can be used in this context to predict job processing times using a regression-based approach. These factors are:

- x_1 – number of functional requirements
- x_2 – number of basic components
- x_3 – number of design interdependencies
- x_4 – number of technologies
- x_5 – number of regulations and standards
- x_6 – number of subsystems
- x_7 – presence of a reference job

Although multiple regression has been used by others such as Salman and Dogru (2004), no such procedure has ever been employed which uses ETO-specific complexity factors (CFs). Once the regression procedure generates a predicted processing time for each job *j*, they can then be used as an input to SPT. The framework for this approach (referred to as RegSPT) is comprised of three parts that are complementary in nature. The first component employs multiple linear regression using a basic CF equation composed of the seven CFs.

$$\mathrm{CF}_j = \beta_1 x_1 + \beta_2 x_2 + \beta_3 x_3 + \beta_4 x_4 + \beta_5 x_5 + \beta_6 x_6 + \beta_7 x_7 \tag{1}$$

where $\mathrm{CF}j > 0$, x_1, x_2, x_3, x_4, x_5, x_6, $x_7 \geq 0$ and initially x_1 through x_6 are defined as discrete integer variables, and initially x_7 is defined as a binary ordinal categorical variable (1, 0).

The second element of the framework is an *estimation algorithm* (EA) which is used to determine the values of the coefficients of the CF equation. This step is necessary for each different ETO firm given that the exact continuous distribution function of the dependent variable data and the exact probability distribution function of the dependent variable data will vary somewhat with each firm. As well, the EA will need to be rerun periodically because it is likely that business conditions will change over time and process improvements will occur. These changes may affect the efficacy of the input data used to build the actual regression equation currently in use. This means every *p* periods, the EA will need to be run using up-to-date historical data for the last *p* periods for both the response variable and the independent variables x_1, x_2, x_3, x_4, x_5, x_6, and x_7. Because not all ETO firms are exactly alike, the length of the update cycle, *p*, and the sample size, *z*, will need to be determined by each firm. The EA is defined as:

Step 1: Gather dependent variable data for the last p periods.

Step 2: Gather independent variable data for the last p periods.

Step 3: Generate updated slope coefficients (β): using Equation (1) as a starting point, generate a statistically valid regression model using the seven aforementioned factors.

The final component of the framework is the RegSPT algorithm (RSA) which uses the resulting CF equation as a part of the sequencing process to generate the complexity-based processing times for each job order. Operationally, when a job is received, the production planning step assembles data that are found in the specific job's packet. The independent variable data (x_i's) are then entered at the engineering step into the regression equation for the job at hand to determine the job unique complexity-driven processing time. Once processing times have been predicted for each job to be sequenced, the SPT sequencing algorithm is then utilized. RSA is defined as:

Let n = the number of jobs.

Step 1: Initialization: Upon arrival of job j, Set $j = 1$.

Step 2: For Equation (1), determine x_1, x_2, x_3, x_4, x_5, x_6, and x_7 using the job information provided by the sales department.

Step 3: Evaluate Equation (1) to generate the predicted process time.

Step 4: Let $j = j + 1$.

Step 5: If $j = n$, Go to step 6. Else, go to Step 1.

Step 6: Sequence job j … n in ascending order of predicted processing times. If ties result, schedule by earliest due date.

Step 7: Stop.

5. Example implementation of the RegSPT method

To illustrate the application of the framework for generating job sequences, actual data from Firm A was chosen because it closely matches the criterion listed in the literature for ETO firms (Hicks et al., 2007). When performing regression analysis, it is important to follow a strategy of solid statistical practice. Therefore, the strategy listed in Kutner, Nachtsheim, and Neter (2004) was followed.

Step 1 – The first step of the EA process is to gather data from the last p periods. At this point, it should be stated that due to differences in software and operational practices, the execution of this task will vary with each firm. Some firms may make use of MS Excel® tables to store these data, while others may store them in their ERP system. At Firm A, response variable data were gathered which represent actual historical engineering labor flow time data in days. These data were resident in their ERP system and consisted of two years' worth (1276 distinct jobs) of data which were exported to a spreadsheet program. Again, for our tests, in order to rule out the impact of sampling bias, the time period from which the data were gathered came from the most recent period of economic stability and represents steady state/normal business conditions. The response variable statistics were mined directly from the data-set. Outliers were identified, and, if determined by Firm A to be inappropriate, were removed.

Step 2 – The independent variable data to be gathered must represent actual historical engineering data used to engineer each of the corresponding jobs identified in Step 1. Once the raw data are obtained, the data need to be separated into the aforementioned seven factors. Again, due to differences in the format in which this data is stored and operational practices, the difficulty of this task will vary with each firm. In this example, similar to some other ETO companies, Firm A makes use of a sophisticated proprietary CAJECSS. With this system, the engineer chooses predefined job characteristics found in the system. Then using logic tables, the CAJECSS generates appropriate wiring diagrams and assembly drawings. In addition, it also generates a top-level Bill of Material which is used in the subsequent production planning step to order materials. For two of the seven factors (number of subsystems and presence of a reference job), the data could easily be extracted directly. However, data categorization for the other five factors was not as simple. These data were discovered to be embedded within multiple levels of tables found within the CAJECSS. Further, given the multilevel nature of the tables, and that factor x_3 is built upon interdependence of design characteristics, the data for each factor could not be extracted using simple sorting techniques. Therefore, in order to accomplish separation of these data into their appropriate factors, a binary mapping table using MS Excel® was developed which took advantage of the logic statements found within the configuration system. Upon reviewing the raw data from the CAJECSS, the amount of raw data was found to be a staggering amount. As a result, the mapping table was then utilized to develop a computerized algorithm to separate the data into the five aforementioned factors. The algorithm was written in the Visual Basic® for Applications programming language. The run time was very quick, requiring less than 30 s. These tools are reusable and once developed, can be used when future updates of the system are required.

Once the data are gathered, the next step is to validate the data and gain an understanding of the data quality. Direct diagnostic plots of the response variable are normally not very important in regression analysis as the values of the observations on the response variable are functions of the levels of the predictor variables. However, in order to ensure that the study is of the highest quality, a cursory look at the data should be performed. The data for each dependent and independent variable were plotted using box plots and reviewed with the SMEs in order to identify and remove gross data errors and outliers.

Step 3 – Next, the regression model is generated using Equation (1) and a standard residual analysis performed to ensure that the model meets the statistical criterion for multiple regressions. Since the distribution of residuals was initially found to be non-normal, a logarithmic transformation was performed on the response, which is referred to in our example as Log PT. Referring to Tables 3 and 4, the model is found to be appropriate. The model itself has a *p*-value of zero, which is well below the normal .050 alpha threshold. Further, all of the factors except for x_2 and x_5 similarly are well

Table 3. Model with factor significance.

$$\text{Log PT} = .635 + .00148\,x_1 + .0422\,x_2 + .143\,x_3 - .0427\,x_4 - .00118\,x_5 + .264\,x_6 - .590\,x_7$$

Predictor	Coef	SE Coef	T	p
Constant	.63492	.03167	20.05	.000
x_1 – # Func. req.	.00148	.00025	6.01	.000
x_2 – # Basic comp.	.04222	.01216	3.47	.001
x_3 – # Design inter.	.14271	.03200	4.46	.000
x_4 – # Technologies	−.04273	.01225	−3.49	.001
x_5 – # Reg. and Std.	−.001185	.00067	−1.77	.076
x_6 – # Subsystems	.26399	.03257	8.11	.000
x_7 – Ref. job?	−.58987	.03055	−19.31	.000
$S = .472984$	$R^2 = 39.8\%$	$R^2_{adj} = 39.4\%$		

Table 4. ANOVA table.

Analysis of variance					
Source	DF	SS	MS	F	p
Regression	7	187.366	26.767	119.65	.000
Residual error	1268	283.669	.224		
Total	1275	471.034			

below the normal .050 alpha threshold and although x_2 and x_5 are marginally above the threshold, investigations with the ETO firms have concluded that these factors are valid. Hence, the model can be used for predictive purposes.

Step 4 – Given that the CF equation with coefficients is available, the RSA can be run. In order to generate the processing times, a second historical data-set consisting of (74) jobs was gathered from Firm A and represents the input data that would be used by the RSA. Recall that the regression formulation is based on p time periods before the current time period. Therefore, the data from the 1276 distinct jobs used in Step 1 iare from a time period preceding this set of 74 jobs. This condition simulates actual operating conditions which would be experienced by practitioners. Equation (1) is embedded in an MS Excel® spreadsheet and is evaluated for each of the 74 jobs to compute the predicted processing times. Examples of the input data and predicted process time values are given in Table 5.

Table 5. Example of input data for RegSPT and output.

Order No.	x_1	x_2	x_3	x_4	x_5	x_6	x_7	LOG predicted process time	Predicted process time
3001774379	128	71	0	68	0	1	1	.59	3.90
3001912763	142	113	1	112	0	1	0	1.24	17.31
3002194681	30	6	0	6	0	1	0	.94	8.72
3002246903	2	0	1	0	0	0	0	.78	6.04

Table 6. Example of the RegSPT schedule.

Order No.	Arrival time hour	Predicted process time (days)
3001774379	0	3.90
3002267611	0	8.27
3002194681	0	8.72
3002341332	0	12.36
3001912763	0	17.31
3002380038	7	2.35
3002400367	7	4.32
3002390220	7	8.03
3002454575	14	8.16
3002452423	14	8.50
3002447003	14	8.58
3002450199	14	11.18
3002491099	21	4.35
3002472412	21	6.44
3002488807	21	7.98

Step 5 – Referring to Table 6, the next step is to group the jobs based on their arrival time and then sequence the jobs in each group in ascending order of their processing time (SPT). The run time to sort the 74 jobs is extremely fast and was completed in less than 5 s.

6. Methods

Given the previous section's description of RegSPT as an alternative to current method of sequencing for the ETO engineering environment, it remains to be proven whether the new methodology is better than current approaches typically used by ETO firms and popular literature approaches. Therefore, an analysis using simulation employing data from one of the five aforementioned partner firms, Firm A, was performed to study the performance and behavior of sequencing methods in a real-world ETO engineering department. A discrete simulation model of the engineering process was built, verified, and validated using the approach of Harrell, Ghosh, and Bowden (2004). The output of the model yields queue lengths, utilization, job cycle times, and many other data useful for analysis of the long-term behavior of the engineering environment. In particular, the cycle times were used to determine job completion times. In order to provide a 95% percent confidence interval with a three-day half-length using the standard deviation of lateness (SDL) value based on an initial run of 30 replications, it was determined that 53 replications of the experiment would be required. Due dates for input into the model were calculated using the approach of Grabenstetter and Usher (2014). The process steps in the model follow those shown in Figure 2 and the processing times for those steps utilize industry values. The arrival patterns form the basis for testing the performance of RegSPT against popular sequencing methods.

Three popular sequencing approaches from the literature were chosen for the analysis: SPT, MDD, and EDD. Baker and Trietsch (2009) find that SPT performs quite robustly for a variety of objective functions (see Table 1) and, in particular, in the presence of imperfect information which is an advantage in the ETO environment. In addition, it is easy for practitioners to implement and understand. However, as can be observed, this rule requires foreknowledge of a job processing time. Hence, in order to simulate the operation of SPT in practice at an ETO firm, it is necessary to develop a method that ETO firms would most likely use to predict processing times using the limited data (material costs) that is available at the beginning of step A3. To accomplish this task, it is important to know that the firms will likely only implement a method which is simple and robust (O'Neil, 1991; Stoop & Wiers, 1996). Therefore, it was decided to use a simple material-based approach similar to what is currently utilized in industry. As such, jobs are grouped into three classes based upon perceived difficulty as predicted by the job material content (Grabenstetter & Usher, 2013). To accomplish this task, the 1276 jobs in the data-set were grouped by the three classes and an average process time by class determined. These average process times are then used to help drive SPT. In the literature, the release time r_j, is another variable that normally appears as a part of sequencing formulations. However, in this case study, it is assumed universally that the release time equals zero. Therefore, this term drops out of the formulations.

The MDD method chooses the assigned job due date (d_j) or the soonest time that the job could be completed ($t + p_j$), whichever time is later.

$$d'_j = \max\{d_j, t + p_j\} \tag{2}$$

The same approach for processing time prediction used previously for SPT is employed here to get the values of p_j.

EDD is the method of choice among ETO firms and has been reported to perform robustly. It is simple, easy to understand and implement, and does not require foreknowledge of job processing times. In order to sequence the jobs via the EDD method, within each arrival period, the jobs are simply arranged in ascending order according to their due dates. If ties are encountered, the jobs are ordered arbitrarily.

In order to judge the performance of the sequencing methods, it is necessary to define the performance metrics used. This study employs the criteria used by other published ETO studies such as Hicks and Pongcharoen (2006, 2009), Hicks and Braiden (2000), Hicks et al. (2007), and Song et al. (2002, 2006). These criteria are:

- Mean lateness (ML): lateness is calculated as the job completion time minus the due date. As such it measures the deviation, early or tardy from the due date.
- Mean absolute lateness (MAL): defined as the absolute value of the job lateness.
- SDL: used to add a measure of variability.

7. Effectiveness of the RegSPT method

The test results show the RegSPT method to be provide robust performance for the studied metrics. Referring to the summary statistics in Table 7, RegSPT produces the best performance in all three categories. In particular, the value for MAL and the standard deviation of ML are both statistically lower when compared to the closest competitor: EDD. Since the data for neither of these metrics is normally distributed, Levene's test is used for variance testing and the Mann–Whitney Test is used for median testing. Referring to Table 8, given that the p-value is less than the .05 alpha value, the ratio of the variances are considered to be unequal. Hence, the variance for RegSPT is significantly lower compared to EDD and Table 9 shows MAL for RegSPT to be likewise significantly different (lower) from that of EDD. Since lower values of absolute lateness are important for lean manufacturing (goal of zero), this is an important finding.

8. Implementation issues

RegSPT has been shown to provide more balanced performance compared to the existing sequencing methods and is easier to implement than the MDD and SPT approaches as there is no need to estimate a processing time for each job. In terms of tools required to implement the approach, all a company needs is a computer and standard software applications such as MS Excel® and possibly Minitab® (although Excel® statistical add-ons will work fine). RegSPT can be thought of as having two components: the CF regression equation and the sequencing algorithm. Because business conditions change over time, there is a need to regularly refresh the coefficients of the regression equation.

Table 7. Summary of lateness results for all methods.

Metric	EDD	MDD	SPT	REG SPT
MAL	10.55	10.68	29.83	7.84
SDL	11.33	12.12	24.49	7.11
AL	−7.06	−7.08	27.01	−7.09

Table 8. Statistical variance testing for ML.

Test and CI for Two Variances: EDD AVE lateness, Ave Lateness by

Method

Null hypothesis	Sigma(EDD AVE lateness by job) / Sigma(Ave Lateness by job-REG SPT) = 1
Alternative hypothesis	Sigma(EDD AVE lateness by job) / Sigma(Ave Lateness by job-REG SPT) not = 1
Significance level	Alpha = 0.05

Statistics

Variable	N	StDev	Variance
EDD AVE lateness by job	74	11.331	128.400
Ave Lateness by job-REG SPT	74	7.111	50.573

Ratio of standard deviations = 1.593
Ratio of variances = 2.539

95% Confidence Intervals

Distribution of Data	CI for StDev Ratio	CI for Variance Ratio
Normal	(1.264, 2.008)	(1.599, 4.031)
Continuous	(1.430, 2.123)	(2.045, 4.508)

Tests

Method	DF1	DF2	Statistic	P-Value
F Test (normal)	73	73	2.54	0.000
Levene's Test (any continuous)	**1**	**146**	**32.22**	**0.000**

Table 9. Statistical median testing for MAL.

Mann-Whitney Test and CI: EDD MAL by job, MAL by Job-REG SPT

	N	Median
EDD MAL by job	74	10.397
MAL by Job-REG SPT	74	6.191

Point estimate for ETA1-ETA2 is 1.696
95.0 Percent CI for ETA1-ETA2 is (0.110,3.269)
W = 6058.0
Test of ETA1 = ETA2 vs ETA1 not = ETA2 is significant at 0.0368

This process is conducted offline to the everyday operation of RegSPT and given that most ETO firms use ERP systems, the historical data needed to compute the coefficients should be simple to access and export to a spreadsheet. The amount of time to accomplish this task depends on the firm's systems/data cleanliness and availability, but it should be on the order of several hours.

Next, given that the regression equation is up-to-date, the jobs which are to be run through the engineering step must be sequenced on a regular predetermined basis (typically weekly). In order to generate the input data required for the sequencing algorithm, the predicted process time for each job needs to be generated. In order to accomplish this task, personnel simply need to enter the aforementioned seven pieces of complexity information (x_i's) into the regression equation which can be encoded into a spreadsheet formula. At this point, it needs to be stated that the process of searching for these data is not an extra step that results from the use of RegSPT, but is actually already currently being performed as a normal part of the engineering review process. Because the job package is reviewed as a normal part of the engineering review process, the seven input data values can be quickly gathered and the RSA executed to generate the predicted processing times. The jobs are then sequenced in ascending order of predicted processing times. Once the input data are extracted, based upon discussions with the five firms studied, it is estimated that the entire sequencing process should typically take 15–20 min vs. the roughly one hour that it takes today. Overall, this algorithmic approach is quick, user-friendly, and runs using spreadsheet programs and tools that are well known by ETO practitioners.

9. Summary

The ETO environment drives complexity and variability which, in turn, provides a justification for administrative convenience. Administrative convenience leads personnel to choose methods which are familiar and easy. Unfortunately, per O'Neil (1991), administrative convenience also often pushes in a direction counter to best performance and, in this case, the use of ad hoc approaches that yield poor results. To combat administrative convenience, O'Neil states that simple, easy to understand robust heuristic approaches are needed and more likely to be implemented. In fact, Stoop and Wiers (1996) state that such simple and robust approaches are the most broadly used in practice. Therefore, research is needed which helps practitioners in terms of choosing the correct simple and robust scheduling rules for the ETO engineering environment. Yet, despite this need, there has been little sequencing research focused on the ETO engineering environment. Therefore, this research has focused on a more robust approach for enhancing scheduling in the ETO environment concentrating on what has been called the main bottleneck in the ETO environment, namely the engineering process.

The engineering process is distinct from traditional shop floor processes as its inputs and outputs are purely intellectual and not physical in nature. These and other issues have led to an overriding question: how can scheduling capabilities which co-exist with existing ERP systems be developed for the highly complex ETO engineering environment while, at the same time, remaining realistic, robust, acceptable, and easily implemented? Restated, the literature survey and the results of the study of the partner firms have shown that what is needed is a scheduling policy which focuses on the engineering process, is robust and relatively simple to implement and use, and is complementary to an ERP system.

As has been demonstrated, the RegSPT sequencing approach, as applied to the ETO engineering environment, meets all of the above goals. First, RegSPT is robust and simple. It requires only seven pieces of information in order to accurately predict the job processing time. These data are available in a typical job quotation package which is received from sales. Because the job package is reviewed as a normal part of the engineering review process, the seven data points can be quickly gathered and the RSA executed. In addition, the algorithm runs using a spreadsheet program which are extremely user-friendly and well known by ETO practitioners. Once the input data are extracted, it is estimated that the entire sequencing process should typically take 15–20 min vs. roughly one hour today.

Second, RegSPT is accurate. The industry and literature approaches were rigorously tested against one another and RegSPT exhibited the best performance in all categories compared to its next closest competitor, EDD. In particular, it achieved statistically better results for MAL (7.84 days vs. 10.55 days). Since lower values of absolute lateness are important for lean manufacturing, this is an important finding. In addition, RegSPT achieved a statistically significantly lower SDL compared to the same competitor (7.11 days vs. 11.33 days). Given the high amount of variability found in the ETO engineering environment, these results are fairly impressive. Lastly, it is known that ERP systems have the capability to export and import data to and from the spreadsheet software. Hence, since RegSPT can easily be implemented using spreadsheet programs, it is compatible with typical ERP systems. Therefore, there is no need to jettison an existing costly ERP system.

In terms of next steps, while RegSPT has been shown to be an improved method of sequencing in the ETO engineering environment, admittedly there are other possible approaches which can take advantage of the REG approach and perhaps the use of a complexity-driven regression might yield in an incremental improvement in performance. Also, an important part of the complexity-driven regressive approach is the determination of p, the number of time periods needed before data refresh. In this study, this is left up to the practitioner to determine, but further studies on an analytic method to determine the parameter p would be beneficial to the industrial community.

ORCID

John M. Usher http://orcid.org/0000-0002-4762-6989

References

Baker, K. R. (1984). Sequencing rules and due-date assignments in a job shop. *Management Science, 30*, 1093–1104.

Baker, K. R., & Bertrand, J. V. M. (1981). A comparison of due date selection rules. *AIIE Transactions, 13*, 123–131.

Baker, K. R., & Trietsch, D. (2009). *Principles of sequencing and scheduling*. New York, NY: Wiley.

Brucker, P., Drexl, A., Moehring, R., Neumann, K., & Pesch, E. (1999). Resource-constrained project scheduling: Notation, classification, models, and methods. *European Journal of Operational Research, 112*, 3–41.

Carroll, D. C. (1965). *Heuristic sequencing of single and multiple component jobs* (Unpublished Ph.D. dissertation). Sloan School of Management, MIT, Cambridge, MA.

Coates, G., Ritchey, I., Duffy, H. B., Hills, W., & Whitfield, R. I. (2000). Integrated engineering environments for large complex products. *Concurrent Engineering: Research and Applications, 8*, 171–182.

Conway, R. W. (1965a). Priority dispatching and work-in-process inventory in a job shop. *Journal of Industrial Engineering, 16*, 123–130.

Conway, R. W. (1965b). Priority dispatching and job lateness in a job shop. *Journal of Industrial Engineering, 16*, 228–237.

Conway, R. W., & Maxwell, W. L. (1962). Network dispatching by shortest operation discipline. *Operations Research, 10*, 51–73.

Conway, R. W., Maxwell, W. L., & Miller, L. W. (1967). *Theory of Scheduling*. Reading, MA: Addison-Wesley.

Cutler, T. R. (2005). Are you ETO? The increasing importance of custom (engineer-to-order) manufacturers. *In MFG*, pp. 1–3.

Cutler, T. R. (2009). Special orders, engineer to order challenges for the industrial engineer. *Industrial Engineer, 41*, 36–38.
Elvers, D. A. (1973). Job shop dispatching using various due date setting criteria. *Production and Inventory Management, 14*, 62–69.
Elvers, D. A., & Taube, L. (1983). Time completion for various dispatching rules in a job shop. *Omega, 11*, 81–89.
Gere, W. S. (1966). Heuristics in job shop scheduling. *Management Science, 13*, 167–190.
Gosling, J., & Naim, M. M. (2009). Engineer to order supply chain management: A literature review and research agenda. *International Journal of Production Economics, 122*, 741–754.
Grabenstetter, D. H., & Usher, J. M. (2013). Determining job complexity in an engineer to order environment for due date estimation using a proposed framework. *International Journal of Production Research, 51*, 5728–5740.
Grabenstetter, D. H., & Usher, J. M. (2014). Developing due dates in an engineer-to-order engineering environment. *International Journal of Production Research, 52*, 6349–6361.
Harrell, C., Ghosh, B. K., & Bowden, R. O. (2004). *Simulation using ProModel* (2nd ed.). New York, NY: McGraw-Hill.
Hicks, C., & Braiden, P. M. (2000). Computer-aided production management issues in the engineer-to-order production of complex capital goods explored using a simulation approach. *International Journal of Production Research, 38*, 4783–4810.
Hicks, C., & Pongcharoen, P. (2006). Dispatching rules for production scheduling in the capital goods industry. *International Journal of Production Economics, 104*, 154–163.
Hicks, C., & Pongcharoen, P. (2009). Applying different control approaches for resources with high and low utilisation: A case study of the production of complex products with stochastic processing times. *International Journal of Technology Management, 48*, 202–218.
Hicks, C., McGovern, T., & Earl, C. F. (2000). Supply chain management: A strategic issue in engineer-to-order manufacturing. *International Journal of Production Economics, 65*, 179–190.
Hicks, C., McGovern, T., & Earl, C. F. (2001). A typology of UK engineer-to-order companies. *International Journal of Logistics: Research and Applications, 4*, 43–56.
Hicks, C., Song, D. P., & Earl, C. F. (2007). Dynamic scheduling for complex engineer to order products. *International Journal of Production Research, 45*, 3477–3503.
Kanet, J. J., & Hayya, J. C. (1982). Priority dispatching with operation due dates in a job shop. *Journal of Operations Management, 2*, 167–175.
Kutner, M. H., Nachtsheim, C. J., & Neter, J. (2004). *Applied linear regression models* (4th ed.). New York, NY: McGraw-Hill Irwin.
Little, D., Rollins, R., Peck, M., & Porter, J. K. (2000). Integrated planning and scheduling in the engineer to order sector. *International Journal of Computer Integrated Manufacturing, 13*, 545–554.
Miyazaki, S. (1981). Combined scheduling system for reducing job tardiness in a job shop. *International Journal of Production Research, 19*, 201–211.
Muhlemann, A. P., Lockett, A. G., & Farn, C. K. (1982). Job shop scheduling heuristics and frequency of scheduling. *International Journal of Production Research, 20*, 227–241.
O'Neil, P. (1991). *Performance of lot dispatching and scheduling algorithms through discrete event simulation*. IEEE/SEMI International Semiconductor Manufacturing Science Symposium (pp. 21–24). San Francisco, CA.
Pandit, A., & Zhu, Y. (2009). An ontology-based approach to support decision-making for the design of ETO (engineer to order) products. *Automation in Construction, 16*, 759–770.
Pongcharoen, P., Hicks, C., Braiden, P. M., & Stewardson, D. J. (2002). Determining optimum genetic algorithm parameters for scheduling the manufacturing and assembly of complex products. *International Journal of Production Economics, 78*, 311–322.
Pongcharoen, P., Hicks, C., & Braiden, P. M. (2004). The development of genetic algorithms for the finite capacity scheduling of complex products, with multiple levels of product structure. *European Journal of Operational Research, 152*, 215–225.
Qi, X., & Tu, F. S. (1998). Scheduling a single machine to minimize earliness penalties subject to the SLK due-date determination method. *European Journal of Operational Research, 105*, 502–508.
Salman, N., & Dogru, A. (2004). Design effort estimation using complexity metrics. *Journal of Integrated Design and Process Science, 8*, 83–88.

Song, D. P., Hicks, C., & Earl, C. F. (2001). Stage due date planning for multistage assembly systems. *International Journal of Production Research, 39*, 1943–1954.

Song, D. P., Hicks, C., & Earl, C. F. (2002). Product due date assignments for complex assemblies. *International Journal of Production Research, 76*, 4877–4895.

Song, D. P., Hicks, C., & Earl, C. F. (2006). An ordinal optimization based evolution strategy to schedule complex make to order products. *International Journal of Production Research, 44*, 4877–4895.

Stevenson, M., Hendry, L. C., & Kingsman, B. G. (2005). A review of production planning and control: The applicability of key concepts to the make-to-order industry. *International Journal of Production Research, 43*, 869–898.

Stoop, P. P. M., & Wiers, V. C. S. (1996). The complexity of scheduling in practice. *International Journal of Operations and Production Management, 16*, 37–53.

Weeks, J. K. (1979). A simulation study of predictable due dates. *Management Science, 25*, 363–373.

Worthen, B. (2002). Nestlé's ERP Odyssey. *CIO, 15*, 1–9.

14

Lean thinking for a maintenance process

Sherif Mostafa[a]*, Sang-Heon Lee[b], Jantanee Dumrak[c], Nicholas Chileshe[a] and Hassan Soltan[d]

[a]School of Natural and Built Environments, University of South Australia, Adelaide 5000, Australia; [b]School of Engineering, University of South Australia, Mawson Lakes 5095, Australia; [c]Global Management Program, Torrens University Australia, Adelaide, SA 5000, Australia; [d]Production Engineering and Mechanical Design Department, Faculty of Engineering Mansoura University, Mansoura 35516, Egypt

The maintenance process shares significant operating costs in an organisation. Lean thinking can be incorporated into maintenance activities through applying its principles and practices/tools. Lean maintenance (LM) is a prerequisite for lean manufacturing systems. This research proposes a new structure for LM process based on a systematic literature review of a significant number of related articles that were published on LM. The process structure is designed based on the five lean principles to guide and support organisations to pursue maintenance excellence. This study establishes a scheme for LM tools that are structured into 2 level 4 bundles and 26 lean practices/tools and develops a House of Waste (HoW) to demonstrate the association between maintenance wastes and the LM tools. With a successful accomplishment of the proposed scheme, the performance of a maintenance department can create more improvement opportunities over time to reach the maintenance excellence status.

Keywords: total productive maintenance (TPM); lean maintenance; maintenance wastes and value stream mapping; lean maintenance tools; House of Waste (HoW)

1. Introduction

Maintenance function becomes a significant contributor towards to achieve strategic objectives of an organisation in today's competitive markets (Fraser, 2014). The maintenance process is to serve the production facilities of high productivity. It comprises planned and unplanned actions carried out to retain a physical asset to the acceptable operating conditions (Faccio, Persona, Sgarbossa, & Zanin, 2014). Maintenance aims at increasing the value of the reliability, safety, availability, and quality of an asset (e.g. production plant, equipment, or building) with acceptable economical costs (Márquez, 2007). Over the last decades, the maintenance has been considered as a necessary evil from the organisational management as the maintenance operation is limited to corrective functions that are usually executed under the emergency situations such as machine breakdown. However, this practice is no longer acceptable since the role of maintenance has been recognised as a strategic element of revenue generation for organisations. The maintenance with role creates a significant impact on some critical elements in production plants such as product quality, safety requirements, and operating budget levels (Khazraei & Deuse, 2011).

*Corresponding author. Email: Sherif.Mostafa@mymail.unisa.edu.au

The cost of maintenance activities could be ranged from 15 to 70% of the total production costs (Fraser, 2014; Pinjala, Pintelon, & Vereecke, 2006). This is the second largest part, after energy costs, of the operational budget. In the United States, the estimated cost of maintenance increased from $200 billion in 1979 to $600 billion in 1989 (Bevilacqua & Braglia, 2000). Maintenance activities account for an average 28 per cent of the total cost of finished goods (Blanchard, 1997). One of the reasons for such significant portion of maintenance of the total operation cost is that the machinery has become highly automated and technologically very complex. For example, usually, the modern operation systems depend on sensor-driven management systems that provide alerts, alarms, and indicators. Consequently, maintenance costs are expected to be even higher in future. Generally, the maintenance costs are proportional to the downtime (DT). The DT is the time interval when equipment/system is down for maintenance until it is back to the normal working conditions (Tinga, 2013). The increased DT is caused by the non-value added (NVA) activities or wastes within the entire maintenance process. One of waste elimination strategies is the application of lean thinking in all activities between suppliers and customers (value stream). Integrating lean thinking in maintenance is known as lean maintenance (LM). Baluch, Abdullah, and Mohtar (2012) emphasised that LM is prerequisite for success of a lean manufacturer as it provides a holistic approach to the function of maintenance. In general, the lean integration in any process is carried through adopting lean principles which begins with specifying the customer value (Bhasin, 2015). In the maintenance environment, any maintenance service could be considered as a final intangible product. The service is provided to a customer which, in this case, could be assumed as an asset (e.g. production line). Therefore, it is essential to identify the value from the asset perspective which can be improving its availability and reliability through efficient maintenance. Then, mapping the maintenance value stream which fundamentally consists of all the collective activities to deliver the maintenance service. Later, improving the maintenance value stream by abolishing the waste which assist in minimising the lead time (in this case is DT).

Investigation into the applicability of lean principles in maintenance in the existing research is still at a marginal level. Davies and Greenough (2010) emphasis on the necessity for more researches to apply lean principles to maintenance operations. Ghayebloo and Shahanaghi (2010) formulate a multi-objective decision-making (MODM) model that can determine the minimal level of maintenance requirements (i.e. labour and spare parts) which satisfies expected reliability level with the use of the lean concept. Tendayi and Fourie (2013) use a combined approach of quality function deployment (QFD) and analytic hierarchy process (AHP) to evaluate the importance of a set of maintenance excellence criteria and prioritise the lean tools against these criteria. Soltan and Mostafa (2014) introduce a framework for measuring maintenance strategies based on lean and agile components, i.e. waste removal and responsiveness. The study of McCarthy and Rich (2004) discussed lean total productive maintenance system (or lean TPM) which conceptualises the application of lean-specific techniques in TPM. The system is focused on maintaining equipment in its optimal operational state and continually improving its productivity. However, an integrative structure of lean thinking or lean TPM (e.g. principles, practices/tools, waste identification, and value stream mapping (VSM)) within the maintenance activities has not been fully established. This provides an opportunity for this study to develop and propose a process for lean thinking to be integrated to the maintenance operation. The process is formulated according to the hypothesis of Womack and Jones (2003) stating that lean principles can be applied to any sector.

This study carried out a systematic literature review following Denyer and Tranfield (2009) approach. The review was conducted to identify and understand the existing literature on LM and evaluate contributions and summarise knowledge, thereby, identifying potential directions of future research. The grand electronic databases were explored to gather the literature on LM. A total of 43 related articles published between 2000 and 2014 have been included in this study.

This study is organised as follows. Section 2 provides an overview on maintenance and lean manufacturing concepts. Sections 3 and 4 demonstrate the aim, objectives, and methodology of the proposed research. Section 5 discusses the LM and LM practices/tools. In Section 6, a LM process is proposed with the discussion of the five stages of the lean process in maintenance and introduction of the LM scheme including four bundles and 26 lean tools. The section also demonstrates the relationships between maintenance wastes and lean tools using House of Waste (HoW). Section 7 highlights benefits of the proposed LM process and Section 8 concludes the study with some directions of future research.

2. Maintenance and lean manufacturing

2.1. Lean manufacturing

The word lean in manufacturing means the efficient use of the available resources by cutting the NVA activities or wastes (Carrasqueira & Machado, 2008). Lean manufacturing represents a collection of tools that work together synergistically to create a streamlined, high-quality system that produces finished products at the same pace of the customer demand (Shah & Ward, 2007). Waste in lean manufacturing is defined as any activities that add cost to a product/service without adding values from a customer's perspective. It may be classified to three major types: unobvious waste, less obvious waste, and obvious waste (Hopp & Spearman, 2004). De Treville and Antonakis (2006) list examples of obvious waste as unnecessary inventory, unneeded processes, excessive set-up times, unreliable machines, and rework. They also argue that the less obvious waste occurs due to the various reasons such as process times, delivery times, yield rates, staffing levels, and demand rates.

In recent decades, lean system is gaining a momentum across different industrial sectors. It has been originally started as Toyota production system which describes the manufacturing philosophy of Toyota Motor Corporation (Dombrowski & Malorny, 2014; Holweg, 2007; Womack & Jones, 2003). This lean system has been successfully extended to service industries, such as maintenance service, retail banking, airlines, restaurants, public sector, education, food, and hospitals (Burgess & Radnor, 2013; George, 2003; Pedersen & Huniche, 2011; Resta, Powell, Gaiardelli, & Dotti, 2015; Smith & Hawkins, 2004; Thirkell & Ashman, 2014; Zarei, Fakhrzad, & Jamali Paghaleh, 2011). As an example, in the health care sector, hospitals achieved high-quality health care outcomes (e.g. recorded lower 30-day mortality rates) by applying lean management practices (McConnell, Lindrooth, Wholey, Maddox, & Bloom, 2013). Clearly, the implementation of lean concept in diverse sectors proves to be true Womack and Jones (2003) hypothesis stating that lean principles can be transferred to any organisation. They coined this process as lean thinking which refers to the thinking process of lean inside an organisation and its extended supply chain. This means that within the same organisation, lean thinking should be extended from the shop floor or production level to other areas such as maintenance department. The key behind this lean thinking is that service/maintenance departments and production processes are inseparable and complement each other to sustain the competitive edge of an organisation.

2.2. *Maintenance strategies*

Maintenance includes all activities required to keep an asset at maximum operating condition. The activities are usually carried out according to a certain maintenance strategy. The maintenance strategies may have developed accordingly with the development of manufacturing systems (Shahin, Shirouyehzad, & Pourjavad, 2012). In the early days, maintenance had been mainly concentrated around corrective maintenance, for example, repairs and replacements were conducted when needed with no optimisation strategy and there was no or little consideration of the DT. More recently, maintenance became a full-scale function, instead of a sub-function of the whole production operation. Today, maintenance management becomes a complex function, encompassing technical and management skills, while still requiring flexibility to cope with the dynamic business environment (Lee & Wang, 2008). Maintenance strategies have gradually changed from preventive maintenance (PM) (including condition-based maintenance (CBM) and time-based maintenance (TBM) to design-out maintenance (DOM) and total productive maintenance (TPM) as demonstrated in Figure 1.

The classification of the maintenance strategies shown in Figure 1 is based on the time of maintenance activities and failure that requires maintenance (Potes Ruiz, Kamsu Foguem, & Grabot, 2014). The maintenance activities are only performed after the failure occurrence in the corrective maintenance strategy. Whereas, in the PM, the intervention of maintenance activities is conducted before the failure occurrence. Maintenance strategies have been diversely used in the existing literature using similar terms such as preventive, predictive, planned, corrective, and TPM. The most common three maintenance strategies are discussed below.

2.2.1. *Corrective maintenance*

Corrective maintenance is known as failure-based maintenance, emergency maintenance, fire-fighting maintenance, or breakdown maintenance as the concept of corrective

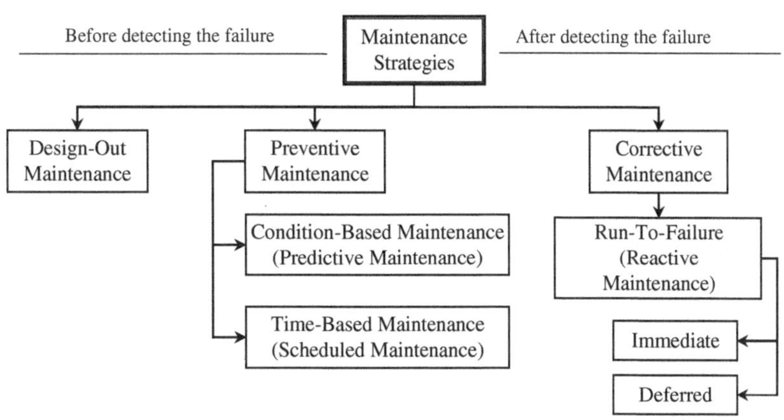

Figure 1. Types of maintenance strategies.

maintenance strategy is based on fix-it when broke (Márquez, 2007). Corrective maintenance is a conventional maintenance strategy appeared early in the industry. It has been employed in maintenance operations due to knowledge shortage on the equipment failure behaviours (Waeyenbergh & Pintelon, 2002). Corrective maintenance can be carried out immediately or deferred by appropriate maintenance technicians whom are contracted to assess the situation and fix the repairs. In situations where the failure is not critical (i.e. plenty of DT is available) and the values of the assets are not of a great concern, the corrective mode of maintenance may prove to be an acceptable option. However, the market competition, environmental and safety issues force the maintenance managers to search for more efficient maintenance strategies besides the corrective maintenance (Shahin et al., 2012).

2.2.2. *Preventive maintenance*

PM is carried out according to prescribed criteria. It intends to reduce the probability of failure or the degradation of the functioning of an item (Fouladgar, Yazdani-Chamzini, Lashgari, Zavadskas, & Turskis, 2012). PM can be divided into TBM and CBM. In TBM, the maintenance activities are performed based on fixed operating time interval or number of output units without considering the current condition state of the item. On the other hand, CBM is based on performance and/or parameter monitoring (e.g. corrosion and electric current monitoring, lubricant and vibration analysis, leak and crack detection, and ultrasonic testing) (Al-Najjar & Alsyouf, 2003; Khazraei & Deuse, 2011). CBM could be described as a process that integrates technology and human skills using a combination of all available diagnostic and performance data, maintenance history, operator logs, and design data to determine the likelihood of a potential failure. As a result, CBM requires a high initial cost to acquire and install the necessary sensors as well as to monitor technology (Hellingrath & Cordes, 2014; Nezami & Yildirim, 2013).

DOM focuses on improving the design of a product in order to eliminate the cause to maintenance. DOM makes maintenance easier during the life cycle of a product (Waeyenbergh & Pintelon, 2004). DOM is based on the successive design corrections derived from the maintenance knowledge. It is appropriate for items with high maintenance cost, which arises because of defective design or operation outside design specifications. The DOM concept is used in some parts of motor vehicles such as permanent bearing (bearing using solid lubricant and permanently sealed) (Gopalakrishnan & banerji, 2013).

2.3. *TPM and lean thinking*

TPM is a concept developed in Japan, which could be defined as a productive maintenance that involves total participations meaning all employees across the whole levels of the operational hierarchy (Cua, McKone, & Schroeder, 2001; Jasiulewicz-Kaczmarek, 2013). It attempts to eliminate any losses in equipment and production efficiency through active team-based participation. Waeyenbergh and Pintelon (2002) identified the six categories of losses in TPM: (1) breakdown losses; (2) set-up and adjustment losses; (3) minor stoppage/idling losses; (4) reduced speed losses; (5) defects/rework losses; (6) start-up losses. The first two losses refer to time losses and influence equipment availability. The third and fourth losses denote speed losses and measure the equipment performance efficiency. The last two losses designate the quality of production process and are used to calculate the quality rate of the equipment (Chan, Lau, Ip, Chan, & Kong, 2005). The effectiveness of TPM strategy with regard to these six losses is measured using overall equipment effectiveness (OEE). The measurement of OEE is a function of the availability, performance efficiency, and quality. TPM entails eight main elements/pillars that can be considered as principles/tools of TPM in an organisation.

- Autonomous maintenance
- Performance improvement
- Early equipment management
- Planned maintenance
- Environment health and safety
- Office TPM
- Education and training
- Quality maintenance

The detailed description of these pillars mentioned above is presented in Ahmed, Hassan, and Taha (2005) and Chong, Chin, and Hamzah (2012). The TPM pillars can be collectively used to improve equipment availability and reliability. However, according to Baluch et al. (2012), removing or missing application of any pillars could lead to unattainable results. Likewise, the lack of comprehensive approach for the TPM implementation has resulted in a decrease of 50% TPM initiatives in US. Moreover, there are some deficiencies related to the TPM strategy. TPM concerns the operational issues that handle equipment failures rather than long-term strategic business issues. Levitt (2008) stated that TPM is no longer lean as it focuses primarily on the operational issues of equipment and it would take a long time to implement. Moreover, EPA (2015) highlighted that TPM failed to consider the environmental aspects during equipment efficiency improvement. This leaves potential waste minimisation and pollution prevention opportunities to be more researched.

While referring to major literature on TPM, it was observed that there is a link between TPM and lean as well as LM emerged as an effective maintenance strategy. Levitt (2008) mentioned that TPM and lean are broadly linked. This is clear in some of the literature that considered TPM as a subsection of lean system which mainly focused on enhancing the efficiency and availability of the manufacturing facilities. For instance, Shah and Ward (2003) postulated 21 manufacturing practices as the key features of lean system. They grouped the inter-related and internally consistent practices into bundles. TPM was one bundle which contained four practices: PM, maintenance optimisation, new process equipment, and safety improvement. These four practices sound very generic and not provide practitioners what tools to undertake the maintenance optimisation, or new process equipment. Another study by Mostafa (2011) extended the TPM practices, suggested by Shah and Ward (2003), to include housekeeping, cross-training and teams of maintenance technicians, operator involvement (autonomous maintenance), and information tracking of the work orders. Clearly, these suggested practices indicated another dimension of TPM in lean system by considering human resource perspective. Abdulmalek, Rajgopal, and Needy (2006) gathered 14 lean tools into three areas: quality, production process, and methods. TPM was located under quality area which can be considered as another perspective for TPM in lean environment. It can be concluded that TPM is considered as the foundation for the maintenance process in lean environment which must be supplemented with some lean practices. That is why some studies coined relatively new terms like lean TPM (Georgescu, 2010; McCarthy & Rich, 2004) and LM (Baluch et al., 2012; Okhovat, Ariffin, Nehzati, & Hosseini, 2012; Romano, Murino, Asta, & Costagliola, 2013; Smith & Hawkins, 2004).

LM principles take its lead from lean manufacturing through applying some new techniques to TPM concepts to render a more structured implementation path (Levitt, 2008; McCarthy & Rich, 2004; Romano et al., 2013; Smith & Hawkins, 2004). It is a prerequisite for success as a lean manufacturer that provides a holistic approach to the function of maintenance (Baluch et al., 2012; Soltan & Mostafa, 2014). As lean concept has been taking hold in the manufacturing sector, there is an increasing insight that maintenance must not be seen only from narrow operational perspective dealing with equipment failures and their consequences. Instead, maintenance must be viewed in the long-term strategic perspective and must integrate the different technical and commercial issues in an effective manner. However, LM approach cannot just be a mirror image of a lean manufacturing approach because the business dynamics of asset maintenance and those of production are fundamentally different (Baluch et al., 2012; Brown, Collins, & McCombs, 2006; Clarke, Mulryan, & Liggan, 2010). Therefore, it is clear that there is a need to develop an effective process to collectively integrate lean thinking into the maintenance with long-term strategic perspective. This study addresses such an issue to cover the major deficiencies in the existing literature.

3. Research aim and objectives

This research aims to adopt lean principles and practices/tools and collectively integrate them into the maintenance process. It has adopted recommendations of previous research works such as a study of Davies and Greenough (2010) that emphasised on the necessity of conducting more research in application of lean manufacturing principles in maintenance operations. To achieve the main aim of this research, four main objectives have been established:

(1) To review the LM concept.
(2) To develop a process for adopting lean principles into maintenance processes.
(3) To document the maintenance wastes and establish a scheme of lean practices/tools applied to maintenance.
(4) To develop the HoW that demonstrates the association of lean tools with the maintenance wastes.

4. Research methodology

To achieve the research objectives, this study employed a systematic review of the literature in order to explore major publications related to lean manufacturing and maintenance concepts, especially research works published from 2001 to 2014. Okoli and Schabram (2010) indicate that a systematic literature review is a systematic, explicit, comprehensive, and reproducible method for identifying, evaluating, and synthesising the existing body of completed and recorded work produced by researchers, scholars, and practitioners. According to Denyer and Tranfield (2009), the systematic literature review has become an essential scientific activity.

From the literature review, it has been found that exploration of the applicability of lean principles in maintenance in the existing research works is minimal. Davies and Greenough (2010) and Soltan and Mostafa (2014) state the necessity of conducting more research on applying lean principles in maintenance operations. As an initial step of this research, a seven-step systematic literature is conducted as summarised in Figure 2 to demonstrate development of the proposed LM process. It is noted that these research steps are adapted from several academic sources, such as Joffe (2011), Okoli and Schabram (2010), and Tranfield, Denyer, and Smart (2003).

4.1. Research selection criteria

The research inclusion and exclusion criteria are very critical to the quality assessment of papers (Booth, Papaioannou, & Sutton, 2012). Okoli and Schabram (2010) indicated that simplifying research using these criteria (by reviewing the title and then the abstract when needed) saves the researcher time and effort. In this study, the authors examined research articles by title and then abstracts. In the systematic review of this study, the criteria were addressed in order to clarify the selection of the research-related articles. The following criteria have been considered to include/ exclude the articles:

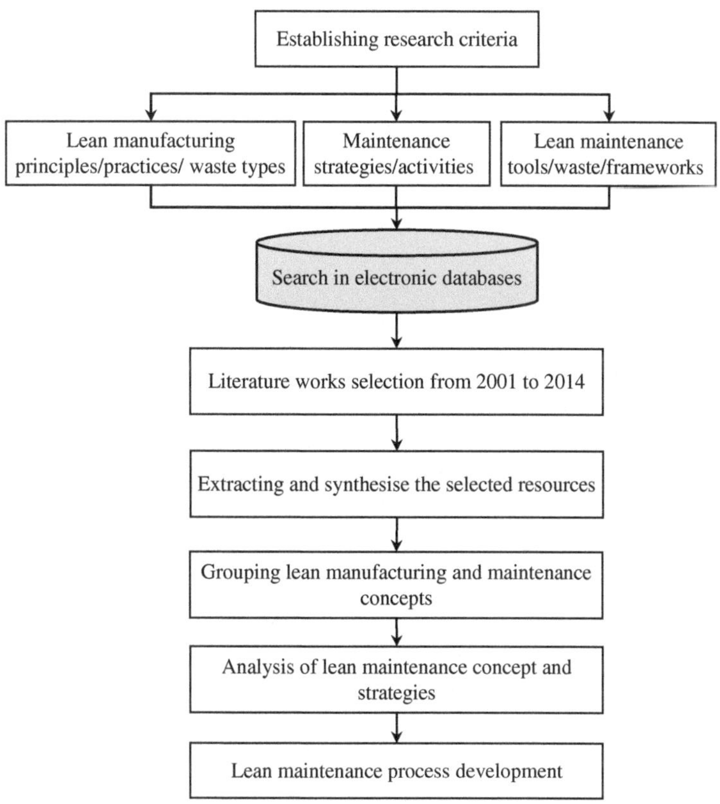

Figure 2. Research methodology framework.

- Articles published between 2001 and 2014 as the rational for considering the year 2001 is that LM as a field of research was firstly addressed by Davies and Greenough (2001)
- Search for articles published in peer-reviewed scientific journals or conference proceedings
- Ensure essential relevance by requiring that selected articles contain at least one of the search keywords in the title or abstract (see Table 1)
- Search well-known online databases which are Taylor & Francis, EBSCO, Emerald, IEEE, Inderscience, ProQuest, Sage, Science Direct, and Springer Link

5. Development of LM in the literature

The screened LM studies are dated from 2001 to 2014 with the total number of 43 retrievable publications as shown in Table 2.

The selection of published LM studies shows that most of the studies were published in the year 2014 and focused on three aspects, namely LM implementation initiative, suggested lean tools, and interrelationship within LM application as illustrated in Figure 3.

From the review of 43 publications, the focus on the LM implementation has been continuously growing since 2011 for general practice and specific industries. The suggested lean tools and interrelationship within the LM initiatives were not in parallel with the LM implementation. The LM tools were found in 18 publications. None of these publications provided a complete set of lean tools. Only two publications in 2012 and 2013 demonstrated the interrelationship within LM application. The low volume of publications indicates that more research in LM is crucially needed, especially in establishing a logical interrelationship between LM concepts and implementation and a complete set of LM tools.

5.1. Lean maintenance

LM term was coined in the last decade of the twentieth century. Smith (2004) defines LM as a proactive maintenance operation employing planned and scheduled maintenance activities through TPM practices using maintenance strategies developed through application of reliability-centred maintenance (RCM) decision logic and practiced by

Table 1. Research primary and secondary keywords.

Time period considered	Search engines used	Primary keywords	Secondary keywords
2001–2014	Taylor & Francis, EBSCO, Emerald, IEEE, Inderscience, ProQuest, Sage, Science Direct, and Springer Link	• Lean maintenance • Lean TPM	• Lean principles AND maintenance • Lean practices AND maintenance • Lean tools AND TPM

Table 2. Review of existing literature on lean maintenance.

S. no.	Author(s) and year	Key aspects covered	Application industry	Suggested lean maintenance implementation	Limitations
1	Davies and Greenough (2001)	(1) Investigate the level of presence of lean thinking within the maintenance domain and its influence and interaction with the maintenance function		No	No
2	Davies and Greenough (2003)	(1) A set of performance measures that can be used to analyse the impact of lean thinking within the maintenance function		No	No
3	Davies (2003)	(1) Explained the contribution of lean thinking to the maintenance of manufacturing system (2) An overall measure of maintenance performance comprised a number of indicators that signify change through maintenance activity has been suggested		No	No
4	Heisler (2003)	(1) Highlight the role of effective planning and scheduling of maintenance process in lean environment	General-planning and scheduling	Lean maintenance process flow	• Present the flow of maintenance processes for the lean maintenance planning and scheduling • No lean tools or principles demonstrated
5	Smith (2004)	(1) Defined the lean maintenance (2) Discussed the elements for lean maintenance success		No	

(Continued)

Table 2. (*Continued*).

S. no.	Author(s) and year	Key aspects covered	Application industry	Suggested lean maintenance implementation	Limitations
6	Smith and Hawkins (2004)	(1) Reviewed the history and evolution of lean (2) Explained the lean maintenance element and TPM (3) Discussed the pre-planning and mobilising lean maintenance (4) Introduced the key elements for sustain lean	General	The maintenance management pyramid	• Limited tools provided • Flow of the implementation process is limited
7	Verma and Ghadmode (2004)	(1) A model for ship fleet repair and maintenance based upon lean principles and current best practices	Ship fleet repair	Proposed implementation model for ship fleet repair	• The model contained 7 steps with no interrelationship or logic to apply them • No feedback or measuring of the performance • No lean tools suggested
8	Zwas (2006)	(1) Investigated how lean techniques can be employed in bus and rail maintenance		No	
9	Colbert (2008)	(1) Benefits of integrating lean into IR predictive maintenance programmes (2) Lean concepts/tools are able to better manage and maintain the real value of their IR programmes while drastically increasing their ROI		No	No
10	Van den Heuvel (2008)	(1) Increasing the efficiency of the activities during regular maintenance shutdowns at steel Sheet Plant using lean and six sigma approaches		No	No

11	Marks (2008)	(1) Created an accurate and enhanced CMMS database and a safe and efficient Rotable Plant storage system (2) Explored and implemented an Effective Rotable Plant Management Programme that fulfils the lean maintenance strategy (3) Demonstrated how all value adding in terms of Effective Rotable Plant Management is maintained using very little resources		No	No
12	Yile, Hang, and Lei (2008)	(1) Based on the value stream analysis (VSA), inherit 'value' relation of TBM, RCM, and remanufacture engineering is established (2) A simple layered technology framework of lean maintenance	Frictional clutch maintenance	Maintenance management pyramid	• The pyramid defined the lean maintenance as an integration of 13 practices • No limitation process flow were provided • The lean principles were not included
13	Georgescu and Militaru (2009)	(1) A lean TPM approach to the current understanding of the TPM system (2) 7Ss as a critical first step in any improvement programme; instant maintenance; improvement set-up operations		No	No
14	Thiruvengadam (2009)	(1) A method for assessing the breakdown maintenance factors using a value stream maintenance map in a clearly and precisely measurable technique		No	No

(Continued)

Table 2. (*Continued*).

S. no.	Author(s) and year	Key aspects covered	Application industry	Suggested lean maintenance implementation	Limitations
15	Clarke et al. (2010)	(1) One of the most important aspects of lean maintenance is developing an understanding of the maintenance processes and applying a risk-based approach in a pharmaceutical asset maintenance. This involves evaluating whether each element of maintenance practice used adds value to the product and benefits the customer		No	No
16	Davies and Greenough (2010)	(1) A lean practice template which represent activities possible within a company's maintenance department (2) Performance indicators to identify the impact of lean thinking within maintenance activities		No	No
17	Georgescu (2010)	(1) A new maintenance system called lean TPM. It is a systematic approach based on three techniques: the 7Ss as a critical first step in any improvement programme; instantaneous maintenance; improvement set-up operations		No	No
18	Ghayebloo and Shahanaghi (2010)	(1) A MODM model to determine minimum level of maintenance requirements and to satisfy expected reliability level		No	No
19	Sheng and Tofoya (2010)	(1) Four core lean rules (activities, connections, flows, and improvements) transferred in the maintenance context		No	No

20	Ayeni, Baines, Lightfoot, and Ball (2011)	(1) Application of lean within MRO in the aviation industry		No	No
21	Kolanjiappan and Maran (2011)	(1) Application of root cause failure analysis (RCFA), RCM, Failure mode and effect analysis, and maintenance procedure effectiveness analysis	Aircraft maintenance	Schematic diagram for servicing time reduction of aircrafts	• The focus of the diagram is to reduce the service time not to implement lean maintenance • process flow is missing • no lean tools provided
22	Qiang, Zhu, and Li (2011)	(1) The characteristics of the military equipment lean maintenance (MELM) and main components for improving the MELM	Military equipment	Components of MELM	• The MELM contains three stages: models of maintenance, components of MELM, and output • No implementation introduced • Lean tools has been suggested
23	Baluch et al. (2012)	(1) Lean maintenance cannot be a mirror of lean production as the business dynamics of asset maintenance and production are fundamentally different (2) TPM and lean initiatives that allowed organisations to focus on improving the efficiency of production processes		No	No
24	De Silva, Ranasinghe, and De Silva (2012)	(1) Improving maintainability of buildings can promote lean maintenance (2) Challenges of lean maintenance in the construction industry		No	No

(Continued)

Table 2. (*Continued*).

S. no.	Author(s) and year	Key aspects covered	Application industry	Suggested lean maintenance implementation	Limitations
25	Huang, Bian, and Cai (2012)	(1) Key technical characteristics of lean maintenance (RCA, FMEA, value stream analysis) (2) Realisation of lean maintenance (LM) through updating maintenance concept, highlighting the academic innovation and integration, and enhancing the culture construction of LM theory (3) A theoretical basis for the equipment maintenance reform	Military equipment maintenance	LM technical supporting frame	• The frame showed the key techniques of LM • Some lean tools have been provided • No implementation of the LM introduced
26	Okhovat et al. (2012)	(1) Developed a strategic framework to provide guidance and support to reach world-class standards both in maintenance and manufacturing processes through continual improvement (2) The framework integrated three process improvement strategies: six sigma, TPM, and lean	General-world-class manufacturing	Framework for world-class standard	• General frame for production and maintenance included TPM, lean, and 6σ • Limited lean tools documented for maintenance process
27	Rastegari (2012)	(1) A maintenance strategy in manufacturing organisation that can be linked to the company's business strategy in lean environment		No	No
28	Tinashe George Tendayi and Fourie (2012)	(1) A framework for implementing a lean in a rolling stock maintenance environment that seeks to eliminate waste, add value, and continuously improve its supply chain	Rolling stock maintenance	Lean maintenance supply chain framework	• Frame for specific case study supply chain • Not incorporated the lean principles • Lean tools are not included

29	Hou, Zhao, Meng, and Ma (2013)	(1) Lean Maintenance System Model for China's hydropower equipment maintenance enterprises by the (2) Application of the lean production theory and ideas, and the model consists of lean culture building, system file structure, maintenance project management, maintenance on-site management and safety management		No	No
30	Jahanbakhsh, Moghaddam, and Samaie (2013)	(1) Benefits of using lean approach in Teen Dairy Industry to reduce the amount of losses and rework of maintenance systems		No	No
31	Romano et al. (2013)	(1) A LRCDA method to reduce scraps and work-in-process in manufacturing context (2) LRCDA merges steps of RCFA, lean maintenance, and TPM	Power cables factory	LRCDA model	• The model is based on the RCDA not lean principles • The model contained limited lean practice within the model
32	Tendayi (2013)	(1) A framework, based on lean thinking tools and relevant performance measures to prove the applicability or otherwise, of lean thinking in an operational maintenance environment outside the traditional domain of manufacturing	Rolling stock maintenance	Framework for lean thinking in maintenance	• Insufficient implementation process flow • Limited lean tools introduced
33	Djurovic and Bulatovic (2014)	(1) Implementation of the 'S5' LEAN methods concept in maintenance, storage, as well as other elements of the business system		No	No
34	Bulatovic and Djurovic (2014)	(1) Lean production cannot be implemented without the use of lean strategy in the equipment maintenance		No	No

(Continued)

Table 2. (*Continued*).

S. no.	Author(s) and year	Key aspects covered	Application industry	Suggested lean maintenance implementation	Limitations
		(2) 5S is one of the most basic tools in the lean maintenance			
35	Dunn (2014)	(1) Focus on key elements within each of the work management framework processes that will produce improvements in maintenance productivity and remove waste (1) Three ways in which continuous improvement approaches can further improve Maintenance Productivity and eliminate waste: • Improving maintenance efficiency • Improving the predictive maintenance programme • Eliminating defects		No	No
36	Irajpour, Fallahian-Najafabadi, Mahbod, and Karimi (2014)	(1) Identify and evaluate the effectiveness of a given maintenance strategy and to rank components of maintenance system (2) Using DEMATEL method on maintenance strategy as a guideline to rank the lean components in maintenance	General-rank component of maintenance	Model of effective maintenance strategy	• The model is used to rank the effective elements of maintenance strategy • Lack of model elements dependency • Limited lean tools provided • Lean principles is not incorporated in the model
37	Jasiulewicz-Kaczmarek (2013)	(1) The common elements of the lean and green paradigms introduced for better understanding of synergy between them and overall improvement of maintenance efficiency	General-production management	Lean & green maintenance results	• The frame demonstrated the requirements and results of using lean and green in maintenance process • No presentation of how to incorporate either lean or green in the maintenance

38	Nima, Hesamodin, and Alireza (2014)	(1) Development of a practical model to define an optimised effective maintenance strategy using lean approach		No	No
39	Önder (2014)	(1) How maintenance may contribute to decreasing the environmental impact of production	General-lean green maintenance concept	Reducing environmental impacts and improving efficiency with lean green maintenance	• Generic framework contained four phases: plan, learn, action, and improve • Not comprehensive nor designed based on lean principles • Very limited lean tools were suggested
40	Shou, Wang, Wang, Hou, and Truijens (2014)	(1) Apply BIM and lean concepts to practical maintenance to improve efficiency		No	No
41	Soltan and Mostafa (2014)	(1) Lean and agile can manage the maintenance processes (2) Lean and agile maintenance should be considered as a prerequisite for any successful lean or agile application (3) A framework to measure the performance of maintenance strategies based on lean and agile factors	General	Lean and agile factors in maintenance operations	• No lean tools provided • No implementation process flow provided

(Continued)

Table 2. (*Continued*).

S. no.	Author(s) and year	Key aspects covered	Application industry	Suggested lean maintenance implementation	Limitations
42	Stuchly and Jasiulewicz-Kaczmarek (2014)	(1) The maintenance aim has been shifted from production paradigm to sustainable development that resulted in a change towards the product life cycle as well as taking into account economic, environmental, and social aspects (2) Maintenance offers numerous opportunities of decreasing influence of business processes on natural environment and more efficient resources utilisation		No	No
43	Sunjka and Murphy (2014)	(1) Determine the status of lean implementation within South African aircraft maintenance organisations (AMOs)		No	No

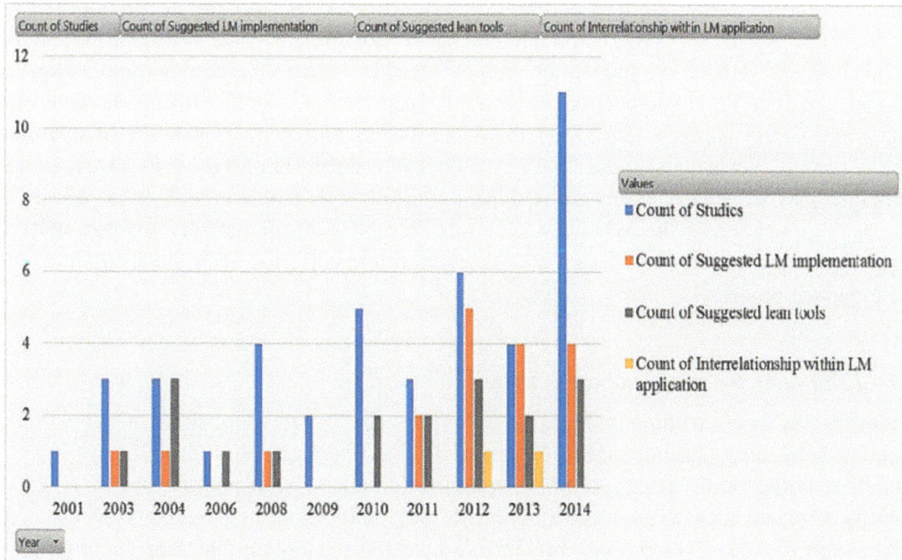

Figure 3. Distribution of lean maintenance studies.

empowered (self-directed) action teams. LM generates a desirable outcome by minimising consumption of inputs (Smith & Hawkins, 2004). LM represents adopting lean principles into the maintenance, repair, and overhaul (MRO) operations. It could reduce unscheduled DT through optimising maintenance support activities and maintenance overhead. The lean tools are representing the lean principles for the implementation process (Mostafa, Dumrak, & Soltan, 2013). To achieve LM improvement effectively, key lean tools such as VSM, 5S, and visual management need to be employed (Smith, 2004; Smith & Hawkins, 2004). A comprehensive lean tools developed for maintenance activities within an organisation include 5S, TPM, OEE, Kaizen, Poka-Yoke, process activity mapping, Kanban, computer managed maintenance system (CMMS), enterprise asset management (EAM) system, and Takt time (Davies & Greenough, 2010; Smith, 2004).

Despite the benefits of LM mentioned earlier, the review of previous studies conducted in this study found that the investigation on the applicability of lean principles into maintenance is marginal. The existing research works have been largely limited to the manufacturing environment where LM is practised as a prerequisite for lean manufacturing (Tendayi, 2013). This proposition has been mentioned in Davies and Greenough (2010) emphasising on the necessity of conducting more research on practical application of lean manufacturing principles in maintenance operations. It was discovered that the previous studies mainly focused on ranking the maintenance strategies based on some specific scope. Moreover, few initiatives have included comprehensive frameworks or models that can integrate lean thinking in operational maintenance environments outside of the manufacturing context. Ghayebloo and Shahanaghi (2010) formulate a model for determining the minimal level of maintenance requirements and satisfying reliability level through the use of the lean concept. Tendayi and Fourie (2013) use a combined approach between QFD and AHP to evaluate the importance of maintenance excellence criteria and prioritise the lean tools upon these criteria. The recent study of Soltan and Mostafa (2014) introduces a framework for measuring maintenance strategies based on lean and agile components, i.e. waste removal and responsiveness. However, the study cannot provide sufficient practical application of lean concept in the maintenance process. Romano et al. (2013) formulate lean root cause and defect analysis (LRCDA) to reduce scraps and work-in-process in manufacturing system. Nevertheless, the LRCDA model introduced is based on RCDA (not lean principles) as well as contains limited lean practice within the model. The paucity of practical application in the existing LM studies provides an opportunity for this study to expand the prevailing knowledge into a new framework for lean integration in the maintenance process.

5.2 Lean tools for maintenance activities

Reducing the NVA within maintenance activities can be accomplished through implementing lean tools (Jasiulewicz-Kaczmarek, 2013). The lean tools that suit the maintenance activities have been stated in previous studies. Smith and Hawkins (2004) identify the key lean tools including VSM, 5S, and visual management. Davies and Greenough (2010) develop a comprehensive lean tools template that represents possible lean activities within the maintenance process within an organisation. The tools are 5S, TPM, OEE, standards, mapping, inventory management, and visual management. Okhovat et al. (2012) suggest six lean tools that fit the maintenance processes of an organisation. These tools include visual control, 5S, seven wastes, single minute exchange of die (SMED), and Poka-Yoke (mistake proofing). Clarke et al. (2010) target eight LM practices as a preparation for delivering lean project objectives in a pharmaceutical organisation. A list of the references including LM tools is demonstrated in Table 3.

6. A proposed LM process

This section demonstrates an attempt to propose a process for adopting lean thinking into the maintenance activities. The process adopts the hypothesis of Womack and Jones (2003) that lean principles can be deployed to all organisations and sectors. Lean principles have being increasingly extended for industrial and service sectors. This is known as lean thinking which refers to the thinking process of lean (Holweg, 2007). The process proposed in this study is designed on the basis of the five lean manufacturing principles stated by Womack and Jones (2003). Some authors including Karim and Arif-Uz-Zaman (2013) develop lean implementation methodology based on the five lean principles for a manufacturing environment. Mostafa et al. (2013) state that lean practices/tools represent lean principles in the implementation process.

The process introduced in this study could be considered as an attempt to pave the way of applying or adopting lean principles to maintenance activities (Figure 4). The process provides guidance and support towards maintenance excellence for an organisation pursuing to extend lean practice to its maintenance department or other organisations starting introducing lean thinking to maintenance department. The process was developed through conceptual integration of five lean principles as they are the backbones of any lean initiatives (Womack & Jones, 2003). The principles specify the value, identify the value stream, flow the value, pull the value, and pursuing perfection. In addition, analysing and addressing limitations of the existing initiatives assist in developing the comprehensive LM process proposed in this study. The process is more flexible and can be adjusted according to any maintenance strategy. It could work

Table 3. Lean maintenance practices/tools reported in previous studies.

	Previous study																	
Lean maintenance tool	1	2	3	4	5	6	7	8	9	10	11	12	13	14	15	16	17	18
TPM	*	*	*	*		*	*			*	*			*		*	*	
5S/CANDO	*	*	*	*	*		*		*		*		*	*	*	*		
Kaizen (continuous improvement)	*	*	*	*						*		*		*	*	*	*	*
CMMS/EAM	*					*	*				*			*			*	
Distributed MRO storerooms	*	*				*								*				
RCFA/FMEA	*		*	*		*		*			*						*	
PdM	*	*				*								*			*	
Autonomous maintenance		*					*			*				*			*	
SMED		*			*		*		*									
Poka-Yoke			*	*	*				*	*	*	*	*		*	*		
PDCA										*	*					*		
OEE				*														
Kanban				*				*					*		*	*	*	
Jidoka													*					*
JIT/inventory management			*	*			*						*		*	*		*
RCM						*	*	*			*			*			*	
Process mapping (VSM)				*						*			*			*		
Maintenance and reliability group				*										*				
Work standardisation	*		*	*				*				*		*	*	*		
Story boarding			*	*														
Visual control			*	*	*				*						*			
Work order system			*											*			*	
Self-audit				*														
Supplier association				*														
Open book management		*		*														
Empowered maintenance team												*		*				
Multi-skilled work team												*						
Maintenance crew training and learning							*					*						
Hoshin planning										*							*	*
A3										*								

Note: 1 – Baluch et al. (2012); 2 – Clarke et al. (2010); 3 – Davies and Greenough (2003); 4 – Davies and Greenough (2010); 5 – Djurovic and Bulatovic (2014); 6 – Huang et al. (2012); 7 – Irajpour et al. (2014); 8 – Kolanjiappan and Maran (2011); 9 – Okhovat et al. (2012); 10 – Önder (2014); 11 – Qiang et al. (2011); 12 – Romano et al. (2013); 13 – Smith and Hawkins (2004); 14 – Smith (2004); 15 – Tendayi (2013); 16 – Verma and Ghadmode (2004); 17 – Yile et al. (2008); 18 Zwas (2006).
*Presence of the lean maintenance practice within the study.

simultaneously and complementary with previous framework developed for shop floor area as it is inclusive for the maintenance processes. The proposed process entails five stages and detailed steps within each stage.

6.1. *Stage one: specify the value*

The first stage focuses on defining an organisational maintenance system including activities, maintenance planning, strategies, and maintenance crew. In this stage, the employees training on LM wastes are assigned. Furthermore, identifying the types of wastes in maintenance processes is included. The core concept of lean manufacturing is eliminating the seven cardinal forms of waste. The seven cardinal types of waste in the maintenance process can be discussed in the same manner as in the eight waste types identified in the production system (Baluch et al., 2012; Clarke et al., 2010; Davies & Greenough, 2010).

(1) Too much maintenance: performing PM and predictive maintenance (PdM) tasks at intervals more often than optimal which results in the overproduction of maintenance work.
(2) Waiting for maintenance resources: production department is waiting for maintenance personnel to perform the maintenance service. It involves waiting for tools, parts documentation, and buys extra tools and stores them near the job location.
(3) Centralised maintenance: the centralisation of the MRO stores that are far from the job, commonly used repetitive parts that have not been kitted, documentation that must be hunted down, and work orders for machines that are not available all cause excess transportation. Therefore, maintenance personnel spend more time in motion and transportation which do not add value to the process.
(4) Non-standard maintenance: maintenance operations are normally conducted to achieve operation as soon as possible with no standard guidelines. This sometimes eliminates an opportunity to perform a higher quality repair.
(5) Excessive stock: the MRO inventory contains needed materials and spares. Additionally, work-in-process inventories may be used to ensure availability of required materials. Inventory for a maintenance operation also includes the work order backlog. Excessive maintenance work inventory results in slow response, unexpected breakdowns, and a high reactive labour percentage.
(6) Double handling: the wasted motion is usually concentrated around PM tasks. Doing inspection monthly on a pump that has not changed status in three years should be extended longer to quarterly, semi-annually, or annually depending upon the criticality of that piece of equipment.
(7) Poor maintenance: performing incorrect repair is a source of poor maintenance. Incorrect maintenance requires several repeated times to complete the repair job correctly. This affects the maintenance cost and the quality of the product. Applying proper training and detailed procedures can assist in poor maintenance elimination.
(8) Under-utilisation of maintenance crew: maintenance technicians do NVA work or do not perform as required/at the best interest of the organisation.

6.2. Stage two: identify the value stream

This stage includes all maintenance-related activities and processes. The stage starts by mapping the maintenance value stream then locating the wastes sources. This stage ends with setting equipment performance measures such as availability, OEE, and meantime between failures. VSM is used for visualising the flows of information and materials within a supply chain. VSM primarily helps an organisational management to recognise different forms of waste and its sources. One key metric of VSM is value added (VA) time percentage which measures VA activities against NVA activities (Monden, 1998). Standard icons for drawing the current and future VSM are available in Sullivan, McDonald, and Van Aken (2002). These icons should be modified to fit the maintenance activities. In this study, a new set of VSM icons fitting all the maintenance activities was introduced (see Figure 5). The new icons were designed using Edraw Max™

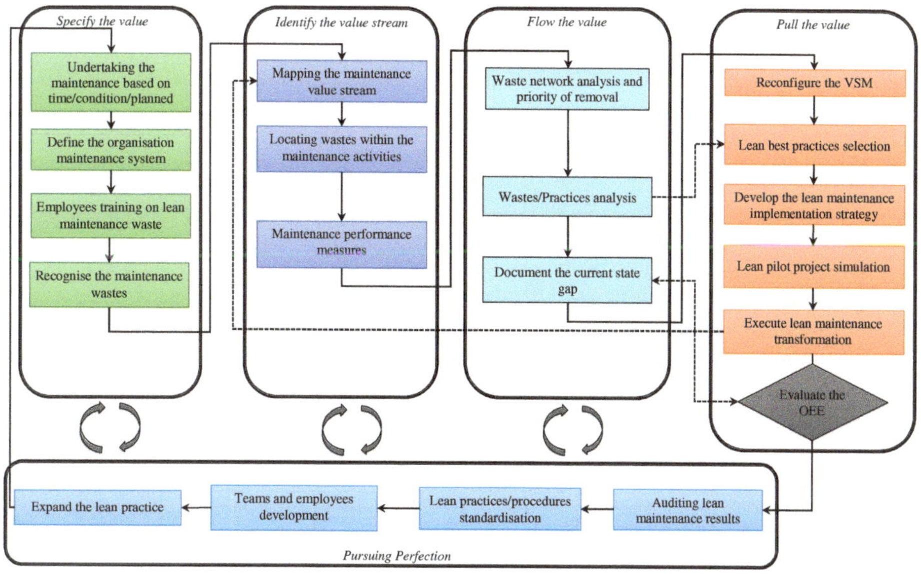

Figure 4. Proposed lean maintenance process.

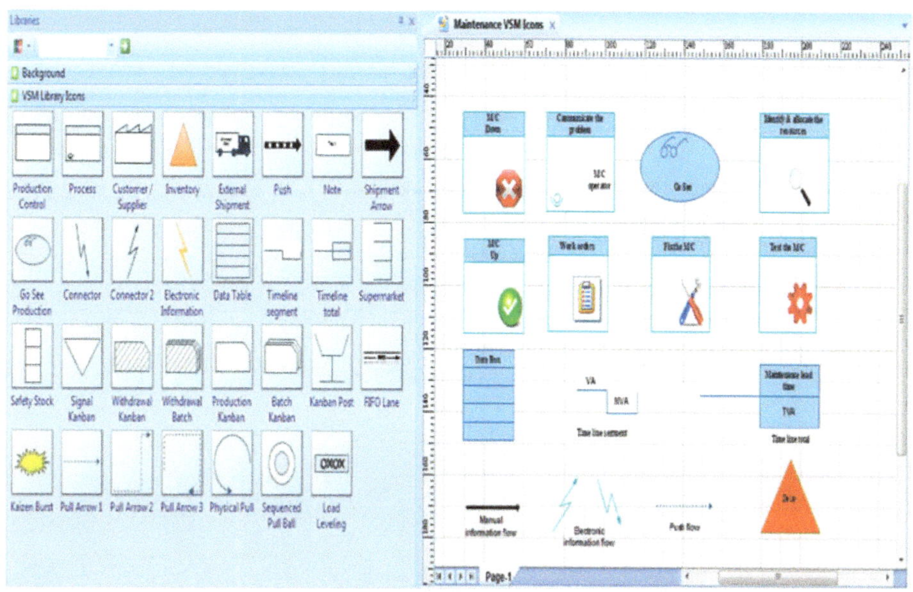

Figure 5. VSM symbols for maintenance activities (developed from Edraw Max software).

professional software package. The discussion of each activity is displayed in Table 4. An example of mapping the maintenance process using these icons for visualising a general current state is displayed in Figure 6.

Table 4. Maintenance VSM symbol description.

	Symbol	Description
(1)	Machine down	Machine need maintenance (e.g. break down, time-based, or condition-based)
(2)	Communicate the problem	Machine operator communicates the maintenance department
(3)	Go see	Maintenance personnel go and check the machine condition and report to the maintenance department
(4)	Identify and allocate resources	Identification and locating of appropriate resources such as tools, spare parts, manpower for fixing an equipment
(5)	Work orders	Generating maintenance work order through the maintenance software
(6)	Machine fix	Steps for fixing the machine
(7)	Test the machine	Testing of machine after repair until first good part is produced
(8)	Delay	Waiting time due to unavailability of resources (e.g. technician, tools, and spare parts)
(9)	Manual information flow	The flow of information from reports
(10)	Electronic information flow	The flow of information from the internet, intranet, local area network, wide area network, and other notes
(11)	Push flow	Represent the physical flow sequence of the maintenance activities
(12)	Time line segment	Represent the VA and NVA time for each activity
(13)	Time line total	Represent the maintenance lead time (down time)
(14)	Data box	Record the information of each maintenance process including cycle time (C/T), changeover time (C/O), and number of employees

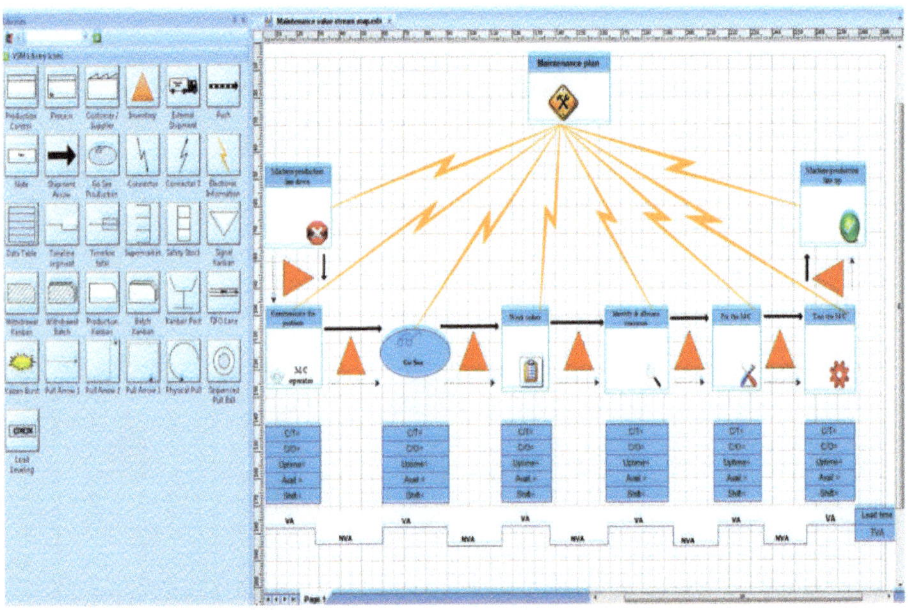

Figure 6. Example of general maintenance state map (developed from Edraw Max software).

6.3. *Stage three: flow the value*

The process of 'flowing the value' starts from waste network and waste/practices analyses before the results of the analyses are documented at the current state gap of maintenance using the calculation of OEE within an organisation. The documentation of lean practices for maintenance was conducted through the literature presented in Table 3 through marking out the most frequently suggested practices in these studies and then constructed the hierarchical scheme for LM practices as demonstrated in Figure 7. The structure is similar to Shah and Ward (2003), however, for the use of this structure is dedicated to the maintenance process not the manufacturing process. The structure consists of two levels and four bundles: just-in-time (JIT), total quality management (TQM), human resource management (HRM), and TPM and practices assigned under each bundle. The scheme can be used to indicate the association between the eight types of maintenance waste and the LM practices. The success of the LM depends on the application of each bundle. Each practice provides some benefits within the maintenance process. As a result, the performance of a whole maintenance department can be improved. The four LM bundles are briefly explained below.

6.3.1. JIT bundle

The JIT bundle encompasses all practices which are designed to reducing and eliminating unnecessary inventory and waiting in maintenance activities. These practices are Kanban, SMED, work standardisation, Takt Time, visual control, distributed MRO storeroom, and CMMS.

6.3.2. TQM bundle

The TQM bundle aims at continuously improving and sustaining the quality of products and processes through the participation of management, workforce, suppliers, and

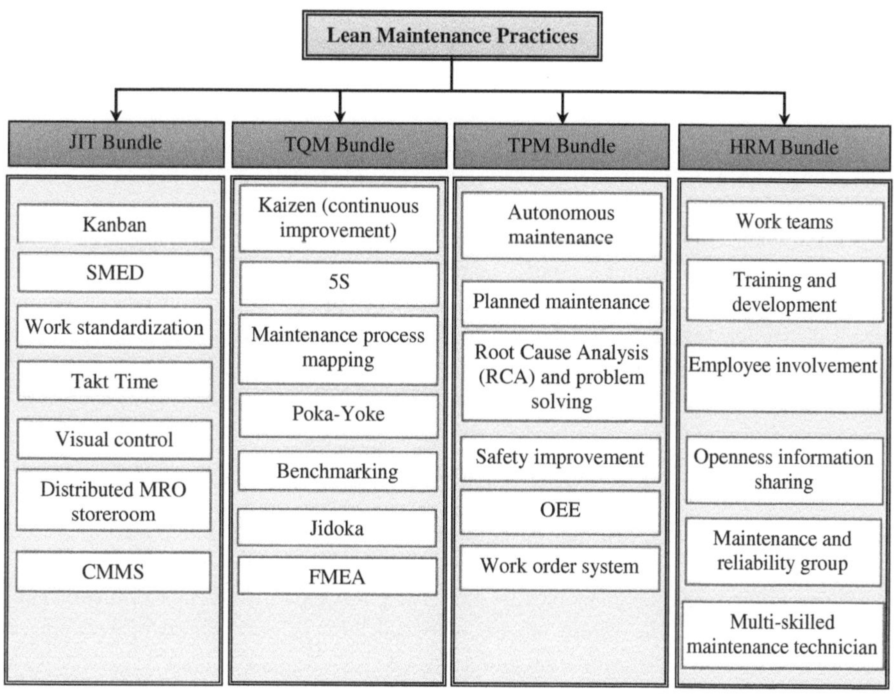

Figure 7. Scheme for lean maintenance practices.

customers, in order to meet or exceed customer expectations (Cua et al., 2001). The practices of TQM are customer focus, leadership, strategic quality planning, use of information and analysis, management of people (participation and partnership), and continuous improvement (Chin, Rao Tummala, & Chan, 2002). This ensures higher customer satisfaction, better quality of products, and higher market share; improving the competitiveness, effectiveness, and flexibility of the whole organisation (Pheng & Chuan, 2001). In this study, the TQM bundle consists of Kaizen, 5S, maintenance process mapping, Poka-Yoke, benchmarking, Jidoka, and FMFA.

6.3.3. TPM bundle

The TPM bundle incorporates all practices that are designed to maximise the equipment effectiveness. The bundle contains autonomous maintenance, planned maintenance, root cause analysis (RCA) and problem solving, safety improvement, OEE, and work order system. The concept of TPM was developed in Japan by Seichii Nakajima aiming at achieving zero losses (e.g. zero breakdowns, zero defects, zero accidents) and attaining the Takt time through improving and maintaining equipment to its highest performance level (Brown et al., 2006). TPM can be integrated with lean in order to distinguishing and attacking six big losses which are reducing the effectiveness of equipment (Rodrigues & Hatakeyama, 2006). These six categories of losses are (1) breakdown losses, (2) set-up losses, (3) minor stoppage/idle losses, (4) reduced speed losses (equipment speed is slow than the designed speed), (5) rework losses, and (6) start-up losses (start-up after periodic repair, start-up after holidays, start-up after lunch breaks, and start-up after suspension) (Waeyenbergh & Pintelon, 2002).

6.3.4. HRM bundle

The HRM bundle encompasses all the practices to ensure that the human resources of an organisation are used in such a way that the employer obtains the greatest possible benefit from their abilities. At the same time, the employees obtain both material and psychological rewards from their work (Hiltrop, 1996). The HRM practices form an organisation performance through increase employees' knowledge, skills, and abilities (KSAs), motivate employees to leverage their KSAs for the firm's benefit and employees empowerment. In addition, work teams, performance appraisal, and information sharing have been recommended to enhance an organisation performance (Huemann, Keegan, & Turner, 2007). In this study, the HRM bundle includes work teams, training and development, employee involvement, openness information sharing, maintenance and reliability group, and multi-skilled maintenance technician.

6.4. Stage four: pulling the value

The fourth stage is to confirm that an equipment (i.e. considered as the customer in the maintenance process) is pulling the value through all maintenance processes. The execution of lean principles takes place at this stage. The stage involves steps in reconfiguring the VSM or designing the future stream map, selecting LM practices, developing the lean transformation strategy, and evaluating the OEE.

6.4.1. Association between maintenance wastes and LM practices

Selecting LM tools is the next step after identifying and discussing the types of waste and lean tools. It assists an organisation to achieve LM targets by associating suitable tools to tackle the spotted waste types. The QFD is suggested in this study to develop the association between these tools and wastes. Mostafa (2011) suggested using QFD to assign the lean practices for each waste. Nonetheless, the study was in a general manufacturing environment rather than a maintenance process. Likewise, Tendayi (2013) applies QFD coupled with AHP to rank the maintenance excellence criteria with some lean tools. The study focused only on a certain case study of rolling stock maintenance. It used limited lean tools and did not include the maintenance wastes. In contrast to Tendayi (2013), this study has identified 26 lean tools for maintenance (see, Figure 7) and eight types of maintenance wastes (discussed in Section 6.1). QFD is a structured tool that identifies important customer expectations and translates them into appropriate technical characteristics which are operational in design, verification, and production. QFD enables resources to be focused on meeting major customer demands. Figure 8 displays the structure of QFD which is frequently called a house of quality (HOQ) because of the shape presented (Bottani & Rizzi, 2006).

The HOQ comprises customer requirements (horizontal axis) and technical characteristics (vertical axis). The customer axis designates what customers require, importance of the requirements, and competitive performance. The customer requirements are often referred as what's. The technical axis describes the technical characteristics that affect customer satisfaction directly for one or more customer expectations. Moreover, on the technical axis are the correlations, importance, and targets of the technical characteristics and technical competitive benchmarking. The technical characteristics are denoted as How's, meaning how to address what's. The technical targets are accordingly called

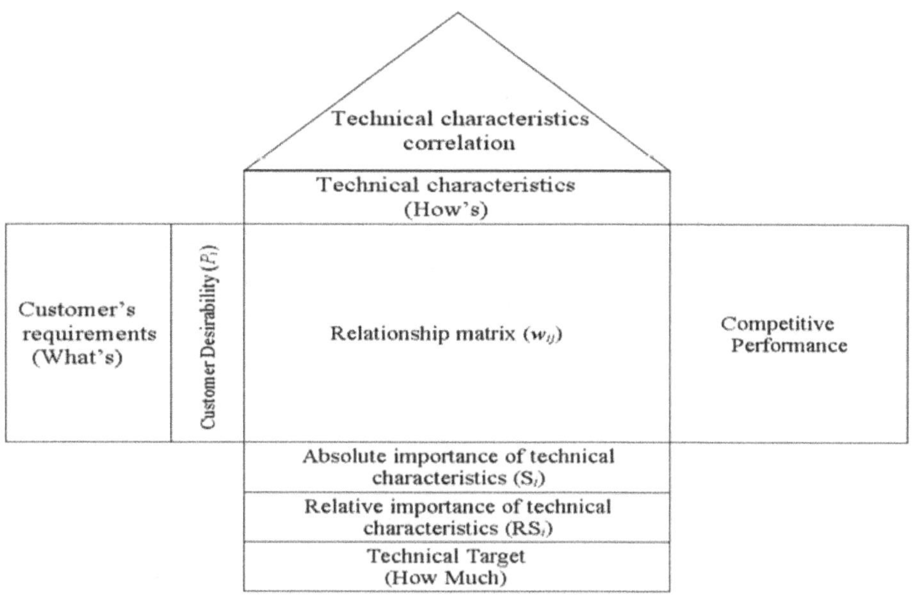

Figure 8. House of quality.

How Much. The interrelationships between customer wants and technical characteristics are evaluated in the relationship matrix.

6.4.2. *House of Waste*

The HoW is developed in this study to demonstrate the maintenance wastes/lean tools association. HoW identifies the importance of each LM tool in the elimination of each waste type as demonstrated in Table 5. It has two main parts: the horizontal part 'What's' which contains information relevant to the waste types and the vertical part 'How's' which comprises corresponding LM tools. The basic process underlying 'How' resides in the centre of the matrix where intersect of maintenance waste types and lean tools provides an opportunity to examine each waste type versus each tool.

Five steps in developing the wastes/tool association using HoW are briefly described as follows:

(1) List the waste types within the maintenance process in the horizontal axis.
(2) Determine each waste priority (*Pi*), which rates the significance of each waste to the maintenance performance. Otherwise, it can be based on the waste occurrence or cost of removal. A scale of 1 to 9 can be used, where 9 is given to extremely important, 7 strongly important, 5 to very important, 3 to important, and 1 to not important.
(3) List LM tools on the vertical axis. An organisation can state suitable LM tools for their own maintenance process in terms of the application cost of LM tools.
(4) Identify the interrelationships between wastes and tools (w_{ij}). The strength of relationship may be classified into three levels, where a rating of 9 is assigned to a strong relation, 3 to a medium relation, and 1 to a weak relation. Each tool must be interrelated to at least one waste; one waste must also be addressed by at least one tool. This ensures that all wastes are concerned in the maintenance removal planning, and all tools are properly established.
(5) Calculate the score (S_j) and (RS_j) to rank each LM tool. The following formula can be used to calculate the score

$$S_j = \sum_{i=1}^{8} w_{ij} P_i, \ \forall j \tag{1}$$

$$RS_j = \frac{S_j}{\sum_{j=1}^{26} S_j}, \ j = 1, \ldots, 26 \tag{2}$$

where

P_i relative priority of *i*th waste

i type of maintenance waste, $i = 1, 2, \ldots, 8$

T_j LM tool, $j = 1, 2, 3, \ldots, 26$

w_{ij} numerical VA to position (i, j) of the HoW matrix. This refers to the weight of eliminating waste *i*th by tool *j*th. Numerical scale can be (9, 3, 1)

S_j score of tool *j*th over the total types of waste

RS_j relative score of tool *j*th over the total score of tools

Table 5. House of Waste.

What's (*maintenance wastes*)	Waste priority	How's (*lean maintenance tools*) Kanban	SMED	Work standardisation	Takt Time	Visual control	Distributed MRO storeroom	CMMS	Kaizen	5S	Maintenance VSM	Poke Yoke	Jidoka	Benchmarking	FMEA	Autonomous maintenance	Planned maintenance	RCA	OEE	Work order system	Work teams	Training and development	Employee involvement	Openness information system	Maintenance & reliability group	Multi-skilled technician	Safety improvement
To eliminate too much maintenance	P_1	w_{11}	w_{12}	w_{13}	w_{14}	…																			…	w_{125}	w_{126}
To eliminate waiting for resources	P_2	w_{21}	w_{22}	w_{23}	w_{24}	…																			…	w_{225}	w_{226}
To eliminate centralised maintenance	P_3	w_{31}	w_{32}	w_{33}	w_{34}	…																			…	w_{325}	w_{326}
To eliminate non-standard maintenance	P_4	w_{41}	w_{42}	w_{43}	w_{44}	…																			…	w_{425}	w_{426}
To eliminate excessive stocks	P_5	w_{51}	w_{52}	w_{53}	w_{54}	…																			…	w_{525}	w_{526}
To eliminate double handling	P_6	w_{61}	w_{62}	w_{63}	w_{64}	…																			…	w_{625}	w_{626}
To eliminate incorrect maintenance	P_7	w_{71}	w_{72}	w_{73}	w_{74}	…																			…	w_{725}	w_{726}
To eliminate underutilisation of maintenance crew	P_8	w_{81}	w_{82}	w_{83}	w_{84}	…																			…	w_{825}	w_{826}
LM score		S_1	S_2	S_3	S_4	S_5	S_6	S_7	S_8	S_9	S_{10}	S_{11}	S_{12}	S_{13}	S_{14}	S_{15}	S_{16}	S_{17}	S_{18}	S_{19}	S_{20}	S_{21}	S_{22}	S_{23}	S_{24}	S_{25}	S_{26}
Relative score		RS_1	RS_2	…																					…	RS_{25}	RS_{26}
Rank																											

The score S_j represents the absolute importance of tool jth over all types of wastes. The relative score RS_j refers to the importance of the tool jth over all tools. According to Bottani and Rizzi (2006), the technical characteristics are usually ranked based on relative importance rather than absolute importance. Therefore, LM tools should be prioritised based on RS_j. The higher the RS_j, the more important the LM tool that should be incorporated in order to eliminate the waste.

6.5. Stage five: pursue perfection

The last stage is to pursue the complete waste elimination from maintenance processes. This could be achieved through auditing the LM results, standardise the lean tools and procedures, teams and employees developments and expand the lean practice.

7. Implications of proposed LM process

Most of literature works on lean implementations fall into the manufacturing environment albeit LM is essential part for the lean manufacturing system (Soltan & Mostafa, 2014; Tendayi & Fourie, 2013). Plethora of studies included and mentioned lean and maintenance concepts (see Table 2). Nevertheless, the integration of lean principles into the maintenance environment has not evidently received full attention. The common aspects among the previous studies were (1) introducing partial lean tools to be applied in the maintenance process; (2) developing diagrammatic structures (model, pyramid, or scheme) with unspecified sequence of LM implementation; and (3) lack of proper structure. Consequently, some researchers including Davies and Greenough (2010) and Soltan and Mostafa (2014) have proposed more research on incorporating lean principles in the maintenance activities. Therefore, this study proposed a LM process to cover most of the limitations or deficiencies of the existing suggested LM initiatives.

The proposed LM process is a straight forward, comprehensive, and easy to understand for maintenance practitioners and maintenance strategy decision makers. The benefits of the proposed process can be summarised as:

(1) The process is straightforward and easy to comprehend. This is because of the simple structure that is combined in the process.
(2) A systematic procedure of applying the five lean principles into maintenance activities can result in better understanding from an organisation's management view.
(3) All advantages of lean are kept as the process focus. These advantages are integrated and systemised in the steps of the process.
(4) Application of the process establishes pursuing a perfectionist culture in an organisation. This is done through auditing the LM results, standardisation, team's development, and expansion of the practice. Hence, with every iteration of the process, an organisation maintenance activities move a further step towards maintenance excellence.
(5) Aligning the human resource aspects (maintenance employees training and development) with other steps can bring high potential for a comprehensive and sustainable LM implementation process as well as overall process improvement in an organisation.
(6) The process promotes the teamwork and problem solving cultures to ensure high-quality outcomes of the implementation process.

(7) The maintenance process can be converted to be a predictive and proactive system that provides the reliable process using lean principles. As a result, it enhances the maintenance excellence and world-class manufacturing as maintenance and manufacturing are inseparable (i.e. reliability and availability of manufacturing facilities).
(8) The process can be applied to all manufacturing and non-manufacturing organisations that are pursuing the world-class status in their maintenance function. The process suits those organisations either in pursuing to transfer lean thinking to their maintenance departments or starting lean transformation from the maintenance department.
(9) Hitherto, some initiatives have been developed for specific larger size organisations such as shipbuilding and military (see Table 3). However, the proposed process can be accepted within all level of skills and organisational sizes. This is due to the fact that the process is more flexible and can be adjusted according to any maintenance strategies.
(10) The proposed process is precisely documented and discusses the maintenance wastes and LM tools. It is to enhance the understanding the waste elimination and lead to successful LM implementation.
(11) The process presented wastes/tools association to assign LM tools for each waste type. This leads to an effective waste elimination process and sustainable outcomes.

8. Conclusion

Maintenance is a critical contributor to progressing towards world-class manufacturing of an organisation. It has rapidly grown into a very complex undertaking as technologies, competition, and product characteristics have evolved. In order to achieve world-class performance, the maintenance strategies should be linked to manufacturing strategies such as the lean concept. Applying an effective maintenance strategy can ensure a high degree of utilisation, reliability, and availability of manufacturing facilities especially in a continuous production process. This study has introduced a process to adopt lean into maintenance activities. The process highlights types of NVA maintenance activities, a package of VSM symbols to capture the maintenance activities, LM tools scheme, and tools/wastes association. Moreover, it promotes the culture of continuous improvement aiming at maintenance excellence. In general, any improvements attained from implementing the suggested process might take time to transform maintenance activities into LM. Nevertheless, commitment and direct involvement of an organisational management along with employee training and teamwork development can be a catalyst to accelerate the transformation process. The proposed LM process can be applied in real conditions to test its validity and reliability of the process. This can be considered as a further suggestion to the proposed process in this research. Hence, application of the process in different industrial sectors and a wide range of manufacturing companies can contribute to additional empirical evidence.

Acknowledgements

The authors would like to express their gratitude to the anonymous reviewers for their constructive comments and suggestions to improve the quality of the paper.

Disclosure statement

No potential conflict of interest was reported by the authors.

ORCID

Sherif Mostafa http://orcid.org/0000-0002-5708-3770
Jantanee Dumrak http://orcid.org/0000-0002-5752-0707
Nicholas Chileshe http://orcid.org/0000-0002-1981-7518

References

Abdulmalek, F. A., Rajgopal, J., & Needy, K. L. (2006). A classification scheme for the process industry to guide the implementation of lean. *Engineering Management Journal, 18*, 15–25.

Ahmed, S., Hassan, M. H., & Taha, Z. (2005). TPM can go beyond maintenance: Excerpt from a case implementation. *Journal of Quality in Maintenance Engineering, 11*, 19–42. doi:10.1108/13552510510589352

Al-Najjar, B., & Alsyouf, I. (2003). Selecting the most efficient maintenance approach using fuzzy multiple criteria decision making. *International Journal of Production Economics, 84*, 85–100. doi:10.1016/S0925-5273(02)00380-8

Ayeni, P., Baines, T. S., Lightfoot, H., & Ball, P. (2011). State-of-the-art of 'Lean' in the aviation maintenance, repair, and overhaul industry. *Proceedings of the Institution of Mechanical Engineers, Part B: Journal of Engineering Manufacture*, 1–16. doi:10.1177/0954405411407122

Baluch, N., Abdullah, C. S., & Mohtar, S. (2012). TPM and lean maintenance – A critical review. *Interdisciplinary Journal of Contemporary Research In Business, 4*, 850–857.

Bevilacqua, M., & Braglia, M. (2000). The analytic hierarchy process applied to maintenance strategy selection. *Reliability Engineering & System Safety, 70*, 71–83. doi:10.1016/S0951-8320(00)00047-8

Bhasin, S. (2015). *Lean management beyond manufacturing*. New York, NY: Springer International Publishing.

Blanchard, B. S. (1997). An enhanced approach for implementing total productive maintenance in the manufacturing environment. *Journal of Quality in Maintenance Engineering, 3*, 69–80. doi:10.1108/13552519710167692

Booth, A., Papaioannou, D., & Sutton, A. (2012). *Systematic approaches to a successful literature review*. Thousand Oaks, CA: Sage.

Bottani, E., & Rizzi, A. (2006). Strategic management of logistics service: A fuzzy QFD approach. *International Journal of Production Economics, 103*, 585–599. doi:10.1016/j.ijpe.2005.11.006

Brown, C. B., Collins, T. R., & McCombs, E. L. (2006). Transformation from batch to lean manufacturing: The performance issues. *Engineering Management Journal, 18*, 3–14.

Bulatovic, M., & Djurovic, D. (2014). Lean concept in the maintenance. *Journal of Trends in the Development of Machinery and Associated Technology, 18*, 135–138. Retrieved from http://tmt.unze.ba/zbornik/TMT2014Journal/30_Journal_TMT_2014.pdf

Burgess, N., & Radnor, Z. (2013). Evaluating lean in healthcare. *International Journal of Health Care Quality Assurance, 26*, 220–235. doi:10.1108/09526861311311418

Carrasqueira, M., & Machado, V. C. (2008). Strategic logistics: Re-designing companies in accordance with Lean Principles. *International Journal of Management Science and Engineering Management, 3*, 294–302. doi:10.1080/17509653.2008.10671056

Chan, F. T. S., Lau, H. C. W., Ip, R. W. L., Chan, H. K., & Kong, S. (2005). Implementation of total productive maintenance: A case study. *International Journal of Production Economics, 95*, 71–94. doi:10.1016/j.ijpe.2003.10.021

Chin, K. S., Rao Tummala, V. M., & Chan, K. M. (2002). Quality management practices based on seven core elements in Hong Kong manufacturing industries. *Technovation, 22*, 213–230. doi:10.1016/S0166-4972(01)00020-7

Chong, M. Y., Chin, J. F., & Hamzah, H. S. (2012). Transfer of total productive maintenance practice to supply chain. *Total Quality Management & Business Excellence, 23*, 467–488. doi:10.1080/14783363.2011.637788

Clarke, G., Mulryan, G., & Liggan, P. (2010). Lean maintenance – A risk-based approach. *Pharmaceutical Engineering, 30*(5), 1–6.

Colbert, F. (2008). The benefits of integrating "lean thinking" concepts into the management of infrared predictive maintenance programs. *Cinde Journal, 29*, 5–12.

Cua, K. O., McKone, K. E., & Schroeder, R. G. (2001). Relationships between implementation of TQM, JIT, and TPM and manufacturing performance. *Journal of Operations Management, 19*, 675–694. doi:10.1016/S0272-6963(01)00066-3

Davies, C. (2003). *The contribution of lean thinking to the maintenance of manufacturing systems* (PhD Thesis). Cranfield University, Cranfield, UK.

Davies, C., & Greenough, R. (2001). *Maintenance survey – Identification of lean thinking within maintenance*. Paper presented at the Proceedings of the 17th National conference on manufacturing research: Advances in manufacturing technology XV, Cardiff, UK.

Davies, C., & Greenough, R. M. (2003). *Testing performance measures within maintenance, through case study research into lean thinking activities*. Retrieved March 24, 2014, from www.plant-maintenance.com/articles/Testing_Performance_Measures.pdf

Davies, C., & Greenough, R. M. (2010). *Measuring the effectiveness of lean thinking activities within maintenance*. Retrieved June 24, 2013, from http://www.plant-maintenance.com/articles/Lean_Maintenance.pdf

De Silva, N., Ranasinghe, M., & De Silva, C. R. (2012). *Maintainability approach for lean maintenance*. Paper presented at the World Construction Conference 2012 – Global Challenges in Construction Industry, Colombo, Sri Lanka. Retrieved from http://www.irbnet.de/daten/iconda/CIB_DC25109.pdf

De Treville, S., & Antonakis, J. (2006). Could lean production job design be intrinsically motivating? Contextual, configurational, and levels-of-analysis issues. *Journal of Operations Management, 24*, 99–123. doi:10.1016/j.jom.2005.04.001

Denyer, D., & Tranfield, D. (2009). Producing a systematic review. In D. A. Buchanan & A. Bryman (Eds.), *The Sage handbook of organizational research methods* (pp. 671–689). Thousand Oaks, CA: Sage.

Djurovic, D., & Bulatovic, M. (2014). *"5 S" as a tool lean concept in the maintenance*. Paper presented at the 8th International Quality Conference, Center for Quality, Faculty of Engineering, University of Kragujevac, Serbia.

Dombrowski, U., & Malorny, C. (2014). Lean after sales service – An opportunity for OEMs to ensure profits. In B. Grabot, B. Vallespir, S. Gomes, A. Bouras, & D. Kiritsis (Eds.), *Advances in production management systems innovative and knowledge-based production management in a global-local world* (Vol. 439, pp. 618–625). Berlin: Springer.

Dunn, S. (2014). 5 keys to lean maintenance and improving maintenance. *Asset Management & Maintenance Journal, 27*, 8–10.

EPA. (2015, 2 February). *Lean thinking and methods: TPM*. Retrieved January 4, 2015, from http://www.epa.gov/lean/environment/methods/tpm.htm

Faccio, M., Persona, A., Sgarbossa, F., & Zanin, G. (2014). Industrial maintenance policy development: A quantitative framework. *International Journal of Production Economics, 147*, 85–93. doi:10.1016/j.ijpe.2012.08.018

Fouladgar, M. M., Yazdani-Chamzini, A., Lashgari, A., Zavadskas, E. K., & Turskis, Z. (2012). Maintenance strategy selection using AHP and COPRAS under fuzzy environment. *International Journal of Strategic Property Management, 16*, 85–104. doi:10.3846/1648715X.2012.666657

Fraser, K. (2014). Facilities management: The strategic selection of a maintenance system. *Journal of Facilities Management, 12*, 18–37. doi:10.1108/JFM-02-2013-0010

George, M. (2003). *Lean six sigma for service: How to use lean speed and six sigma quality to improve services and transactions*. New York, NY: McGraw-Hill.

Georgescu, D. (2010). The maintenance management for lean organization. *Economia. Seria Management, 13*, 437–446.

Georgescu, D., & Militaru, C. (2009, 12–13 November). *7s's System foundation of total productive maintenance house for lean organization*. Paper presented at the The 4th International Conference of the Interdisciplinarity in Engineering, Targu Mures, Romania.

Ghayebloo, S., & Shahanaghi, K. (2010). Determining maintenance system requirements by viewpoint of reliability and lean thinking: A MODM approach. *Journal of Quality in Maintenance Engineering, 16*, 89–106. doi:10.1108/13552511011030345

Gopalakrishnan, P., & banerji, A. (2013). *Maintenance and spare parts management* (2nd ed.). New Delhi: PHI Learning Private Ltd.

Heisler, R. (2003). *Planning and scheduling in a lean maintenance environment.* Retrieved December 20, 2014, from http://www.reliabilityweb.com/index.php/search/results/?q=Randy+Heisler&sa=Go&cx=008700119099951100649%3A5ndeqiorkim&cof=FORID%3A10&ie=UTF-8

Hellingrath, B., & Cordes, A.-K. (2014). Conceptual approach for integrating condition monitoring information and spare parts forecasting methods. *Production & Manufacturing Research, 2*, 725–737. doi:10.1080/21693277.2014.943431

Hiltrop, J.-M. (1996). A framework for diagnosing human resource management practices. *European Management Journal, 14*, 243–254. doi:10.1016/0263-2373(96)00004-7

Holweg, M. (2007). The genealogy of lean production. *Journal of Operations Management, 25*, 420–437. doi:10.1016/j.jom.2006.04.001

Hopp, W. J., & Spearman, M. L. (2004). Commissioned paper to pull or not to pull: What is the question? *Manufacturing & Service Operations Management, 6*, 133–148.

Hou, Y.-H., Zhao, H.-F., Meng, X.-K., & Ma, Y.-X. (2013). Research on lean maintenance system of hydroelectric station equipment. *Industrial Engineering and Management, 3*, 117–121.

Huang, J., Bian, Y., & Cai, W. (2012, 15–18 June). *Weapon equipment lean maintenance strategy research.* Paper presented at the International Conference on Quality, Reliability, Risk, Maintenance, and Safety Engineering (ICQR2MSE), Chengdu, China.

Huemann, M., Keegan, A., & Turner, J. R. (2007). Human resource management in the project-oriented company: A review. *International Journal of Project Management, 25*, 315–323. doi:10.1016/j.ijproman.2006.10.001

Irajpour, A., Fallahian-Najafabadi, A., Mahbod, M. A., & Karimi, M. (2014). A framework to determine the effectiveness of maintenance strategies lean thinking approach. *Mathematical Problems in Engineering, 2014*, 1–11.

Jahanbakhsh, M., Moghaddam, N., & Samaie, H. M. (2013). Lean maintenance (case study: Teen Dairy Industry Co.). *International Research Journal of Applied and Basic Sciences, 4*, 2033–2040.

Jasiulewicz-Kaczmarek, M. (2013). Sustainability: Orientation in maintenance management: Case study. In P. Golinska (Ed.), *EcoProduction and logistics* (pp. 135–154). London: Springer.

Joffe, H. (2011). Thematic analysis. In D. Harper & A. R. Thompson (Eds.), *Qualitative research methods in mental health and psychotherapy* (pp. 209–223). West Sussex: Wiley.

Karim, A., & Arif-Uz-Zaman, K. (2013). A methodology for effective implementation of lean strategies and its performance evaluation in manufacturing organizations. *Business Process Management Journal, 19*, 169–196. doi:10.1108/14637151311294912

Khazraei, K., & Deuse, J. (2011). A strategic standpoint on maintenance taxonomy. *Journal of Facilities Management, 9*, 96–113. doi:10.1108/14725961111128452

Kolanjiappan, S., & Maran, D. K. (2011). Lean practices in aircraft maintenance. *Software Engineering and Technology, 3*, 450–457.

Lee, J., & Wang, H. (2008). New Technologies for Maintenance. In K. A. H. Kobbacy & D. N. P. Murthy (Eds.), *Complex System Maintenance Handbook* (pp. 49–78). London: Springer (eBook).

Levitt, J. (2008). *Lean maintenance.* New York, NY: Industrial Press.

Marks, A. (2008). Effective rotable plant management in a lean maintenance culture. *Asset Management and Maintenance Journal, 21*, 36–43.

Márquez, C. A. (2007). *The maintenance management framework: Models and methods for complex systems maintenance.* London: Springer-Verlag.

McCarthy, D., & Rich, N. (2004). *Lean TPM.* Oxford: Elsevier.

McConnell, K., Lindrooth, R. C., Wholey, D. R., Maddox, T. M., & Bloom, N. (2013). Management practices and the quality of care in cardiac units. *JAMA Internal Medicine, 173*, 684–692. doi:10.1001/jamainternmed.2013.3577

Monden, Y. (1998). *Toyota production system: An integrated approach to just-in-time.* Norcross, GA: Engineering & Management Press.

Mostafa, S. (2011). *A lean framework for maintenance decisions* (Master of Industrial Engineering). Mansoura University, Egypt.

Mostafa, S., Dumrak, J., & Soltan, H. (2013). A framework for lean manufacturing implementation. *Production & Manufacturing Research, 1*, 44–64. doi:10.1080/21693277.2013.862159

Nezami, F. G., & Yildirim, M. B. (2013). A sustainability approach for selecting maintenance strategy. *International Journal of Sustainable Engineering, 6*, 332–343. doi:10.1080/19397038.2013.765928

Nima, R., Hesamodin, N., & Alireza, R. (2014). Presenting an optimal lean maintenance strategy using fuzzy MADM. *International Journal of Management, IT and Engineering, 4*, 263–278.

Okhovat, M. A., Ariffin, M. K. A. M., Nehzati, T., & Hosseini, S. A. (2012). Development of world class manufacturing framework by using six-sigma, total productive maintenance and lean. *Scientific Research and Essays, 7*, 4230–4241. doi:10.5897/SRE11.368

Okoli, C., & Schabram, K. (2010). A guide to conducting a systematic literature review of information systems research. *Sprouts: Working Papers on Information Systems, 10*, 1–50. Retrieved form http://sprouts.aisnet.org/10-26

Önder, M. (2014). *Maintenance as a contributor in green production systems: Interviews with Volvo, Scania, and Dynamate* (Master Thesis). Mälardalen University, Mälardalen, Sweden.

Pedersen, E. R. G., & Huniche, M. (2011). Determinants of lean success and failure in the Danish public sector. *International Journal of Public Sector Management, 24*, 403–420. doi:10.1108/09513551111147141

Pheng, L., & Chuan, C. (2001). Just-in-time management of precast concrete components. *Journal of Construction Engineering and Management, 127*, 494–501. doi:10.1061/(ASCE)0733-9364(2001)127:6(494)

Pinjala, S. K., Pintelon, L., & Vereecke, A. (2006). An empirical investigation on the relationship between business and maintenance strategies. *International Journal of Production Economics, 104*, 214–229. doi:10.1016/j.ijpe.2004.12.024

Potes Ruiz, P., Kamsu Foguem, B., & Grabot, B. (2014). Generating knowledge in maintenance from experience feedback. *Knowledge-Based Systems, 68*, 4–20. doi:10.1016/j.knosys.2014.02.002

Qiang, T., Zhu, B., & Li, L. (2011). *A study on military equipment lean maintenance*. Paper presented at the International Conference on Quality, Reliability, Risk, Maintenance, and Safety Engineering, Xi'an.

Rastegari, A. (2012). *Strategic maintenance management in lean environment* (Master Thesis). Mälardalen University, Västerås, Sweden.

Resta, B., Powell, D., Gaiardelli, P., & Dotti, S. (2015). Towards a framework for lean operations in product-oriented product service systems. *CIRP Journal of Manufacturing Science and Technology, 9*, 12–22. doi:10.1016/j.cirpj.2015.01.008

Rodrigues, M., & Hatakeyama, K. (2006). Analysis of the fall of TPM in companies. *Journal of Materials Processing Technology, 179*, 276–279. doi:10.1016/j.jmatprotec.2006.03.102

Romano, E., Murino, T., Asta, F., & Costagliola, P. (2013). Lean Maintenance model to reduce scraps and WIP in manufacturing system: Case study in power cables factory. *WSEAS Transactions on Systems, 12*, 650–666.

Shah, R., & Ward, P. (2003). Lean manufacturing: Context, practice bundles, and performance. *Journal of Operations Management, 21*, 129–149. doi:10.1016/s0272-6963(02)00108-0

Shah, R., & Ward, P. T. (2007). Defining and developing measures of lean production. *Journal of Operations Management, 25*, 785–805. doi:10.1016/j.jom.2007.01.019

Shahin, A., Shirouyehzad, H., & Pourjavad, E. (2012). Optimum maintenance strategy: A case study in the mining industry. *International Journal of Services and Operations Management, 12*, 368–386. doi:10.1504/IJSOM.2012.047626

Sheng, T. L., & Tofoya, J. (2010, 18–20 October). *The secret of manufacturing excellence: Lean maintenance*. Paper presented at the International Symposium on Semiconductor Manufacturing (ISSM), Tokyo.

Shou, W., Wang, X., Wang, J., Hou, L., & Truijens, M. (2014). Integration of BIM and lean concepts to improve maintenance efficiency: A case study. *Computing in Civil and Building Engineering, 2014*, 373–380.

Smith, R. (2004). What is lean maintenance? Elements that need to be in place for success. *Maintenance Technology, 17*, 15–21.

Smith, R., & Hawkins, B. (2004). *Lean maintenance: Reduce costs, improve quality, and increase market share*. Burlington, MA: Elsevier.

Soltan, H., & Mostafa, S. (2014). Leanness and agility within maintenance process. *International Journal of Engineering Research and Technology, 3*, 553–555.

Stuchly, V., & Jasiulewicz-Kaczmarek, M. (2014). Maintenance in sustainable manufacturing. *Scientific Journal of Logistics, 10*, 273–284. Retrieved from http://www.logforum.net/vol10/issue3/no4

Sullivan, W. G., McDonald, T. N., & Van Aken, E. M. (2002). Equipment replacement decisions and lean manufacturing. *Robotics and Computer-Integrated Manufacturing, 18*, 255–265. doi:10.1016/S0736-5845(02)00016-9

Sunjka, B. P., & Murphy, S. F. (2014). Lean implementation within South African aircraft maintenance organisations. *International Journal of Lean Enterprise Research, 1*, 59–80. doi:10.1504/IJLER.2014.062285

Tendayi, T. G. (2013). *An investigation into the applicability of lean thinking in an operational maintenance environment* (Master of Science in Engineering). Stellenbosch University, South Africa.

Tendayi, T. G., & Fourie, C. J. (2012, July). *A lean maintenance supply chain framework for rolling stock maintenance: A case study.* Paper presented at the The 42nd International Conference on Computers and Industrial Engineering (CIE42), Cape Town, South Africa.

Tendayi, T. G., & Fourie, C. J. (2013). *The combined AHP-QFD approach and its use in lean maintenance.* Paper presented at the 25th Annual Southern African Institute of Industrial Engineering Conference, Stellenbosch.

Thirkell, E., & Ashman, I. (2014). Lean towards learning: Connecting Lean Thinking and human resource management in UK higher education. *The International Journal of Human Resource Management, 25*, 2957–2977. doi:10.1080/09585192.2014.948901

Thiruvengadam, A. (2009). *A practical method for assessing maintenance factors using a value stream maintenance map* (Master of Science). Wichita State University, Wichita, KS.

Tinga, T. (2013). Maintenance concepts. In H. Pham (Ed.), *Principles of loads and failure mechanisms* (pp. 161–186). London: Springer-Verlag.

Tranfield, D., Denyer, D., & Smart, P. (2003). Towards a methodology for developing evidence-informed management knowledge by means of systematic review. *British Journal of Management, 14*, 207–222. doi:10.1111/1467-8551.00375

Van den Heuvel, M. (2008). *Improving maintenance shutdown processes: Reducing delay and increasing work efficiency at Corus's Direct Sheet Plant* (Master Thesis). University of Technology Delft, Delft, The Netherlands.

Verma, A. K., & Ghadmode, A. (2004). An integrated lean implementation model for fleet repair and maintenance. *Naval Engineers Journal, 116*, 79–90. doi:10.1111/j.1559-3584.2004.tb00306.x

Waeyenbergh, G., & Pintelon, L. (2002). A framework for maintenance concept development. *International Journal of Production Economics, 77*, 299–313. doi:10.1016/S0925-5273(01)00156-6

Waeyenbergh, G., & Pintelon, L. (2004). Maintenance concept development: A case study. *International Journal of Production Economics, 89*, 395–405. doi:10.1016/j.ijpe.2003.09.008

Womack, J. P., & Jones, D. T. (2003). *Lean thinking: Banish waste and create wealth in your corporation* (2nd ed.). New York, NY: Free Press, Simon & Schuster.

Yile, L., Hang, X. X., & Lei, Z. (2008). *Lean Maintenance framework and its application in clutch maintenance.* Paper presented at the International Conference on Information Management, Innovation Management and Industrial Engineering, Taipei.

Zarei, M., Fakhrzad, M. B., & Jamali Paghaleh, M. (2011). Food supply chain leanness using a developed QFD model. *Journal of Food Engineering, 102*, 25–33. doi:10.1016/j.jfoodeng.2010.07.026

Zwas, A. (2006). Lean manufacturing techniques in bus and rail maintenance: Study at Chicago Transit Authority in Illinois. *Transportation Research Record: Journal of the Transportation Research Board, 1986*, 54–58. doi:10.3141/1986-09

15

Validating quality productivity improvement framework for sponge iron industry in India

Vinod S. Gorantiwar* and R.L. Shrivastava

Department of Mechanical Engineering, Yeshwantrao Chavan College of Engineering, Nagpur 441 110, India

In this work, an attempt has been made to validate the quality–productivity improvement framework developed for sponge iron (SI) industry, through its implementation in two different SI plants. The study proved that the developed framework is valid and reliable for coal-based SI industry in India. The same framework may be implemented elsewhere in the world with little modification according to the prevailing environment in that country. The concept can also be expanded to other fields.

Keywords: quality; productivity; sponge iron; critical success factors; performance measurement factors

1. Introduction

Identification of critical success factors (CSFs), performance measurement factors (PMFs), and its relation with each other (called as framework) for particular industry helps improving the desired parameters through its implementation. However, developed framework must be validated by implementing it in representative unit. Through the literature review, it was noticed that the quality–productivity improvement framework has not yet been developed for sponge iron (SI) industry anywhere in the world. The authors have developed a framework empirically for coal-based SI industry in India and validated it through its implementation in two different SI plants.

2. Literature review

India accounted for major share of global SI production and has occupied the top slot across the globe. With the country's strong economic environment, the industry prospects are certainly bright. There has been an inspiring track record of growth of SI industry in India during the last decade.

Table 1 shows the data from the ministry of steel about production of gas and coal-based SI in India from 2006–2007 to 2012–2013; while Table 2 shows segment-wise production and capacity utilization for year 2010–2012. Reduced growth rate in the year 2012 demands something to be done towards the improvement in quality and productivity in SI industry. Figure 1 shows the graph of production of SI from 2000 to 2010.

The year 2011–2012 was the year of political turmoil in India. A lot of policy changes have occurred which affected the SI industry, directly or indirectly. Availability

*Corresponding author. Email: principal@ycce.edu

Table 1. Production of SI in India.

Year	Production of SI (in MT)		
	Gas based	Coal based	Total (gas + coal) based (in MT)
2006–2007	5.26	13.08	18.34
2007–2008	5.84	14.53	20.37
2008–2009	5.52	15.57	21.09
2009–2010	6.15	18.18	24.33
2010–2011	5.79	20.92	26.71
2011–2012	5.17	15.20	20.37
2012–2013(Apr–Dec)	3.23*	11.66*	14.89*

Source: Annual Report (2011–2012), and Annual Report (2012–2013), Ministry of Steel.
*Indicates the production for three quarers.

Table 2. Segment-wise production and capacity utilization in India.

Particulars	2010–2011	2011–2012	% Growth
(A) Gas-based units – production (tons)	6,189,917	5,150,018	(−) .83
Annual installed capacity (lakh tons)	96	96	
Capacity utilization (%)	64.47	53.64	
(B) Coal-based units – production (tons)	17,065,432	15,407,645	(−) .90
Annual installed capacity (lakh tons)	253.35	257.9	
Capacity utilization (%)	67.35	59.93	
Grand total (A + B)	23,255,349	20,557,663	(−) .88
Total capacity (lakh tons)	349.35	353.09	
Overall capacity utilization	66.56	58.22	

Source: www.spongeironindia.in/prodfig0607.html, The official website of SIMA India, Retrieved from 10 January 2013.

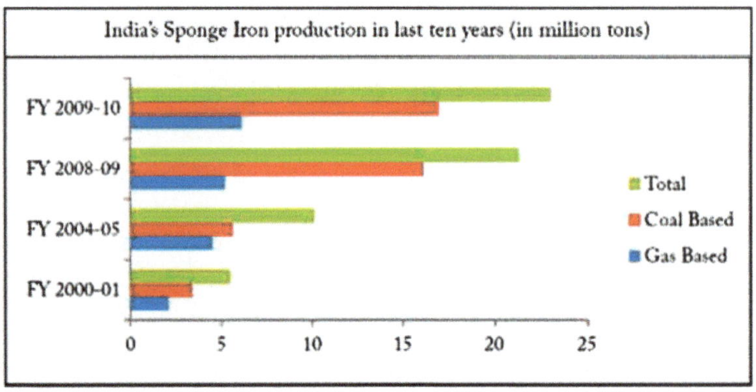

Figure 1. India's SI production.
Source: Ore Team Research (2011).

of iron ore and coal reduced drastically. Many units could not operate to its full capacity. Efforts for quality and productivity improvement were totally at bay. However, few companies (e.g. Tata Sponge Iron Ltd) still managed to earn the profit, resulting from their efforts for quality and productivity improvement through total quality management (TQM) proving the importance and necessity of implementation of TQM in SI industry in India. (www.spongeironindia.in/presentations/PCTATASPONGESIMA Presentation100912.pdf, Retrieved from 22 December 2012; www.tatasponge.com/investor/annualreports/annualreport2010-11.pdf, Retrieved from 10 November 2012; www.tatasponge.com/investor/annualreports/annualreport2011-12.pdf, Retrieved from 10 January 2013.)

During 80s, there was a great demand for SI due to shortage of scrap. Moreover, many new SI plants were established after 1985. Thus, during 80s, quality of SI was not an important factor as there was a huge demand due to the shortage of scrap. Very little attention was paid to the quality and productivity of SI. But as the competition grew, there was a need to focus on the quality and the productivity improvement to survive. After the formation of SIMA i.e. Sponge Iron Manufacturers Association in 1992, the manufacturers had to set standards of quality of SI. Thus, quality of SI gradually began a point of concern in the past.

2.1. *Targeted vs. achieved SI production in India*

Table 3 shows the data for targeted production and achieved production for gas-based and coal-based SI manufacturing in India for the year 2004–2005 to 2011–2012.

The graph drawn from the data from Table 3 is shown in Figure 2. It is clear from the figure that the achieved production is always less than the targeted production in coal-based SI plants. This fact demands the need for identification and validation of critical factors affecting quality and productivity in coal-based SI industry.

2.2. *Problems of SI industry in India*

A panoramic view through the review of Sengupta (2000), Sengupta (2001), Pandit, Sarangi, Babu, and Sheshadri (2003), Sengupta (2005), Raju (2005), Raja and Pal (2006), Kalluraya, Ramchandra, and Sriniwasan (2006), Battacharjee (2007), Equitymaster.com (2009), Sengupta (2009), Khattoi (2010), Reddy (2010), Borwankar (2011), Infocus (2011), Sengupta (2011), Steel Sector Analysis Report (2011), Dhopatkar (2012), Invest India (2012), Kashiva (2012), Scribe.com (2012), and Sengupta (2012a, 2012b), etc. indicates the SI industry with investments of over Rs. 5000 crore contributing more than Rs. 450 crore per annum by way of taxes to the national exchequer, saving sub-

Table 3. Targeted vs. achieved production for gas- and coal-based SI in India.

	Targeted production (MT)			Achieved production (MT)		
Year	Gas-based	Coal-based	Total	Gas-based	Coal-based	Total
2004–2005	6.1	6.0	12.1	4.6	5.5	10.1
2005–2006	6.1	8.5	14.6	5.7	6.5	12.2
2006–2007	7.1	11.0	18.1	7.0	8.5	15.5
2007–2008	7.1	13.0	20.1	7.0	10.0	17.0
2008–2009	7.1	17.0	24.1	7.0	15.3	22.3
2009–2010	7.1	18.0	25.1	7.0	14.0	21.0
2010–2011	9.6	25.34	34.94	6.19	17.07	23.26
2011–2012	9.6	25.34	34.94	5.15	15.41	20.56

Source: Chatterjee (2010) and www.steelguru.com, Retrieved from 10 June 2014.

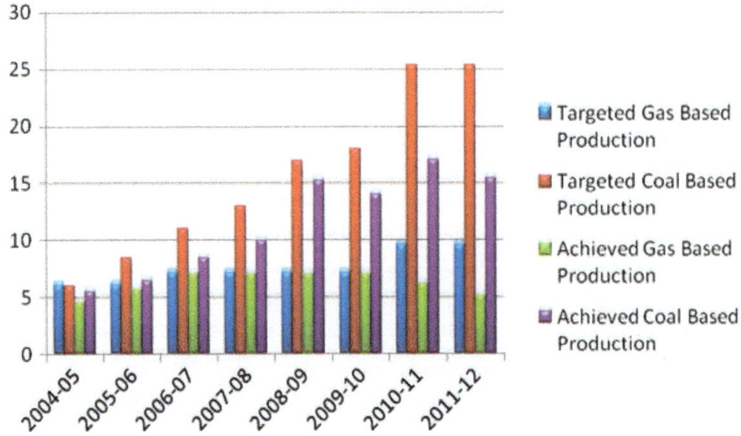

Figure 2. Gas- and coal-based, targeted vs. achieved production of SI in India.

stantial foreign exchange and employing directly and indirectly nearly 1,25,000 people. On the other hand, it is also an industry which is under tremendous viability pressure due to raw material availability and pricing constraints. Selling prices are determined by market forces, whereas all input costs are controlled by the government. The problems of the SI industry are:

- The cost of basic inputs like iron ore, coal, and gas have steeply gone up.
- All input costs (administered pricing) are higher in India than overseas.
- Inputs/services for SI production are mainly in government hands.
- High cost of capital.
- Slowdown of economy, resulting in demand recession.
- Availability of right grade and quality of non-coking coal is a must to optimize cost of production of SI. Indian coal having high-ash content (30–35%) will necessarily have to be blended with imported low-ash content coal (below 12%). Therefore, reduction of custom duty on non-coking coal (below 12% ash content) for metallurgical use is necessary.
- High prices of natural gas in comparison to international prices and cost of production resulting from high input cost and inadequate export incentives.

Table 4. List of CSFs (input factors) and PMFs (output factors).

Critical success factors (CSFs)/input factors	Performance measurement factors (PMFs)/output factors
IF 1: Process management	*OF 1*: Productivity improvement
IF 2: Top management commitment	*OF 2*: Business result
IF 3: Employee involvement	*OF 3*: Employee satisfaction
IF 4: Organization culture	*OF 4*: Time dimension
IF 5: Effective networking	*OF 5*: Customer satisfaction
IF 6: Quality assurance	*OF 6*: Quality performance
IF 7: Competitive assessment	
IF 8: Training	

- The iron ore used as a raw material for SI needs to be of higher quality with more than 62% Fe content. Currently, out of India's estimated reserves of 25 billion tons of iron ore, higher grade constituents are only about 8.7%.

Thus, the various gaps those have been identified are as follows:

(1) Large numbers of SI plants have come up with small capacity of 50–100 TPD due to government's de-licensing policy. They are relatively new and facing many problems. They are coal based due to large availability of non-coking coal and availability of technology. However, for the sake of higher and higher production, Quality and productivity is highly neglected. In the present crisis of closure of many plants, an urgent need has been generated to take care of quality and productivity in SI industry.

(2) Most of the plants are run in traditional ways. Thumb rules and conventional knowledge is employed for getting the work done. Except very few plants, awareness of quality management system is extremely poor and its implementation is the matter of concern.

(3) Exhaustive literature review indicated that CSFs and PMFs for SI industry have not yet been derived. An attempt was made by Gorantiwar, Kshirsagar, and

Table 5. Multiple regression analysis.

	OF					
IF	OF – 1: Productivity improvement	OF – 2: Business result	OF – 3: Employee satisfaction	OF – 4: Time dimension	OF – 5: Customer satisfaction	OF – 6: Quality performance
IF – 1: Process management	.223 (2.641)**	.472 (6.694)**	.270 (3.161)**	.227 (2.967)**	−.039 (−.380)	.305 (3.602)**
IF – 2: Top management commitment	.209 (2.993)**	.147 (2.223)*	.112 (1.403)	−.060 (−.785)	.283 (3.470)**	.184 (2.421)*
IF – 3: Employee involvement	.179 (2.418)*	.138 (1.890)	.272 (3.333)**	.298 (3.824)**	.185 (2.270)*	.099 (1.165)
IF – 4: Organization culture	.136 (2.186)*	.254 (4.196)**	.150 (2.277)*	.054 (.800)	.052 (.710)	.105 (1.561)
IF – 5: Effective networking	−.272 (−5.676)**	−.137 (2.931)**	−.042 (−.767)	.305 (6.198)**	.015 (.247)	.003 (.042)
IF – 6: Quality assurance	.186 (3.391)**	−.038 (−.693)	−.002 (−.036)	−.061 (−1.018)	.067 (.982)	.274 (4.075)**
IF – 7: Competitive assessment	.044 (.746)	.086 (1.624)	.001 (.019)	−.004 (−.061)	−.020 (−.337)	−.110 (−1.975)*
IF – 8: Training	.029 (.514)	.083 (1.481)	.082 (1.308)	.062 (.999)	−.026 (−.385)	.049 (.738)
R	.7463	.7483	.6253	.6656	.4370	.6285
R^2	.557	.560	.391	.443	.191	.395
F-value	53.768**	82.439**	55.577**	69.049**	30.772**	42.230**
Intercept	2.210	2.127	2.392	2.345	3.051	2.632

*Indicates one tailed significance.
**Indicates two tailed significance.

Shrivastava (2011), to identify the factors affecting quality and productivity in direct-reduced iron (DRI) industry; however, it was limited to small region and it lacked validation.

(4) Available literature indicates that no attempt has yet been made to evolve quality–productivity improvement framework for SI industry anywhere in the world.

(5) There is a need to understand the correlation between CSFs and PMFs for appropriate allocation of resources for better performance of SI industry.

To fulfill the gap, authors have developed the framework for coal-based SI industry in India. The framework is represented in the form of models as given in the section below.

3. Models for SI industry

The data collected from the respondents through a specially prepared instrument was analyzed for factor analysis which resulted into eight CSFs & six PMFs for coal-based SI industry in India as listed in Table 4.

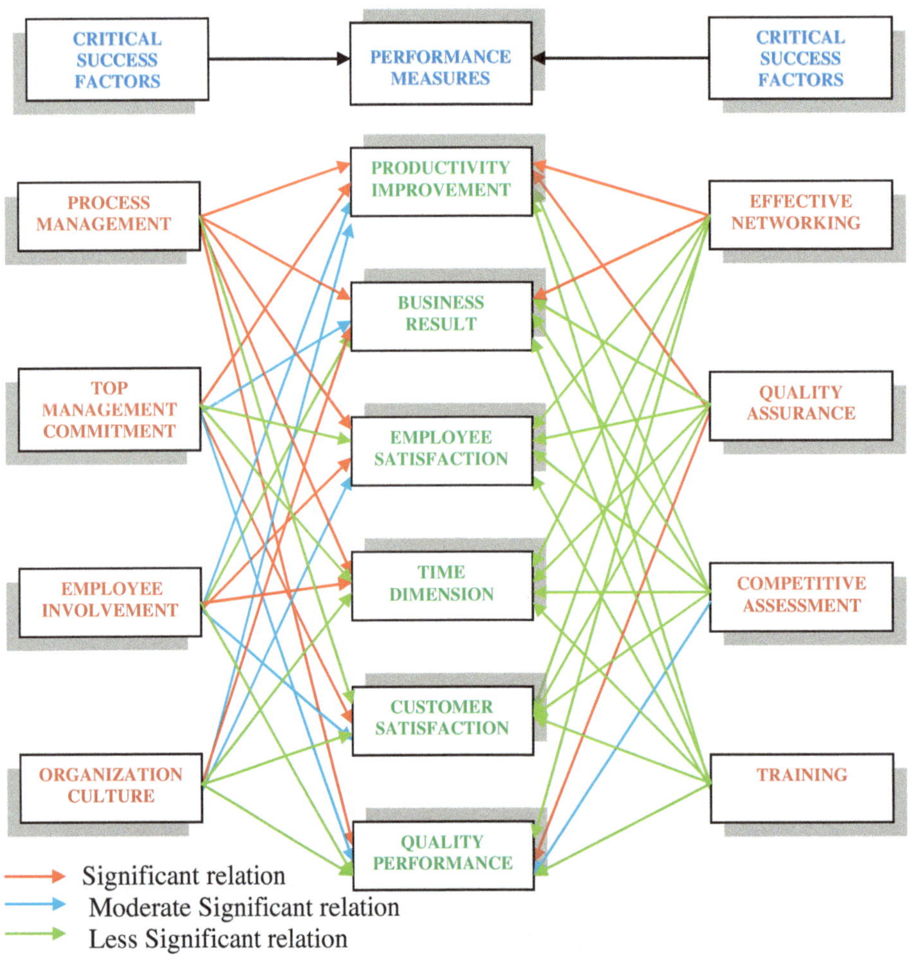

Figure 3. Graphical representation of regression analysis.

The data were regressed to find out the relations between output factors and input factors. Table 5 shows the result of multiple regression analysis. Software SPSS V 20 was used (George & Mallery, 2009; Gupta & Gupta, 2011; IBM SPSS, n.d.; Kerr, Hall, & Kozub, 2002; SPSS online, n.d.). Red color values in the table show significant relations, blue color values show moderate significant relations, and green color values show less significant relations.

The graphical representation of regression analysis is shown in Figure 3.

The regression analysis has resulted into six models for quality–productivity improvement in coal-based SI industry in India as below. They show the relations between output factors and input factors as follows:

(1) *OF 1 (Productivity Improvement)* = 2.210 + .223 (IF1) + .209 (IF2) + .179 (IF3) + .136 (IF4) – .272 (IF5) + .186 (IF6)
(2) *OF 2 (Business Result)* = 2.127 + .472 (IF1) + .147 (IF2) + .254 (IF4) – .137 (IF5)
(3) *OF 3 (Employee Satisfaction)* = 2.392 + .270 (IF1) + .272 (IF3) + .150 (IF4)
(4) *OF 4 (Time Dimension)* = 2.345 + .227 (IF1) + .298 (IF3) + .305 (IF5)
(5) *OF 5 (Customer Satisfaction)* = 3.051 + .283 (IF2) + .185 (IF3)
(6) *OF 6 (Quality Performance)* = 2.632 + .305 (IF1) + .184 (IF2) + .274 (IF6) – .110 (IF7)

Input factors and their contribution in every performance measure can be observed from equations above. It helps managers to allocate the proper share of resources to input factors so as to get desired performance.

4. Validation of models

Model validation is done through model implementation called as case study. According to Zainal (2007), case study is a robust research methodology, when a holistic and in-depth investigation is needed. Case study gives chance to study one aspect of a real-world problem in detail from many different viewpoints. According to Pegram (1999), case study research is a single, in-depth investigation. Yin (1984) has defined the case study method as 'an empirical inquiry that investigates a contemporary phenomenon within its real life context, when the boundaries between phenomenon and context are not clearly evident; and in which multiple sources of evidence are used'. According to Tellis (1997), case studies are designed to bring out the details from the viewpoint of participants by using multiple sources of data. In the opinion of Zainal (2007), case studies are considered useful in research as they enable researchers to examine data at the micro-level. Sroka and Silic (n.d.) found the case studies to be very solid and widely used method by researchers.

Validation of the developed model was done in two different SI manufacturing units. The selection was done in such a way that both the plants differ in many aspects such as production capacity, production method, year of establishment, number of employees, location, ownership, etc.

4.1. Company A (Case study – I)

The plant is equipped with state-of-the-art machineries and the latest technology, making the plant one of the best in India. The usage of most advanced technology helps in keeping emission low and optimize every process in the plants to be energy efficient, thus making the product eco-friendly. The company has pioneered the industry with two induction furnaces of 40 metric tons each along with a common four-strand castor which produces high-quality steel billets. The same is rolled in the most advanced rolling mills to produce TMT bars. With world's largest medium frequency induction melting furnace at its steel plant which is one of the large steel manufacturing units with the capacity to produce .5 million tone steel per annum.

4.1.1. Quality policy

The company has always strived to produce the finest steel in the world. The committed excellence of company in delivering quality products ensures that the integrated steel plant delivers the highest quality steel. The consistent mechanical and chemical properties of steel are attributed to the usage of high-quality virgin steel that is rolled with superior technology. Stringent quality control at every step makes sure that each product comes out with a stamp of consistency. Company adheres to extremely stringent norms to ensure that the higher quality steel reaches the end users. The fully equipped labs with some of the most ultra modern testing and constant upgradation and research make the company best in its class. The company is certified for international quality standard ISO 9001:2008 and ISO 9001:14000 from JNS-ANZ Australia. The company is highly conscious about maintaining the high quality of its product always.

4.2. Company B (Case study – II)

The plant of Company B has three different units in the same premises. SI is produced in 4 kilns of 100 TPD capacities each. The produced SI is supplied as raw material to steel melting shop (SMS) in the same premises for billet making. The waste heat, along with coal is used for power generation. Thus, the three units are:

(1) DRI unit to producc SI
(2) SMS to produce billets
(3) Power plant to produce electricity

4.2.1. Quality policy

The basic principle of the company is, 'there is no survival in long term without quality and services.'

4.3. Data analysis

In order to evaluate the companies' TQM implementation practices and organizational performance, an instrument was prepared as given in Appendix 1. The instrument was administered to evaluate the current situation of TQM implementation and performance. The responses were collected from employees of the plant having different designations and different experiences. The evaluation results are presented in subsequent sections which are translated into score according to scoring methods. It also highlights the extent of relationship between various TQM implementation practices and overall performance.

Table 6. Reliability analysis of instrument.

	Cronbach's alpha
Company A	.916
Company B	.946

4.3.1. *Reliability of research instrument*

The data received from the respondents of Company A as well as from Company B were analyzed for the reliability of research instrument using the software SPSS V 20. The Cronbach's Alpha was found much above .7, indicating the reliability of instrument. Thus, it permits the further analysis of data (Table 6).

4.3.2. *Assessment of company*

The responses in the form of data were analyzed, mean of each item was estimated and items were grouped as respective factors. Further, grand mean of all factors were obtained. The mean of responses were proportionally marked in consultation with the experts. Number '1' indicates that the organization is extremely weak in the practice and number '8' indicates that the organization is extremely strong in the practice. During the process of assessment, the strength and weaknesses of the items of the respective factors were pointed out. Score above and below indicates strong and weak items, respectively. If the company is neither strong nor weak in the particular item, it is indicated as 'Average'. It is relative judgment. Table 7 explains the marking scheme and grading scheme. Based on the total score of factors, the overall grading criterion was decided. The total score is indicated as X.

Table 8 gives the assessment result after detailed analysis of the degree of emphasis of TQM implementation.

Based on above analysis, overall grade of the company for emphasis of TQM implementation and its impact on organizational performance was estimated as follows: (Table 9).

Thus, by using the instrument developed for case study, the degree of emphasis of TQM implementation and its impact on organizational performance was measured. The overall grade achieved as per grading criteria is 'A'. Through the assessment of the companies' quality management practices and overall business performance, most of the established relationship between implementation factors and the performance measures of the companies found to be strong to very strong. Two relations such as Top Management Commitment with Productivity and Quality assurance with Customer Satisfaction

Table 7. Mean and proportionate score and grading criteria.

Mean (response)	Score (strength)	Total score range	Grading
1.00–1.50	1 (Extremely weak)	$X > 90\%$	A^+
1.51–2.00	2 (Very weak)	$80\% < X < 90\%$	A
2.01–2.50	3 (Weak)	$70\% < X < 80\%$	B
2.51–3.00	4 (Average)	$60\% < X < 70\%$	C
3.01–3.50	5 (Above average)	$50\% < X < 60\%$	D
3.51–4.00	6 (Strong)	$X < 50\%$	E
4.01–4.50	7 (Very strong)		
4.51–5.00	8 (Extremely strong)		

Table 8. Assessment results for Company A and Company B.

S. No	Critical success factors	Company – A score (out of 8)	Company – B score (out of 8)
1	Process management	7	7
2	Top management commitment	7	7
3	Employees involvement	6	7
4	Organization culture	7	7
5	Effective networking	7	6
6	Quality assurance	7	6
7	Competitive assessment	7	6
8	Training	7	7

Table 9. Estimation of overall grade of company.

S. No	Particulars	Company A	Company B
1	Maximum numerical weightage assigned to each factor	8	8
2	Total number of factors	8	8
3	Maximum score	$8 \times 8 = 64$	$8 \times 8 = 64$
4	Total score achieved	54	53
5	% of scores achieved	84.375%	82.81%
6	*Overall grade*	*A*	*A*

are found extremely strong. Thus, company is doing well on the front of quality–productivity improvement. This has also validated the instrument used for case study.

4.3.3. Impact of quality practices on the organizational performance

As a result of competitiveness and to strive for better performance, the company has implemented various performance improvement approaches. Many steps have been taken by the company during past to enhance overall performance. The aim of the organization is to achieve better result than the past.

4.3.3.1. TQM implementation. Companies have taken many initiatives for the implementation of QMS for the quality–productivity improvement.

(1) *Process management*: every care is taken for the process management at both the companies. All departments are involved in quality improvement; there are robust feedback mechanisms; all employees audit the process regularly; there are faster adoptions of new technology for continuous improvement; automation is promoted in production; employees' suggestions are considered favorably; the process is monitored continuously for improvement; R&D and innovation in the product development are promoted; employees are motivated for QMS implementation.
(2) *Top management commitment*: top management of both the companies are committed in delivering quality products through quality practices. They have clear vision, mission, and strategic direction. Top management is able to take the responsibility for continuous quality improvement. Adequate budgetary allocations are made for quality improvement initiatives. Constructive efforts of

Table 10. Mean and *F*-value of ANOVA for Company – A.

	1-Productivity	2-Business results	3-Employee satisfaction	4-Time dimension	5-Customer satisfaction	6-Quality performance
1-Process management	4.67 (.310)	3.85 (1.308)	3.81 (2.603)	4.15 (1.219)	4.15 (.579)	4.18 (.441)
2-Top management commitment	4.59 (.691)	4.11 (2.639)	3.81 (.463)	4.00 (1.570)	3.70 (.654)	4.04 (.873)
3-Employee involvement	4.26 (1.149)	3.96 (1.503)	4.11 (1.484)	3.89 (4.191)*	3.59 (2.634)	4.00 (1.549)
4-Organization culture	4.26 (3.465)*	4.07 (1.080)	3.85 (.886)	4.07 (2.365)	4.11 (1.020)	4.22 (1.318)
5-Effective networking	4.26 (.221)	4.44 (.024)	3.89 (2.730)	3.74 (.924)	4.04 (2.147)	4.04 (2.648)
6-Quality assurance	4.26 (.993)	3.96 (.242)	4.15 (.388)	4.18 (1.279)	4.55 (1.096)	4.48 (.765)
7-Competitive assessment	4.33 (.135)	4.3 (.298)	3.18 (2.097)	3.67 (2.233)	3.89 (.756)	4.18 (.233)
8-Training	4.41 (5.201)**	4.15 (7.653)**	4.30 (.584)	4.26 (2.109)	4.00 (.265)	4.48 (2.886)

*Indicates one tailed significance.
**Indicates two tailed significance.

Table 11. Mean and *F*-value of ANOVA for Company – B.

	1-Productivity improvement	2-Business results	3-Employee satisfaction	4-Time dimension	5-Customer satisfaction	6-Quality performance
1-Process management	4.06 (.954)	4.18 (1.044)	3.82 (.557)	3.53 (.566)	4.41 (.684)	4.47 (.366)
2-Top management commitment	4.29 (6.898)**	4.06 (.195)	3.94 (.275)	3.65 (2.464)	4.35 (.765)	4.35 (.131)
3-Employee involvement	4.35 (3.524)*	4.06 (2.079)	3.41 (.777)	3.94 (.908)	4.00 (2.569)	4.41 (3.704)*
4-Organization culture	4.23 (1.753)	3.94 (.153)	3.94 (.430)	3.88 (1.058)	3.94 (3.366)	4.12 (2.737)
5-Effective networking	4.29 (1.842)	4.18 (1.488)	3.88 (.698)	3.76 (.801)	3.88 (5.134)*	3.82 (8.125)**
6-Quality assurance	4.00 (.603)	4.06 (.858)	3.70 (.535)	3.29 (.592)	4.06 (.384)	4.29 (.744)
7-Competitive assessment	3.94 (.369)	3.76 (.398)	3.59 (.040)	3.59 (1.034)	3.70 (.986)	3.65 (1.463)
8-Training	4.47 (.566)	4.23 (3.526)*	4.00 (2.208)	3.94 (1.256)	3.70 (3.183)	4.06 (1.824)

*Indicates one tailed significance.
**Indicates two tailed significance.

employees are recognized. Congenial work environment is provided along with fair compensation package to the employees.

(3) *Employee involvement*: both the companies put the efforts for capacity building of employees. Innovation in the work is always encouraged. Good coordination

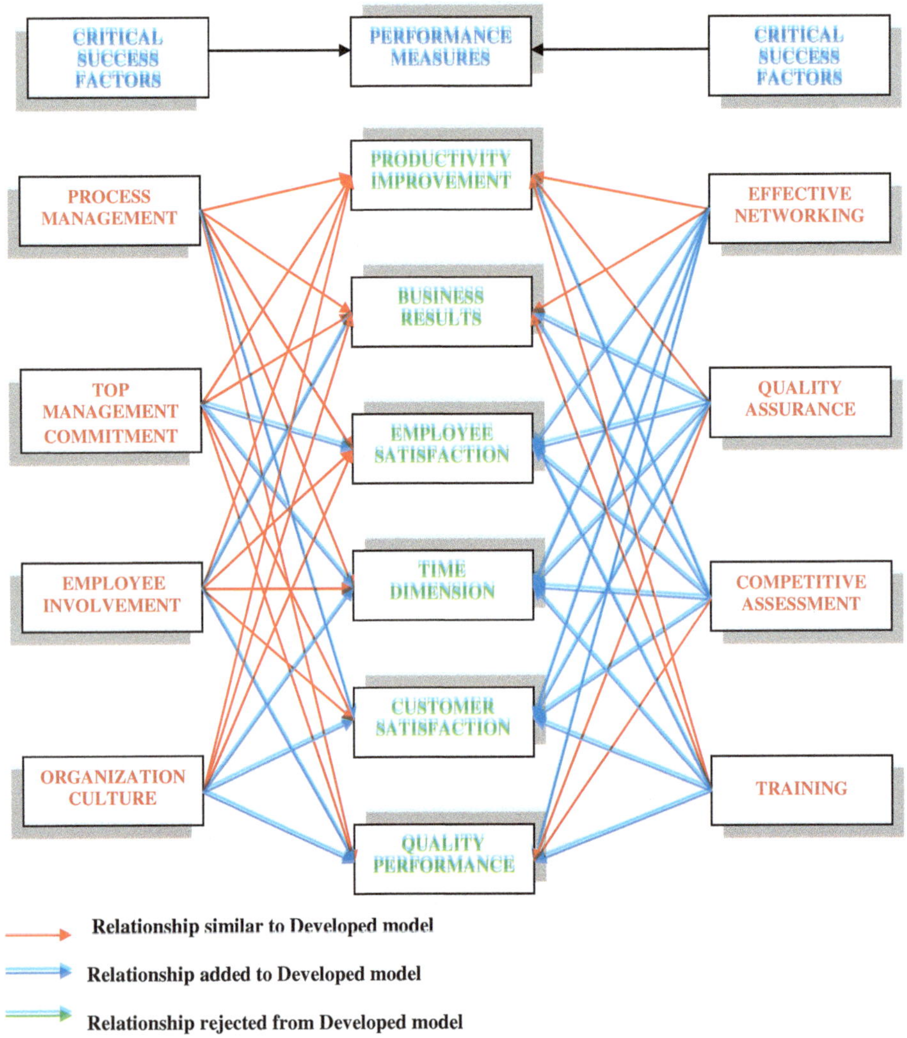

Figure 4. Graphical representation of relations for Company A.

and cooperation is maintained in the plant. All these help in more and more involvement of employees.

(4) *Organization culture*: companies have well-defined organizational structure and governance system. Through the well-defined rules, regulations, and operating procedures and team spirit among the employees, companies get better performance.

(5) *Effective networking*: it is observed that effective networking has been managed by the companies through various ways such as honest and two-way communication between management and employees, regular monitoring of internal and external customer satisfaction; providing technical guidance to suppliers etc.

(6) *Quality assurance*: quality is assured and improved continuously through various ways. Every effort is directed towards the achievement of better and better quality through integration of prevention and correction in production process.

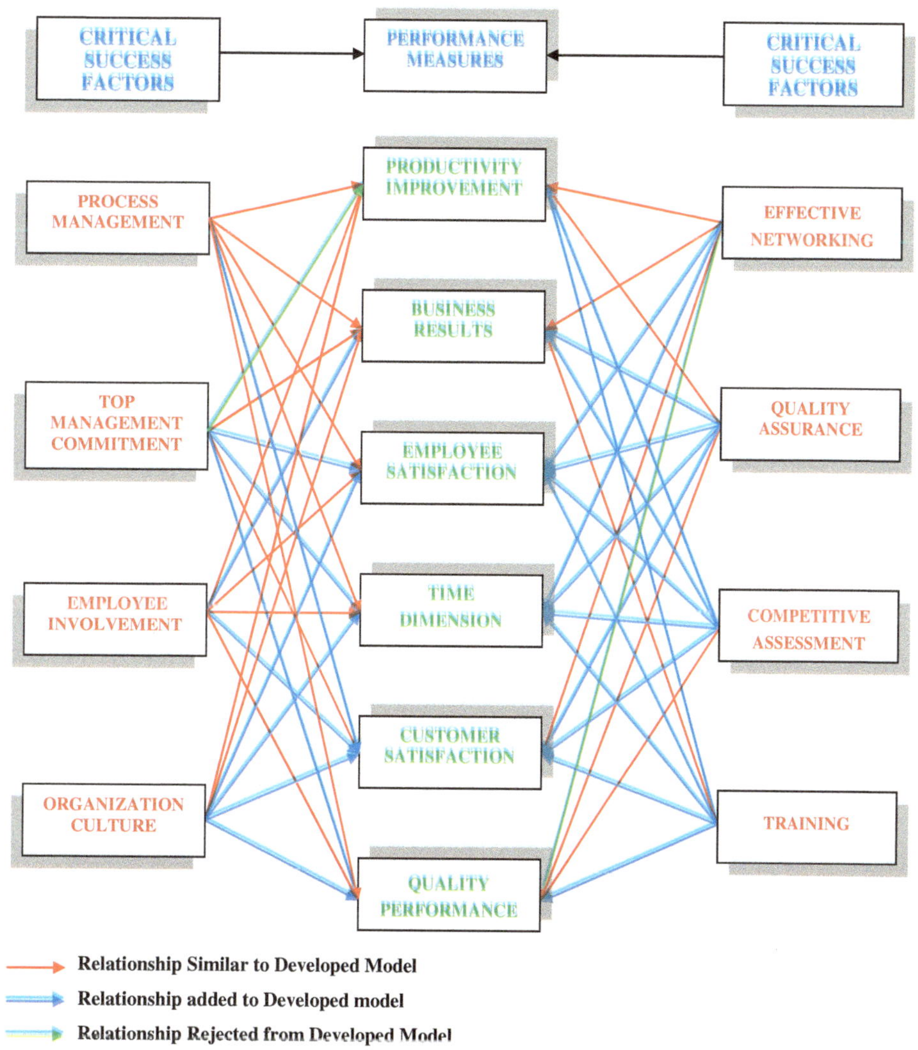

Figure 5. Graphical representation of relations for Company B.

(7) *Competitive assessment*: companies maintain the internal supply chain and manages the relationship with suppliers. Based on the current performance, the goal is set for future.

(8) *Training*: the employees are given on job training to understand the practical aspects of SI manufacturing. Top management of both the companies provides guidance and ensures the effectiveness of training.

Thus, by using the instrument developed for case study, the degree of emphasis of TQM implementation and its impact on organizational performance was measured. The overall grade achieved as per grading criteria is ‘A’. Through the assessment of the companies’ quality management practices and overall business performance, most of the established relationship between implementation factors and the performance measures of the company found to be strong to very strong. One relation i.e. Training with Time Dimension is found extremely strong, while two relations i.e. Employee Involvement with Employee Satisfaction and Quality Assurance with Time Dimension found above average. Thus, company is doing well on the front of quality–productivity improvement. This has also validated the instrument used for case study.

5. Result

For the validation of the models developed for coal-based SI industry in India as given in section 3 and to find out the significance of developed relations, the collected data for company A and Company B was analyzed using the software SPSS V 20 to carry out ANOVA through mean and F-value. ANOVA is useful to identify the differences between the perceptions of employees based on some parameter. Tables 10 and 11 show the mean and F-value of ANOVA for Company A and Company B, respectively.

The Figure 4 shows similarity of relations with the earlier developed relations for Company A.

The Figure 5 shows similarity of relations with the earlier developed relations for Company B.

6. Conclusion

From the data given in Tables 10 and 11 and subsequently drawn into Figures 4 and 5, it is observed that there are many relations similar to developed model for coal-based SI industry in India, shown in Figure 3.

The relations those are observed similar to the earlier developed relations are:

(1) Process management and productivity improvement
(2) Process management and business result
(3) Process management and employee satisfaction
(4) Process management and time dimension
(5) Process management and quality performance
(6) Top management commitment and business results
(7) Top management commitment and customer satisfaction
(8) Employee involvement and productivity improvement
(9) Employee involvement and employee satisfaction
(10) Organization culture and productivity improvement
(11) Organization culture and business results
(12) Effective networking and productivity improvement
(13) Effective networking and business results
(14) Quality assurance and productivity improvement
(15) Quality assurance and quality performance
(16) Competitive assessment and quality performance
(17) Training and productivity improvement

In both the companies, some relationships have been added to earlier relations as follows –

(18) Process management and customer satisfaction
(19) Top management commitment and time dimension
(20) Organization culture and time dimension
(21) Organization culture and customer satisfaction

(22) Organization culture and quality performance
(23) Effective networking and employee satisfaction
(24) Effective networking and time dimension
(25) Effective networking and quality performance
(26) Quality assurance and business results
(27) Quality assurance and employee satisfaction
(28) Quality assurance and customer satisfaction
(29) Competitive assessment and productivity improvement
(30) Competitive assessment and business results
(31) Competitive assessment and employee satisfaction
(32) Competitive assessment and customer satisfaction
(33) Training and employee satisfaction
(34) Training and time dimension
(35) Training and customer satisfaction
(36) Training and quality performance

Company A does not reject any relation, while only two relations are rejected by Company B. Since almost all the relations indicated by model developed for coal-based SI industry are endorsed by both case studies, *both the case studies validate the model developed for coal-based SI industry in India.*

7. Discussion

Validation of the model, developed for coal-based SI industry in India has been done through the two case studies. Both the plants have different capacity, location, manufacturing process, and number of employees. This has enabled to get the quality–productivity status from the SI manufacturers of different nature. It is found that significant relationships exist between implementation factors and performance measures of companies. It was noticed that various performance indicators have remarkable improvement over the years. The companies have achieved both tangible and intangible benefits by practicing quality management.

8. Future scope

The validated model may be implemented to all SI plants in India for quality–productivity improvement. Ultimately, it will help improve the bottom line of the company – the very essence of any business. The concept can also be expanded to other fields.

Disclosure statement

No potential conflict of interest was reported by the author(s).

References

Battacharjee, S. (2007). *Sponge iron industry: Current scenario* [online]. Retrieved November 20, 2012, from http://www.steelworld.com/report0707.pdf

Borwankar, R. S. (2011, January). Gas based sponge iron industry – Outlook. *DRI Update, A Journal of Sponge Iron Manufacturing Association,* 14–19.

Chatterjee, A. (2010). *Sponge iron industry in India – Scope for clean technology.* Retrieved January 10, 2013, from http://ebookbrowse.com/amit-chatterjee-sponge-iron-industry-in-india-scope-for-clean-technolog

Dhopatkar, V. (2012, February). Coal sector in India. *Steelworld*, pp. 16–20.

Equitymaster.com. (2009). *Steel*. Retrieved November 4, 2012, from www.equitymaster.com/research-it/sector-info/steel/index-jul09.html

George, D., & Mallery, P. (2009). *SPSS for windows step by step* (8th ed.). Pearson Education.

Gorantiwar, V. S., Kshirsagar, V. P., & Shrivastava, R. L. (2011). Quality & productivity in direct reduced iron industry and factors affecting it. In *Proceedings of the international conference on sustainable manufacturing: issues, trends and practices (ICSM – 2011)* (pp. 303–312). November 10–12, 2011, BITS, Pilani, India.

Gupta, S. L., & Gupta, H. (2011). *SPSS 17.0 for researchers* (1st ed.). International Book House.

IBM SPSS. (n.d.). *Statistics 20 brief guide* [online]. Retrieved October 10, 2012, from ftp://public.dhe.ibm.com/software/analytics/spss/documentation/statistics/20.0/en/client/Manuals/IBM_SPSS_Statistics_Brief_Guide.pdf

Infocus. (2011). *Sponge iron industry: An analysis* [online]. Retrieved November 10, 2011, from http://www.dhanbank.com

Invest India. (2012). *A joint venture of Ministry of Commerce & Industry, government of India and Federation of Indian Chambers of Commerce & Industry, 'steel – Investment opportunity'*. Retrieved December 30, 2012, from www.investindia.gov.in/?q=steel-sector

Kalluraya, A. S., Ramchandra, G. K., & Sriniwasan, T. M. (2006, March). Iron ore pellets as feed to sponge iron plants. *Steelworld*, pp. 17–21.

Kashiva, D. (2012, March). Sponge iron industry faces raw material crisis. *Steelworld*, pp. 18–20.

Kerr, A. W., Hall, H. K., & Kozub, S. A. (2002). *Doing statistics with SPSS* (1st ed.). New Delhi: Sage.

Khattoi, S. C. (2010, April). Changing scenario of coal based sponge iron. *DRI Updates, Journal of Sponge Iron Manufacturer's Association of India,* 16–18.

Ministry of Steel, Government of India. (2011–2012). *Annual report*. Retrieved January 5, 2013, from www.steel.gov.in

Ministry of Steel, Government of India. (2012–2013). *Annual report*. Retrieved June 10, 2013, from www.steel.gov.in

Ore Team Research. (2011, January). Sponge iron industry – Indian outlook. *DRI Update, Journal of Sponge Iron Manufacturing Association,* 39–40.

Pandit, A., Sarangi, B. M., Babu, K., & Sheshadri, M. K. (2003, December). Coal based sponge iron industry – A prime movers to enhance steel making capacities in India. *Steelworld*.

Pegram, A. (1999). What is case study research? *Nurse Researcher, 7*, 5–16.

Raja, B. V. R., & Pal, N. (2006, April). India sponge iron industry – Status, potential & prospects. *IIM Metal News, 9*, 11–13.

Raju, P. R. K. (2005, July). Sponge iron industry – An overview of problems & solutions. *Steelworld*, p. 20.

Reddy, J. M. (2010, August). Role of raw materials for production of sponge iron. *Steelworld*, pp. 40–42.

Scribe.com. (2012). *Sponge iron report*. Retrieved October 22, 2012, from http://www.scribd.com/doc/7278475/Sponge-Iron-Report

Sengupta, S. (2000, September). Indian sponge iron industry: Growth despite constraints. *Steelworld*.

Sengupta, S. (2001, October). Indian sponge iron industry in global perspective. *Steelworld*.

Sengupta, S. (2005, July). Gearing up to meet growing metallics requirement. *Steelworld*, pp. 11–18.

Sengupta, S. (2009, June). An outline of Indian ferro alloy industry. *Steelworld*, pp. 14–21.

Sengupta, S. (2011, April). Indian steel today & tomorrow. *Steelworld*, pp. 16–24.

Sengupta, S. (2012a, April). Indian steel: Big challenge ahead. *Steelworld*, pp. 26–42.

Sengupta, S. (2012b, June). Indian steel: Performance in FY'12 and its future prospects. *Steelworld*, pp. 34–46.

SPSS. (n.d.) [online]. Retrieved October 10, 2012, from http://en.wikipedia.org/wiki/SPSS

Sroka, A., & Silic, M. (n.d.). *Which requirements do case studies have to meet and how can they be analyzed?* Report on Ph.D. Seminar. St. Gallen: University of St. Gallen.

Steel Sector Analysis Report. (2011). Retrieved from www.equitymaster.com/research-it/sector-info/steel/Steel-Sector-Analysis-Report.asp

Tellis, W. (1997). Application of a case study methodology. *The Qualitative Report, 3*(3), 1–18.

Yin, R. K. (1984). *Case study research: Design and methods* (1st ed.). Beverly Hills, CA: Sage.

Zainal, Z. (2007). Case study as a research method. *Jurnal Kemanusiaan bil,* (9), 1–6.

Appendix 1

Questionnaire for case study validation

Identification of critical success factors and performance measures for sponge iron industry – model validation (The information provided here will be kept strictly confidential).

Name of the respondent:
Qualification: Designation:
E-mail: Experience: _______ yrs.
Mobile No.:
Industry name & address:

Critical success factors and performance measurement factors

Based on the analysis of responses, eight CSFs and six PMFs were derived. More focus on CSFs (also called as input factors, TQM factors or significant factors) will lead to the improvement in PMFs (also called as output factors). Detailed descriptions of factors are –

(A) Critical success factors

(1) *Process management* includes all those attributes that suggest the need of process management in sponge iron industry, such as involvement of all departments in quality improvement processes; regular measurement and monitoring of quality cost; consideration of employee's suggestions/feedback; robust feedback mechanism; regular process auditing by all employees; cross-functional team participation; collective responsibility; training programs for quality techniques; advanced statistical techniques training programs; continuous process monitoring and improvement; data-driven process improvement; faster adoption of new technology for continuous improvement; promoting R&D and innovation for product development; promoting automation in production; conducive environment for implementation of QMS; and motivation to the employees for implementing QMS etc.

(2) *Top management commitment* signifies an important role of top management's clarity of vision, mission, and strategic direction; top management's ability to take responsibility for continuous quality improvement; adequate budgetary allocation for quality improvement initiatives; recognition of constructive efforts of employees by management; management's commitment for providing good work environment; fair compensation package to employees; and congenial work environment

(3) *Employee involvement* has loading of regular mapping and monitoring of employee's skills and competencies; capacity building of employees; use of bench marking for process improvement; encouraging innovation; good coordination and cooperation; and measurement of employee commitment levels.

(4) *Organization culture* accounts for team spirit; well-defined organizational structure and governance system; well-defined rules, regulations and operating procedures.

(5) *Effective networking* is based on the important role of honest and two-way communication between management and employees; fulfillment of legal, ethical, and social responsibilities by the company; establishment of management information system; providing technical guidance to suppliers; selection of suppliers on the basis of raw material quality; regular monitoring of customer satisfaction; standardization of customer care systems; customer relationship management.

(6) *Quality assurance* is an indicator of use of statistical process control for process management; integration of prevention and correction in production process.

(7) *Competitive assessment* is based on the attributes such as goal setting on the basis of current performance; finalizing strategy after SWOT analysis; supplier relationship management; internal supply chain.
(8) *Training* includes communication skill improvement training programs; leadership development training programs.

(B) Performance measurement factors

(1) *Productivity improvement*: has loading of high capital productivity; high productivity of each employee; high productivity of assets; equipment availability for longer duration; optimum human resource utilization; lower product development time; low machine line downtime; low supplier lead time; etc.
(2) *Business result*: includes high return on investment; top-line growth; sustainable increase in market share; bottom line growth; high processes efficiency; less variation in process; higher process accuracy rate.
(3) *Employee satisfaction*: accounts for higher value addition per employee; high job satisfaction of employees; employee involvement; better environmental compliance; better safety of employees; positive change in corporate culture; better quality of work life.
(4) *Time dimension*: signify an important role of reduced material procurement time; low-order process time; low process through time; high inventory turnover (material productivity); high flexibility in operations / system; etc.
(5) *Customer's satisfaction*: is based on customer retention; low customer complaints; higher customer satisfaction (value for the money spent); etc.
(6) *Quality performance*: accounts for less number of defects in the finished product; less rework and waste; optimum inventory utilization; low cost of quality; etc.

Relationship between critical success factors and performance measures

Based on your experience, please rate the importance of following elements in achieving successful implementation of TQM model in sponge iron industry.

S. No.	Attribute/variables	Very low IMP 1	Low IMP 2	Med IMP 3	High IMP 4	Very high IMP 5
1.	Process management leads to improvement in productivity	☐	☐	☐	☐	☐
2.	Process management leads to improvement in business results	☐	☐	☐	☐	☐
3.	Process management leads to improvement in employee satisfaction	☐	☐	☐	☐	☐
4.	Process management leads to improvement in time dimension	☐	☐	☐	☐	☐
5.	Process management leads to improvement in customer's satisfaction	☐	☐	☐	☐	☐
6.	Process management leads to improvement in quality performance	☐	☐	☐	☐	☐
7.	Top management commitment leads to improvement in productivity	☐	☐	☐	☐	☐

(*Continued*)

Table (*Continued*).

S. No.	Attribute/variables	Very low IMP 1	Low IMP 2	Med IMP 3	High IMP 4	Very high IMP 5
8.	Top management commitment leads to improvement in business results	☐	☐	☐	☐	☐
9.	Top management commitment leads to improvement in employee satisfaction	☐	☐	☐	☐	☐
10.	Top management commitment leads to improvement in time dimension	☐	☐	☐	☐	☐
11.	Top Management Commitment leads to improvement in Customer's Satisfaction	☐	☐	☐	☐	☐
12.	Top management commitment leads to improvement in quality performance	☐	☐	☐	☐	☐
13.	Employees involvement leads to improvement in productivity	☐	☐	☐	☐	☐
14.	Employee involvement leads to improvement in business results	☐	☐	☐	☐	☐
15.	Employee involvement leads to improvement in employee satisfaction	☐	☐	☐	☐	☐
16.	Employee involvement leads to improvement in time dimension	☐	☐	☐	☐	☐
17.	Employee involvement leads to improvement in customer's satisfaction	☐	☐	☐	☐	☐
18.	Employee involvement leads to improvement in quality performance	☐	☐	☐	☐	☐
19.	Organization culture leads to improvement in productivity	☐	☐	☐	☐	☐
20.	Organization culture leads to improvement in business results	☐	☐	☐	☐	☐
21.	Organization culture leads to improvement in employee satisfaction	☐	☐	☐	☐	☐
22.	Organization culture leads to improvement in time dimension	☐	☐	☐	☐	☐
23.	Organization culture leads to improvement in customer's satisfaction	☐	☐	☐	☐	☐
24.	Organization culture leads to improvement in quality performance	☐	☐	☐	☐	☐
25.	Effective networking leads to improvement in productivity	☐	☐	☐	☐	☐
26.	Effective networking leads to improvement in business results	☐	☐	☐	☐	☐
27.	Effective networking leads to improvement in employee satisfaction	☐	☐	☐	☐	☐
28.	Effective networking leads to improvement in time dimension	☐	☐	☐	☐	☐
29.	Effective networking leads to improvement in customer's satisfaction	☐	☐	☐	☐	☐
30.	Effective networking leads to improvement in quality performance	☐	☐	☐	☐	☐
31.	Quality assurance leads to improvement in productivity	☐	☐	☐	☐	☐
32.	Quality assurance leads to improvement in business results	☐	☐	☐	☐	☐

(*Continued*)

Table (*Continued*).

S. No.	Attribute/variables	Very low IMP 1	Low IMP 2	Med IMP 3	High IMP 4	Very high IMP 5
33.	Quality assurance leads to improvement in employee satisfaction	☐	☐	☐	☐	☐
34.	Quality assurance leads to improvement in time dimension	☐	☐	☐	☐	☐
35.	Quality assurance leads to improvement in customer's satisfaction	☐	☐	☐	☐	☐
36.	Quality assurance leads to improvement in quality performance	☐	☐	☐	☐	☐
37.	Competitive assessment leads to improvement in productivity	☐	☐	☐	☐	☐
38.	Competitive assessment leads to improvement in business results	☐	☐	☐	☐	☐
39.	Competitive assessment leads to improvement in employee satisfaction	☐	☐	☐	☐	☐
40.	Competitive assessment leads to improvement in time dimension	☐	☐	☐	☐	☐
41.	Competitive assessment leads to improvement in customer's satisfaction	☐	☐	☐	☐	☐
42.	Competitive assessment leads to improvement in quality performance	☐	☐	☐	☐	☐
43.	Training leads to improvement in productivity	☐	☐	☐	☐	☐
44.	Training leads to improvement in business results	☐	☐	☐	☐	☐
45.	Training leads to improvement in employee satisfaction	☐	☐	☐	☐	☐
46.	Training leads to improvement in time dimension	☐	☐	☐	☐	☐
47.	Training leads to improvement in customer's satisfaction	☐	☐	☐	☐	☐
48.	Training leads to improvement in quality performance	☐	☐	☐	☐	☐

1. In your opinion, what are the areas in Sponge Iron Industry which require immediate attention for quality & productivity improvement?

Ans:- ____

____

2. What solutions you suggest for improvement in quality & productivity in Sponge Iron Industry ?

Ans:- ____

____

3. Kindly suggest, Critical Factors, that you feel are important, but not covered in my study.

Ans:-____

____

4. How important is Industry – Institute partnership for growth of both ?

Ans :-____

____

5. Any other aspects / issues, you would like to mention for improvement / excellence of Sponge Iron Industry.

Ans:- ____

____

Permissions

All chapters in this book were first published in PMR, by Taylor & Francis; hereby published with permission under the Creative Commons Attribution License or equivalent. Every chapter published in this book has been scrutinized by our experts. Their significance has been extensively debated. The topics covered herein carry significant findings which will fuel the growth of the discipline. They may even be implemented as practical applications or may be referred to as a beginning point for another development.

The contributors of this book come from diverse backgrounds, making this book a truly international effort. This book will bring forth new frontiers with its revolutionizing research information and detailed analysis of the nascent developments around the world.

We would like to thank all the contributing authors for lending their expertise to make the book truly unique. They have played a crucial role in the development of this book. Without their invaluable contributions this book wouldn't have been possible. They have made vital efforts to compile up to date information on the varied aspects of this subject to make this book a valuable addition to the collection of many professionals and students.

This book was conceptualized with the vision of imparting up-to-date information and advanced data in this field. To ensure the same, a matchless editorial board was set up. Every individual on the board went through rigorous rounds of assessment to prove their worth. After which they invested a large part of their time researching and compiling the most relevant data for our readers.

The editorial board has been involved in producing this book since its inception. They have spent rigorous hours researching and exploring the diverse topics which have resulted in the successful publishing of this book. They have passed on their knowledge of decades through this book. To expedite this challenging task, the publisher supported the team at every step. A small team of assistant editors was also appointed to further simplify the editing procedure and attain best results for the readers.

Apart from the editorial board, the designing team has also invested a significant amount of their time in understanding the subject and creating the most relevant covers. They scrutinized every image to scout for the most suitable representation of the subject and create an appropriate cover for the book.

The publishing team has been an ardent support to the editorial, designing and production team. Their endless efforts to recruit the best for this project, has resulted in the accomplishment of this book. They are a veteran in the field of academics and their pool of knowledge is as vast as their experience in printing. Their expertise and guidance has proved useful at every step. Their uncompromising quality standards have made this book an exceptional effort. Their encouragement from time to time has been an inspiration for everyone.

The publisher and the editorial board hope that this book will prove to be a valuable piece of knowledge for researchers, students, practitioners and scholars across the globe.

List of Contributors

Yadong Fang
Department of Mechanical and Electrical Engineering, Xi'an Technological University, Xi'an, China

Fabian G. Beck, Eric H. Grosse and Ruben Teßmann
Carlo and Karin Giersch Endowed Chair "Business Management: Industrial Management", Technische Universität Darmstadt, Hochschulstr. 1, 64289 Darmstadt, Germany

Morteza Zangeneh, Asadolah Akram, and Alireza Keyhani
Department of Agricultural Machinery Engineering, Faculty of Agricultural Engineering and Technology, School of Agriculture & Natural Resources, University of Tehran, Karaj, Iran

Peter Nielsenb
Department of Mechanical and Manufacturing Engineering, Aalborg University, Aalborg, Denmark

B. Yuce
School of Engineering, BRE Institute of Sustainable Construction, Cardiff University, Newport Road, The Parade, Queen's Building, Cardiff CF24 3AA, UK

D.T. Phamb
School of Mechanical Engineering, University of Birmingham, Edgbaston, Birmingham B15 2TT, UK

M.S. Packianather
Institute of Mechanical and Manufacturing Engineering, Cardiff University, Queen's Buildings, The Parade, Cardiff CF24 3AA, UK

E. Mastrocinque
School of Management, Royal Holloway, University of London, Egham Hill, Egham, Surrey TW20 0EX, UK

Joakim Wikner and Eva Johansson
Department of Industrial Engineering and Management, School of Engineering, Jönköping University, Jönköping, Sweden

Amjad Hussaina and Aftab Ahmada
Department of Industrial and Manufacturing Engineering, University of Engineering and Technology, Lahore 54890, Pakistan

Keith Case
Mechanical and Manufacturing Engineering, Loughborough University, Loughborough, Leicestershire LE11 3TU, UK

Muhammed Nafis Osman Zahid, Keith Case and Darren Watts
Mechanical and Manufacturing Engineering, Loughborough University, Loughborough, Leicestershire LE11 3TU, UK

Joseph Achebo
Department of Production Engineering, University of Benin, Benin City, Nigeria

Tzong-Ru (Jiun-Shen) Lee and Szu-Hai Lin
Department of Marketing, National Chung Hsing University, Taichung, Taiwan

Yang Liub
Department of Production, University of Vaasa, Vaasa, Finland

Shpend Imeri
Department of Production, LEORON Professional Development Institute, Dubai, United Arab Emirates

Ihsan K. Al Naimi and Kasim M. Daws
Mechanical Department, College of Engineering, Baghdad University, Baghdad, Iraq

Moneer H. Al Saadi
Welding Department, University of Kerbala, Baghdad, Iraq

Niels Bay
Department of Mechanical Engineering, Technical University of Denmark, Lyngby, Denmark

Narges Banaeian, Hossein Mobli and Mahmoud Omid
Department of Agricultural Machinery Engineering, Faculty of Engineering and Technology, University of Tehran, Karaj, Iran

Izabela Ewa Nielsen
Department of Mechanical and Manufacturing Engineering, Aalborg University, Aalborg, Denmark

Panagiotis H. Tsarouhas
Department of Standardization & Transportation of Products - Logistics, Technological Educational Institute of Central Macedonia, Kanellopoulou 2, Katerini 60100, Greece;

Ioannis S. Arvanitoyannis
School of Agricultural Sciences, University of Thessaly, Fytokou Street, 38446 Nea Ionia Magnesias, Volos, Greece

Douglas H. Grabenstetter and John M. Usher
Department of Industrial and Systems Engineering, Mississippi State University, MS 39762, USA

Sherif Mostafa and Nicholas Chileshe
School of Natural and Built Environments, University of South Australia, Adelaide 5000, Australia

Sang-Heon Lee
School of Engineering, University of South Australia, Mawson Lakes 5095, Australia

Jantanee Dumrakc
Global Management Program, Torrens University Australia, Adelaide, SA 5000, Australia

Hassan Soltan
Production Engineering and Mechanical Design Department, Faculty of Engineering Mansoura University, Mansoura 35516, Egypt

Vinod S. Gorantiwar and R.L. Shrivastava
Department of Mechanical Engineering, Yeshwantrao Chavan College of Engineering, Nagpur 441 110, India